Sustainable Rural Development Projects: Diversity and Opportunities from Principles of Responsible Investment in Agriculture and Food Systems

Contents

Guest Editors
Adolfo Cazorla Montero
Department of Agroforest Engineering, Planning and Sustainable Management of Rural-Local Development (GESPLAN)
Universidad Politécnica de Madrid
Madrid
Spain

Ignacio De los Ríos Carmenado
Department of Agroforest Engineering, Planning and Sustainable Management of Rural-Local Development (GESPLAN)
Universidad Politécnica de Madrid
Madrid
Spain

Editorial Office
MDPI AG
Grosspeteranlage 5
4052 Basel, Switzerland

This is a reprint of the Special Issue, published open access by the journal *Sustainability* (ISSN 2071-1050), freely accessible at: https://www.mdpi.com/journal/sustainability/special_issues/5W4ARG79S8.

For citation purposes, cite each article independently as indicated on the article page online and as indicated below:

Lastname, A.A.; Lastname, B.B. Article Title. *Journal Name* **Year**, *Volume Number*, Page Range.

ISBN 978-3-7258-7338-8 (Hbk)
ISBN 978-3-7258-7339-5 (PDF)
https://doi.org/10.3390/books978-3-7258-7339-5

Sustainable Rural Development Projects: Diversity and Opportunities from Principles of Responsible Investment in Agriculture and Food Systems

Guest Editors

Adolfo Cazorla Montero
Ignacio De los Ríos Carmenado

MDPI

Basel • Beijing • Wuhan • Barcelona • Belgrade • Novi Sad • Cluj • Manchester

About the Editors

Adolfo Cazorla Montero

Adolfo Cazorla Montero is an Agricultural Engineer, obtaining a PhD with honors cum laude at the Universidad Politécnica de Madrid, and Full Professor in Engineering Projects at the same university, with four recognized six-year research periods. He is certified as Senior Project Manager Level B by the International Project Management Association and has coordinated since 2010 the Master and Doctoral Program in Planning of Rural Development Projects and Sustainable Management at UPM. He is Director of the Metauniversity for Sustainable Development, an international academic network that integrates universities, enterprises, and rural communities under the Principles for Responsible Investment in Agriculture and Food Systems. Through this role, he has promoted an innovative model that connects research, postgraduate education, public governance, and societal engagement. He has directed more than forty-five international studies and projects in the United States and Latin America, supervised fourteen doctoral theses, and authored numerous books and scientific articles. He has held senior public positions in regional and national administrations in Spain, including Director General of the National Institute for Agricultural and Food Research and Technology, and has represented Spain in European and international research coordination bodies, including CGIAR. His career combines academic excellence, public policy leadership, and international cooperation for sustainable rural development.

Ignacio De los Ríos Carmenado

Ignacio De los Ríos Carmenado is Full Professor in Engineering Project Management at the Universidad Politécnica de Madrid, with four recognized six-year research periods. He holds a PhD in Projects from UPM with honors cum laude and received an Award for the best doctoral thesis. He is Certified Project Manager Level C at the International Project Management Association and serves as Vice President of the Spanish Association of Project Management and Engineering AEIPRO IPMA Spain, as well as Assessor of the International Project Management Certification Body. He is Director of the GESPLAN Research Group, Planning of rural development projects and sustainable management at UPM. He directs the International PhD Program in Project Planning for Rural Development and Sustainable Management and leads the Food Bank Chair. He serves as research coordinator of the Metauniversity for Sustainable Development. He is also a consultant to the Food and Agriculture Organization of the United Nations as an expert in the CSA IRA principles. For more than three decades, he has collaborated with governments, universities, and international organizations contributing to sustainable rural development in Europe and Latin America.

Preface

This Reprint addresses one of the central challenges of our time: how to translate global sustainability principles into concrete actions that improve rural territories and agrifood systems. Its subject is the integration of the Principles for Responsible Investment in Agriculture and Food Systems into university teaching, applied research, and collaboration with enterprises and civil society. Its scope spans Latin America, the Caribbean, and Spain, bringing together academic institutions, agrifood companies, associations, cooperatives, and rural communities within a shared framework for sustainable development.

The aim of this Reprint is to document how the Metauniversity has emerged as a pioneering yet practical response to contemporary development challenges. Grounded in the conceptual framework of Working with People, the model connects political contextual, ethical social, and technical business dimensions, enabling a progression from dissemination of principles to shared knowledge and joint action. This approach places people, institutions, and territories at the center of sustainable transformation.

The Metauniversity demonstrates that research, teaching, governance, and societal engagement can be coherently integrated within an international collaborative structure. Through postgraduate programs, applied research, executive training, and living labs, responsible investment principles are translated into concrete actions in rural territories.

This Reprint is addressed to researchers, university leaders, policymakers, and development practitioners seeking replicable models that combine academic rigor with social responsibility. The Guest Editors express their profound gratitude to the Food and Agriculture Organization of the United Nations for its vision and sustained support, and to all academics, entrepreneurs, public officials, students, and especially the people of rural communities whose trust and commitment give true meaning to working with people in building a more sustainable and just world.

Adolfo Cazorla Montero and Ignacio De los Ríos Carmenado
Guest Editors

sustainability

MDPI

Article

Toward a Metauniversity for Sustainable Development: Responsible Agriculture Investment and Food Systems

Adolfo Cazorla [1,*], Adhemir Cáceres [2] and Carlos Lavalle [3]

1 Planificación y Gestión Sostenible del Desarrollo Rural-Local (GESPLAN) Research Group, School of Agricultural, Food and Biosystems Engineering, Universidad Politécnica de Madrid (UPM), Avda Puerta de Hierro, nº 2-4, 28040 Madrid, Spain

2 Área de Dirección General, PAD Escuela de Dirección, Universidad de Piura, Calle Aldebarán 160, Santiago de Surco, Lima 15023, Peru; adhemir.caceres.v@pad.edu

3 Campus Lima, Universidad de Piura, Calle Mártir José Olaya 162, Miraflores, Lima 15074, Peru; carlos.lavalle@udep.edu.pe

* Correspondence: adolfo.cazorla@upm.es

Abstract

The sustainable development of agrifood systems is a pressing global challenge, highlighting the need for frameworks that guide responsible investment and community engagement. The Principles for Responsible Investment in Agriculture and Food Systems (CSA-IRA), approved by the Food Security Council in 2014, provide such a framework. Recognizing this opportunity, the FAO selected the Gesplan Research Group of the Polytechnic University of Madrid in 2016 to promote these principles in Latin America, the Caribbean, and Spain, leveraging the expertise of PhD graduates in Projects and Planning for Sustainable Rural Development. The main objective of this research was to explore how teaching, research, and civil society engagement can be integrated to operationalize CSA-IRA principles and foster sustainable development. To achieve this, the study applied the "Working with People" model across multiple countries and contexts, using university–business collaborations to implement practical, socially responsible initiatives. Over nine years, the approach generated a network of 46 universities and 52 agrifood companies across 12 countries, demonstrating effective multi-stakeholder collaboration. The accumulated experience led to the proposal of the Metauniversity—a "university of universities"—as an innovative instrument to scale knowledge transfer, research, and community engagement. These findings highlight that structured, collaborative networks can translate CSA-IRA principles into tangible actions, offering a replicable model for sustainable agrifood development globally

Keywords: planning; sustainable development; research universities; working with people

1. Introduction

Higher education is currently undergoing a critical transformation, driven by global demands for sustainable development and the need for more effective responses to complex social problems. Traditional university models, although historically effective, are increasingly showing limitations in addressing the multifaceted challenges of the 21st century. This situation has sparked a profound debate about the need for new paradigms in higher education that enable greater integration among research, teaching, and social engagement [1]. The complexity of current problems, ranging from climate change to food security, requires approaches that transcend traditional institutional and geographical

https://doi.org/10.3390/su17219698

boundaries. In this context, there is a need to develop more adaptable and collaborative university models, capable of integrating diverse knowledge and experience on a global scale [2]. The concept of the Metauniversity [3] emerges as a potential response to these challenges, suggesting a structure that facilitates international collaboration and knowledge exchange while maintaining high academic standards. The Metauniversity is emerging as an innovative educational paradigm that redefines higher education through the integration of emerging technologies, artificial intelligence and digital twins [4]. This concept represents an evolution of the traditional university towards a flexible, inclusive and globally connected learning ecosystem.

The term "Metauniversity" was introduced by Charles Vest in 2006 [3], who envisioned a model of international higher education that transcended the physical barriers of traditional universities. Vest envisioned a system where open educational resources and digital tools would make education accessible to all, fostering global collaboration in research and access to quality resources no matter the student's location or financial situation.

The evolution of this concept has been driven by the development of cloud-based learning platforms, Massive Open Online Courses (MOOCs) and digital knowledge networks [4]. The original idea, rooted in the Greek prefix "meta" (meaning "beyond"), suggests a "model of models" or a "university of universities", a concept that, as will be seen below, goes beyond the abstraction of a traditional institution to encompass a global collaborative network of universities and companies [4] (Cazorla).

In the era of globalization and digitalization, the Metauniversity emerges, as well as a response to the need for new, more adaptable and collaborative university models. Its relevance is based on several key points such as: accessibility and inclusion by eliminating geographical, economic and social barriers; learning-based by doing for the acquisition of skills fostering critical thinking and innovation; personalization and flexibility promoting cultural diversity and lifelong learning; response to the problems and needs of society by focusing on practical skills [2].

The Metauniversity model is intrinsically linked to the United Nations Sustainable Development Goals (SDGs) and is seen as a tool to achieve multiple SDG [5]. Its approach, based on collaboration, innovation and equity, contributes directly to several targets: SDG 4 (Quality Education), SDG 9 (Industry, Innovation and Infrastructure), SDG 10 (Reduced Inequalities) and especially SDG 17 (Partnerships for the Goals) since the concept of MetaUniversity is based on global collaboration between educational institutions, business and civil society, embodying the spirit of partnerships to achieve sustainable development [5,6].

The Metauniversity is therefore not only a technological evolution of education, but a new paradigm that aligns higher education with the global needs of sustainability, equity and innovation. By creating collaborative and accessible learning ecosystems, it has the potential to transform education and empower new generations to build a more sustainable future [2,7–9].

To guide actions towards truly sustainable development, the Principles for Responsible Investment in Agriculture and Food Systems (CFS-IRA), adopted by the Committee on World Food Security (CFS) in 2014, provide a fundamental framework [10]. These principles, aligned with the Sustainable Development Goals (SDGs), seek to guide the actors involved—governments, the private sector, civil society and academia—towards development that promotes food security, gender equality, resource conservation and inclusive governance [11,12]. The need to transform agri-food systems towards more sustainable and inclusive models has put the CFS-RAI Principles in the international debate [13–17]. These principles constitute a voluntary framework that guides States, the private sector, civil society, and international organizations in mobilizing investments that contribute to food security, inclusive rural development, and environmental sustainability [18].

In Latin America, the application of the RAI principles is connected to the structural challenges of rural development such as access to land, land grabbing, market concentration, vulnerability to climate change, and the historical exclusion of peasant and indigenous communities in rural development [14]; the institutional weakness to guarantee transparency and accountability in projects [15]; and the difficulties in generating financial instruments accessible to small producers [16,17]

The CFS–RAI principles provide a widely accepted normative framework to guide responsible investments in agriculture and food systems in Latin America. Its implementation is now a field of research and priority action, especially on how to ensure that investments contribute to food security and sustainable rural development [13].

This work is framed in the confluence of two necessary trends: the evolution towards more collaborative university models in the field of sustainable development planning [19] and the growing need to apply guiding principles, such as the CSA-IRA Principles, in real contexts [18]. The Metauniversity's proposal, promoted by the GESPLAN Research Group of the Polytechnic University of Madrid, is based on the "Working with People" (WWP) metamodel [20]. The WWP model is a participatory, multidimensional approach that connects knowledge and action through social learning, emphasizing the integration of human values, expert knowledge, and local expertise. It is structured in three interconnected dimensions: ethical-social, technical-business and political-contextual, which have proven to be effective in solving rural problems, improving governance and promoting sustainable development in various contexts in Europe and Latin America [13].

Previous research has validated the application of the CFS-IRA Principles in various contexts, demonstrating their ability to drive sustainable economic growth in rural areas, support the transformation of regions affected by illicit crops [21], and strengthen indigenous women's cooperative systems [22]. Similarly, the WWP model has been validated as an effective framework for development planning, the creation of Local Action Groups [23,24], the strengthening of agri-food chains [17,25], and the promotion of rural entrepreneurship [26]. The synergy between the CFS-IRA Principles and the WWP model has been explored in projects that demonstrate improvements in producers' livelihoods and successful access to international markets [13,27].

However, despite these advances, the joint application of these frameworks in a structure of inter-university, innovative teaching governance and business collaboration [28], such as the one proposed by the Metauniversity, has been scarcely explored. This article seeks to address this gap, analyzing how the structure of the Metauniversity, operating from the CSA-IRA principles and the WWP methodology, connects with current trends in sustainable development and aligns social innovation, participatory governance and the creation of multi-stakeholder networks.

This research aims to provide a robust theoretical and practical framework that illustrates how higher education institutions can evolve to lead the transition to a more sustainable and equitable future. For this purpose, the research is based on the outcomes of the FAO-UPM project, developed between 2016 and 2024, which, according to Cazorla et al. [29,30], enabled the implementation of a Metauniversity model specifically oriented toward sustainable development by the end of this period. According to Jiménez Aliaga et al., [25] this project has provided a unique laboratory to examine how higher education institutions can evolve to meet contemporary demands while preserving their core mission of knowledge creation and dissemination. As highlighted by Requelme and Afonso [10], the integration of the CSA-IRA Principles in this process has been essential to ensure that institutional transformation aligns with the goals of sustainable development.

As we have seen, the theoretical framework of this research is based on three interrelated conceptual pillars. First, the CSA-IRA Principles (see Table 1), approved by the

Committee on World Food Security in 2014, provide a fundamental basis for Responsible Investment in Agriculture and Food Systems [10]. These principles align with the Sustainable Development Goals through five key dimensions, Planet, People, Prosperity, Peace, and Partnerships [31], establishing a comprehensive framework for sustainable development.

Table 1. CSA-IRA Principles. Source: Own elaboration based on [18].

Principle	Dimension
P1 Contribute to food security and nutrition	Economic
P2 Contribute to sustainable and inclusive economic development and poverty eradication	Economic
P3 Promote gender equality and the empowerment of women	Social
P4 Strengthen youth participation and empowerment	Social
P5 Respect tenure of land, fisheries, and forests, and access to water	Environmental
P6 Sustainably manage and conserve natural resources, enhance resilience, and reduce disaster risk	Environmental
P7 Respect cultural heritage and traditional knowledge, and support diversity and innovation	Economic
P8 Promote safe and healthy agricultural and food systems	Environmental
P9 Incorporate inclusive and transparent governance structures, processes, and grievance mechanisms	Governance
P10 Assess and address impacts, and promote accountability	Governance

Secondly, the paradigm of research universities, according to San Martin and Cazorla [1], evolves beyond traditional rankings by integrating three essential elements: research as a generator of knowledge, effective engagement with society, and relevant teaching, especially at the postgraduate level. This model is complemented by an efficient administrative structure and strategic leadership, recovering the original concept of the university as an institution for the generation and dissemination of knowledge [32].

Finally, the concept of the Metauniversity emerges as an innovative response to the contemporary challenges of higher education. As noted by Vest [3] and Couturier [4], this model facilitates knowledge transfer among academics, researchers, and society, leveraging digital technologies and promoting global cooperation [33]. This approach, according to Costanza et al., [2] represents a significant evolution in how higher education institutions can collaborate and share resources to address complex global challenges.

The integration of these three elements provides a robust theoretical framework for analyzing and understanding the evolution of higher education toward more collaborative and sustainable development-oriented models.

1.1. FAO, SDGs, and CSA-IRA Principles for Responsible Investment

The CSA-IRA Principles, approved by the Committee on World Food Security in October 2014, establish a comprehensive framework for promoting Responsible Investment in Agriculture and Food Systems [18]. As Requelme and Afonso [10] point out, these principles emerged in response to the need to guide stakeholders involved in food systems toward achieving sustainable development.

The conceptual framework links the SDGs with the CSA-IRA Principles through five core pillars, known as the 5 Ps: People, Planet, Prosperity, Peace, and Partnership [8]. Afonso et al. [10] emphasize that this structure enables the classification of research and experience according to different dimensions of sustainability:

The social dimension (People), as noted by De los Ríos et al., [13,24] focuses on eradicating poverty and hunger, relating to SDGs 1–6. These goals are connected to CSA-IRA Principles 3 and 4, which center on gender equality and youth empowerment [10,25].

The environmental dimension (Planet), according to Castañeda et al. [34] and Cachipuendo et al. [16], focuses on protecting natural resources and combating climate change. SDGs 11–15 are associated with this dimension, and their connection to CSA-IRA Principles 5, 6, and 8 concerns respecting land tenure and conserving natural resources [35].

The economic dimension (Prosperity) refers to aspects related to sustainable economic growth, decent employment, and the reduction in economic inequalities. The most relevant SDGs for this dimension are 7, 8, 9, 10, and 11, and they align with CSA-IRA Principles 1, 2, and 7, which focus on economic growth [36–38]. From the governance dimension (Peace and Partnership), the goal is to achieve strong alliances and well-structured governance systems [39]. This dimension is linked to SDGs 16 and 17 and to CSA-IRA Principles 9 and 10, which specifically emphasize the incorporation of solid institutional structures, accountability, and transparency in management [30,35]. Figure 1 summarizes these dimensions of the CSA-IRA Principles and SDO.

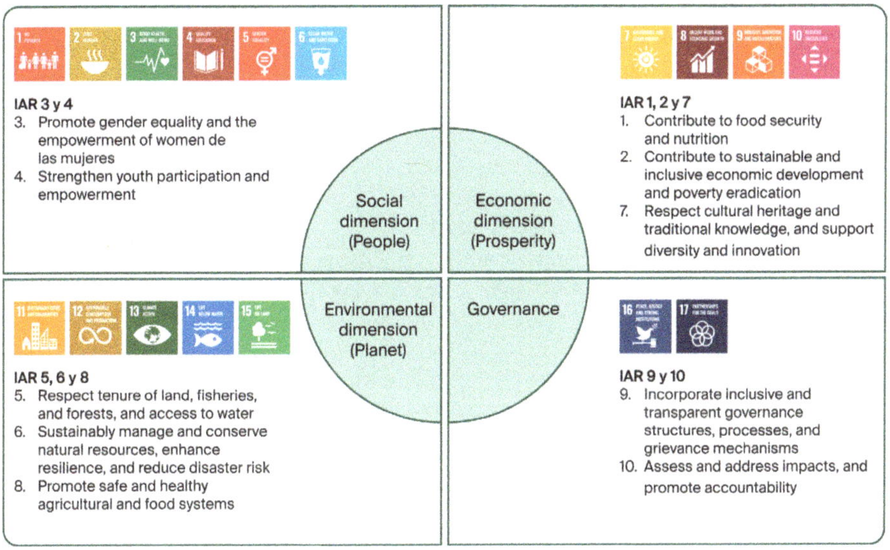

Figure 1. CSA-IRA Principles and SDO. Source: Own elaboration based on [8].

Between 2012 and 2014, FAO's general policy was enriched during the approval process of the CSA-IRA Principles. In addition to its traditional engagement with governments as a United Nations multilateral agency, the FAO began to actively incorporate long-term collaboration with civil society, represented by universities, associations, cooperatives, and enterprises operating in rural areas. This shift reinforced the role of FAO's Partnerships and UN Collaboration Division [34]. As a result, universities began to play an increasingly relevant role, thanks to their dynamic research–action relationships with businesses and associations aiming to achieve sustainable development [19,40].

1.2. The New Paradigm of Research Universities and Their Relationship with Society: The Polytechnic University of Madrid

Until the end of the 20th century, much attention was given to world-class universities and their characteristics to be included in this elite group [40]. These components closely aligned with the indicators that the most prestigious global rankings would begin to use a few years later to classify universities, considering the quality and scope of research as a key factor [1]. However, this key role of research, though always important, was not clearly connected to other crucial aspects of good university governance, such as faculty development, links to teaching, or the real-world needs of the society that universities should serve and help improve.

This was the context in which, in 2003 and 2004, three global university rankings were published: the Shanghai Ranking from Jiao Tong University, the Webometrics Ranking by the Higher Council for Scientific Research (CSIC), and The Times Higher Education Ranking. A fourth, Quacquarelli Symonds (QS), was added in 2012. Starting in 2005, the term "research" became increasingly attached to the concept of "university," not merely to describe institutions that prioritize research—a clear and longstanding trait—but to introduce a new concept with broader ambitions [41].

Cazorla & Stratta [40] identify three fundamental pillars of the research university paradigm: research as a driver of knowledge aligned with societal needs, engagement with society, and relevant teaching, especially in master's programs. Drawing from their experience as vice-rectors of research universities, San Martin and Cazorla [1] complete this approach with two additional and less clear elements necessary for the effective governance of a university: an administrative structure that supports this triad of excellence, and a high-level leadership capable of understanding and meeting the demands of such a challenge.

Figure 2 schematically presents this framework, illustrating this new paradigm of the research university, a model that defines it as a true engine of social transformation [40]. The model is based on five fundamental pillars that, together, create a dynamic institution that generates and disseminates knowledge to respond to the needs of society: (1) Research: it is the central pillar, conceived as an engine of knowledge aligned with the real problems and opportunities of Society; (2) Engagement with society: It reflects the real link between the university and its environment, ensuring that the knowledge generated has a positive impact; (3) Relevant teaching: Focused especially on postgraduate studies (master's and doctorate), to train graduates with high intellectual, creative and innovative preparation; (4) Administrative support structure: An efficient governance support that is essential to sustain excellence in the other pillars; and (5) High-level leadership: strategic leadership capable of guiding the institution to meet the complex demands of this model [42].

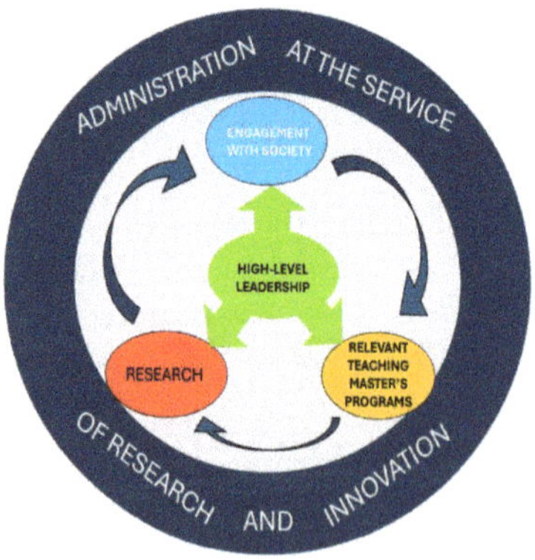

Figure 2. The five pillars of research universities. Source: Own elaboration.

These five pillars define the essential elements for a university to be considered a true engine of social transformation, as Cazorla & Stratta [40] put it, integrating knowledge generation with sustainable development. A university that addresses societal problems or opportunities through innovative knowledge—technological, legal, economic, humanistic, etc.—is a living institution that reclaims the original purpose of the university: not only as a guardian but also as a generator and disseminator of knowledge [10]. This is achieved by faculty, typically working in teams, through their publications and the design of post-graduate programs—master's and doctorate—ensuring that graduates are intellectually, creatively, and innovatively prepared [23,43].

The link between the emergence of global rankings and the new paradigm of research universities has already been noted. Although not developed here, the literature shows that Webometrics, through its indicators, transparency, and openness, best reflects the reality of research universities. It offers more than just a static metric: it provides a path for governance improvement for any university aspiring to become a research university [44–46]. In this way, research universities emerge as highly suitable systems and key instruments to address the complex challenge of sustainable development.

The Polytechnic University of Madrid (UPM) elected a new rector in 2004 who, along with his team, undertook as a governance program the goal of transforming the institution into a research university within the eight-year duration of his mandate. The governance measures and their impact on rankings were significant—the creation of research structures, master's programs, and stronger engagement with society, among many others—which positioned UPM among the top 300 universities in the world [1]. One of the research groups at this university, known as Gesplan (Research in Planning and Sustainable Management of Rural/Local Development) [19], had been working for years under this approach—one of its members had been trained at UC Los Angeles—establishing a network of PhD holders in Latin America (LA) and Spain, many of whom are faculty at universities in those countries. This drew the attention of the FAO in 2016 to undertake a very innovative program—the first of its kind—to connect the United Nations multilateral organization with universities, the business sector, and associative and cooperative networks to promote the dissemination of the CSA-IRA Principles in Latin America and the Caribbean (LAC).

On 27 September 2016, UPM and FAO signed a Framework Agreement, renewed in subsequent editions, to promote a policy that was defined as follows: "Working on the analysis and formulation of policy proposals and projects for strengthening governance for rural development and food and nutrition security in countries of LA and the Caribbean, the geographic area initially targeted by this project. On behalf of UPM, the Gesplan Group and on behalf of FAO, the Agricultural and Development Economics Division (ESA) and FAO's Regional Office for LA and the Caribbean (RLC)" [47].

Since 2016, successive Letters of Agreement (LoAs) have been developed, launching new projects—described later in this document—with increasingly complex and ambitious activities based on the annual results achieved. In the terminology of the field of Project Management, this could be synthesized as a system of projects operating as a process, [23] in which the achievements—new ideas—are articulated through new projects, making it, as Negrillo insists, [48] more appropriate to speak of a development *process* rather than merely development *projects*.

1.3. *Metauniversity in World-Class Universities: Rethinking the Future*

Within the context of American World-Class Universities, the concept of the "Metauniversity" emerges as a way of rethinking the future of higher education through the transfer of knowledge among academics, researchers, and society [3,4,48–50].

The term "meta" gained significance centuries ago and originates from Hellenic culture. In a certain library, some of Aristotle's books were shelved behind works on Physics, and the subject matter therein was referred to as "Metaphysics", meaning "beyond Physics". From that point on, the prefix "meta" enriched the accompanying concept with its original sense of "beyond." In 1984, in the context of a research project conducted at the Chair of Physical Planning and Projects at UPM, Ángel Ramos, a prominent figure, incorporated that prefix into the "model" for the regionalization of a territory, titling it "Metamodel for the Regionalization of a Territory," meaning "beyond" and suggesting that the notion of 'model' as an abstraction, when extended with the prefix 'meta', leads us to another concept: one that completes and adds to the original, as in a "model of models" [51]. More recently, other authors have applied the term "metamodel" as a structure of entities, associations, and constraints that allow for the representation of conceptual models. [24]

With this meaning and linked to technological change, the term "meta" extends to the university domain, giving rise to the concept of the Metauniversity [3,4,25]. The OpenCourseWare initiative developed in the USA, Asia, and Europe enabled the opening and sharing of intellectual resources and educational materials, promoting a dominant spirit of global cooperation: informing, sharing knowledge, and teaching internationally [3].

Cazorla and De los Ríos [19] emphasizes the fundamental role of the professor as a guide and mentor to students, encouraging them to use, adapt, or expand materials to meet their needs, based on the teaching strategy and local context. Within this new conceptual framework, Costanza et al. [2] note that knowledge generation and resource mobilization will take place through digital platforms for shared information, increasing quality and global access.

Thus, we continue to be enriched by a concept that remains highly relevant in the world's leading universities: the university as "a relationship between masters and disciples in search of the unum necessarium: knowledge" [19].

Just as it was previously argued that research universities are "more than" simple universities, likewise, the authors of this publication believe that a concept as complex and multifaceted as sustainable development requires an operational structure suited to such a challenge. The experience gained conceptually "through action" invites us to believe that it is indeed possible.

1.4. Objectives of This Research

Within the new context of change at FAO, the approval of the CSA-IRA Principles, and the development of research universities—particularly the role of UPM over a nine-year period—this research is framed around the challenge of rethinking new models that facilitate the articulation of business relationships with the generation of new knowledge, oriented toward policy proposals and project development within the framework of sustainable development.

Based on everything presented so far, and following the development of the framework, the main objective of this research is to demonstrate the Metauniversity as an effective instrument to promote sustainable development through the integration of the CSA-IRA Principles, from the WWP model, and within an international university–business network in Latin America, Spain, and the Caribbean.

The following section presents and justifies the methodological process undertaken.

2. Materials and Methods

Since 2016, the methodology developed in the successive Letters of Agreement between UPM and FAO has been based on the Working with People model. This WWP model is the result of 25 years of experience by the Gesplan Group in working and publishing in the field of rural development project management in various international contexts. The model is understood as a professional practice developed in cooperation that seeks to connect knowledge and action through joint projects that integrate the learning and values of the people involved.

This WWP conceptual framework is synthesized into three components—ethical-social, technical-business, and political-contextual—which interact to reinforce and enhance one another through processes of social learning. Figure 3 summarizes these three dimensions of the WWP model.

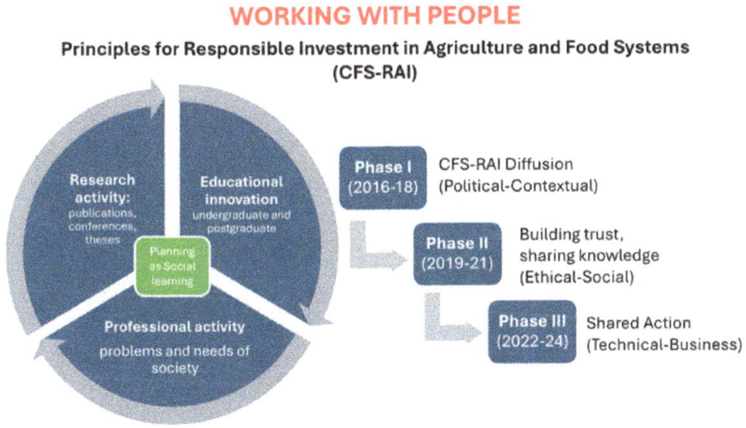

Figure 3. Working with People model [20]. Source: Own elaboration.

The WWP model, [20] in the context of this particular case study —process for the creation of a Metauniversity for Sustainable Development, consolidating a strategic alliance between 46 universities and 52 companies from 12 countries—, is characterized by the following aspects:

1. Participatory approach: This involves different actors (universities, companies, civil society) in a joint learning process.
2. Social learning: This focuses on generating shared knowledge through interaction among participants.
3. Quantitative and qualitative analysis: This presents both descriptive and numerical results of the activities and outcomes of the process.
4. Longitudinal case study: This is quite rarely achieved as, in this case, a nine-year continuity allows for an in-depth analysis of its evolution and results.

Based on the premise that "if there are no ideas, there are no projects or plans", this model, applied to our case study, integrates in its principles the thinking of John Friedmann, a global figure in Planning, particularly through his seminal publication detailed below:

Planning in the Public Domain, where one can extract from "Planning Action" the creative thoughts that ultimately shape a "scientific breakthrough" once, through contrast, something new emerges [29]. In our case, following the emergence of a "new concept" with the CSA-IRA Principles, approved in 2014 by 191 countries around the world, a dissemination project for this concept was proposed in 2016 to a university (UPM), targeting an extensive territorial area: Latin America and the Caribbean (LAC).

The following section describes the phases that make up the model and the work carried out over nine years, in such a way that dissemination evolved into shared knowledge and, in a third stage, into joint actions. Naturally, not all universities or companies have evolved at the same pace, as some have been involved since 2016 while others joined over the subsequent years, the most recent being in 2024. Nevertheless, the body of knowledge that has been developed has enabled these newer institutions to progress more rapidly, for obvious reasons.

2.1. Phases of the Methodological Process

From the political–contextual dimension of the WWP model (Phase I), dissemination actions played a fundamental role in initiating the process and ensuring a level of understanding necessary to guarantee the effectiveness of the subsequent phases. The CSA-IRA Principles require an understanding of contextual and political conditions in order to initiate processes within university–business relationships.

The ethical–social dimension of the WWP model (Phase II) allows for the generation of trust and commitment to advance toward the integration of the CSA-IRA Principles into academic, research, and community engagement programs, thereby fostering shared knowledge.

The technical–business dimension of the model (Phase III) plays an essential role in moving toward "Shared Action" through the development of joint university–business projects, involving stakeholders who have previously been trained and have established mutual trust. These projects would not be possible without having gone through the earlier phases.

2.2. Learning Processes and Scope of Participation

The scope of this research encompasses the nine-year period (2016–2024) of the program, during which the CSA-IRA Principles have been disseminated and integrated into educational programs, applied research, and business engagement—something that was unthinkable just a few years ago—given the challenges multilateral organizations face in reaching that level [29].

From the WWP approach, various actions—such as conferences, workshops, and seminars—were implemented to exchange experiences, generating "living" documents used as "case studies" for the application of the CSA-IRA Principles in education and training efforts across universities, associations, and companies. These were accompanied by well-established methodologies such as project-based learning [19,29,33], question-based learning [28,52,53], and service-learning [54,55].

These joint learning processes help in identifying problems and reflecting on decision-making processes in business, associative, or cooperative environments regarding the Principles for Responsible Investment in Agriculture (IRA) and have been systematized for research and dissemination through a web repository [47].

Over the years, as detailed later in the results section, workshops and training programs have been conducted across various countries in Latin America and the Caribbean (LAC), fostering collaboration between universities and agro-food companies.

In total, 25 workshops and 36 training programs were carried out, with 11,299 participants, distributed as follows: 441 from companies and associations, 764 from universities (faculty and researchers), 151 from public administrations (national, regional, local), 448 from organized civil society, and 9495 students (8868 undergraduate, 601 postgraduate, and 26 in research).

As of 30 June 2024, the initiative had reached 98 committed entities—52 companies and 46 universities—from 12 countries: Argentina, Bolivia, Chile, Paraguay, Ecuador, Peru, Colombia, Venezuela, Mexico, Panama, the Dominican Republic, and Spain.

The following Table 2 quantifies in each of the phases the scope of participation, the objectives and their relationship with the CSA-IRA Principles in each of the phases of the WWP metamodel to evolve towards the creation of a Meta-university oriented to sustainable development [20,30].

Table 2. Phases of research from Working with People" (WWP) model: Scope of participation, objectives and relationship with CSA-IRA Principles.

WWP Dimension: Phases	Scope of Participation Number of Actors Involved	Objectives and Their Relationship to the CFS-IRA Principles
Political-Contextual Dimension *Phase I: Dissemination (2016–2018)*	Total: 398 Academics: 45.5% Companies: 34.7% Public Administration: 9.3% Students: 10.5%	Establish the political and governance foundations for university-business cooperation to create a favorable ecosystem (P9 and P10, Governance). Knowledge for the adaptation of materials to the territorial, environmental (P5) and cultural (P7) context of the countries.
Ethical-Social Dimension *Phase II: Knowledge sharing (2019–2021)*	Total: 10,234 Students 91.5% Civil Society: 4.4% Academics: 3.1% Enterprises: 0.7% Public Administration: 0.3%	Build trust and knowledge by prioritizing gender equality (P3) and youth empowerment (P4). Design of training programs and development of case studies on how the principles impact food security and economic development (P1, P2 and P7)
Technical-Business Dimension *Phase III: shared action (2022–2024)*	Total: 667 Academics: 39.9% Enterprise: 34.9% Public Administration: 12.6% Students: 12.6%	Consolidation of the Meta-University. Transform knowledge into action through joint university-business projects ("Living Labs") by transversally applying all the CSA-IRA principles in the activities of the Metauniversity.

Each phase of the WWP model is aligned with specific objectives, concrete activities and outcomes which, in turn, are linked to the CSA-IRA principles from the interconnected dimensions (ethical-social, technical-business and political-contextual) that reinforce each other through social learning processes. These WWP dimensions were developed in three sequential and progressive phases, designed for the dissemination of the conceptual framework—the CFS-IRA Principles—into shared, tangible actions.

The main novelty of the methodology lies in the application of the WWP metamodel as a structured and sequential framework to operationalize the CSA-IRA Principles through an innovative university paradigm: the Metauniversity. Unlike traditional models, the WWP approach is not limited to the simple transfer of knowledge but articulates a development process in three logical and progressive stages. Phase I ensures strategic dissemination so that a complex concept (the CFS-IRA Principles) is understood and contextualized by all actors (contextual type dimension). Phase II of building social capital is crucial, as it recognizes that effective and sustainable collaboration cannot occur without a foundation of trust and mutual commitment (ethical-social dimension of the model). Finally, in Phase III, the action based on trust is built, which allows the implementation of successful technical and business projects (*living labs*) that would not be viable without the previous phases.

This WWP methodological progression ensures that the Metauniversity is not only a collaborative network, but a social learning ecosystem where research, teaching, and engagement with society are synergistically integrated to address real challenges of sustainable development.

The results of the Metauniversity—with its pillars of research, linkage, teaching, governance and administration—from this nine-year methodological process are shown below.

3. Results

The implementation of the Metauniversity model as an outcome of the FAO-UPM project has produced significant results across multiple dimensions, demonstrating the feasibility and transformative potential of this approach in higher education. The findings are presented according to the methodological dimensions employed, allowing for a comprehensive understanding of the achievements.

The success of the Metauniversity model in creating an international network of universities and companies for sustainable development has been guided by the integration of the three interconnected WWP dimensions [20,30] in its three phases.

The Political-Contextual dimension (Phase I) is crucial to establish the essential framework that guarantees political support, resource mobilization and a collaborative environment. It focuses on diffusion and the ability of actors to build relationships. The Ethical-Social dimension (phase II) its actions are aimed at promoting active participation, developing skills and sharing knowledge. Finally, the Technical-Business dimension (phase III) allows knowledge to be transformed into "Shared Action" with practical actions aimed at sustainable development in the territories.

The results in the process of creation of the Metauniversity are shown, operating under the WWP model, which has demonstrated its viability, by promoting a balance between the three dimensions, to be an instrument for institutional transformation integrating the CFS-RAI Principles.

3.1. Results of the Political–Contextual Dimension (Phase I, 2016–2018): Dissemination

During this initial phase, grounded in the political–contextual dimension of the WWP model, strategic actions were carried out over three years to disseminate the CSA-IRA Principles, establishing the necessary political and contextual foundations to foster a favorable ecosystem. Below are the most notable activities and their specific outcomes.

3.1.1. Activities for the Dissemination of the CSA-IRA Principles

International Multi-stakeholder Workshops on Responsible Investment in Agriculture (2017): Three CSA-IRA high-impact workshops were held in Peru (Lima, 27–28 June 2017), Ecuador (Cuenca, 21–22 September 2017), Colombia (Bogotá 26–27 September 2017), bringing together 195 participants from universities, agrifood companies, associations and public administrations. An additional seminar in Madrid with 65 attendees integrated Spanish actors into the network. The novelty of this approach was the creation of a space for intensive mutual learning, which made it possible to incorporate the productive sector from the beginning and overcome the previous fragmentation of efforts [29].

Facilitator Training Workshop (2018), in the Dominican Republic, 24 professors were trained as multiplying agents of the CSA-IRA principles in their institutions. This strategy ensured local appropriation of knowledge and the sustainability of the model, decentralizing the initiative and guaranteeing its continuity.

3.1.2. Results and Relation to the Principles

The result of this phase was the creation of an initial and committed network of 27 universities in 7 countries, laying the institutional foundations and social capital necessary for the future Meta-university. Specifically, the following instruments were designed for the University:

(A) Methodological Guide "Application of the CSA-IRA Principles in the University Context" (2017): A foundational document translated into two languages (Spanish and English) used as introductory material to present the principles in various academic settings.

(B) Collection of 10 Infographics on the CSA-IRA Principles (2018): Visual materials adapted to facilitate the understanding of each principle with contextualized examples for Latin America and the Caribbean (LAC).
(C) Documentary Video "Investing Responsibly in Agriculture" (2018): A 15 min audiovisual production featuring testimonials from academics and producers on the importance of the principles, used as a didactic resource in the early years.
(D) Colombia Case: In the context of post-conflict Colombia, emphasis was placed on Principles 5 (land tenure) and 10 (impact assessment), in collaboration with the National University of Colombia, adapting materials to this territorial reality.
(E) Mexico Case: In collaboration with the Colegio de Posgraduados, an adapted version of the principles was developed for regions with a strong presence of traditional agriculture, emphasizing Principle 7 (respect for cultural heritage).

The analysis of the political–institutional context initially revealed significant fragmentation in collaboration efforts between universities and businesses in LAC and Spain. However, the results achieved enabled the establishment of effective and lasting cooperation structures during the first phase of the project (2016–2018), including the formation of an initial network of 27 universities in 7 countries, 15 of which committed through an institutional signing event in October 2018, held at the Universidad Nacional Mayor de San Marcos, in the presence of a regional FAO representative. As shown later, this network expanded to 46 institutions in 12 countries by June 2024, becoming the backbone of the Metauniversity proposal approved in September 2024.

The creation of institutional collaboration frameworks has been particularly successful, as evidenced by the signing of a formal cooperation agreement and the establishment of shared working protocols thanks to the preparatory efforts described above. A significant milestone was the integration of the CSA-IRA Principles into the institutional policies of participating universities such as Universidad Politécnica Salesiana and Universidad Nacional Mayor de San Marcos, demonstrating a tangible commitment to sustainable development.

The results of this Phase I especially relate to the CSA-IRA Principles 9 and 10, focused on governance, to establish the necessary political and governance foundations to foster an ecosystem favorable to university-business cooperation. Initially, a fragmentation in collaboration efforts between universities and companies was detected. Through the workshops and the work with the people, it was possible to build trust among the participants to establish effective and lasting cooperation structures. A fundamental milestone was the creation of the initial network of 27 universities from 7 countries, which became the "backbone" to move towards the future Meta-university. In this process, priority was given to the application of Principle 9, to incorporate governance structures in the collaboration network (universities-companies, public sector) with formal agreements and shared work protocols. In addition, in this phase, the principles were not only disseminated in a generic way, but also adapted to specific territorial realities, to assess and address contextual impacts, in line with Principle 10 [17,21]. The creation of a "Methodological Guide for the Application of the CSA-IRA Principles in the University Context" and the collection of 10 contextualized infographics for LAC, a concrete example, contributed to promoting and materializing principles 9 and 10 [23].

In short, this phase was not limited to a simple dissemination of concepts but actively built the governance structures (P9) and the mechanisms of responsibility and adaptation (P10) that allowed the successful development of the subsequent phases [14]. By creating a formal network and adapting content to local contexts, this stage laid the institutional and political foundations indispensable for transforming fragmented collaboration into a robust and functional ecosystem geared towards sustainable development [13].

3.2. Results of the Ethical–Social Dimension (Phase II, 2019–2021): Building Trust and Shared Knowledge

The construction of social capital and trust among participants has been one of the project's most remarkable achievements. It is important to note that this period was marked by the COVID-19 pandemic, which prevented in-person meetings that had previously been the usual method of collaboration. This second phase, grounded in the ethical–social dimension of the WWP model, marked a qualitative leap forward, reflected in several emblematic activities.

3.2.1. Trust-Building and Knowledge-Sharing Activities

The activities developed show an innovative character in terms of their orientation towards building trust and the creation of shared knowledge between traditionally fragmented sectors. The **virtual seminars** organized during the pandemic were not only a mechanism for academic continuity but also provided a space for horizontal interaction between universities and business actors, where explicit commitments were generated to promote the CSA-IRA principles. This aspect is especially relevant, since inter-institutional trust was based on the joint deliberation of challenges and solutions in a context of crisis. Two Seminars hosted jointly by UPM, Universidad Politécnica Salesiana and the Executive Leadership Program of Universidad de Piura in 2020, during the pandemic. These online events gathered 317 faculty and university administrators and 70 business leaders and association executives who began committing to promote the CSA-IRA Principles in their sectors.

Similarly, **legislative research** in the Latin American and Caribbean region provided a common ground of normative reference, contributing to a shared language that links academia with legal and political practice. Research on Legislation in Latin America and the Caribbean (LAC) to identify laws that, even if not explicitly, reflected the CSA-IRA Principles: This effort led to the identification of three relevant legal cases in Peru, Colombia and Chile [47].

The **elaboration of case studies**, based on diverse national realities, functioned as a pedagogical and scientific resource to share transferable learning and strengthen epistemological confidence in the validity of the principles. Preparation of 5 completed Case Studies, which—at FAO's suggestion—were aligned with the CSA-IRA Principles and proposed as research materials. These cases came from Argentina, Peru, Ecuador, Mexico, and Colombia.

Finally, **curricular integration** at multiple educational levels shows a long-term commitment to institutionalizing collective knowledge, promoting not only the transmission of content, but also a culture of cooperation between students, teachers, and sectoral leaders. Following the commitments made by universities in 2018, the CSA-IRA Principles were integrated into academic programs (undergraduate, postgraduate, and research) at each institution, reaching more than 9000 students across the 15 participating universities.

3.2.2. Results and Relation to the Principles

The results of Phase II (2019–2021) allowed the building of trust and shared knowledge around the CSA-IRA Principles related to the social dimension (P3, promote gender equality and the empowerment of women; P4 (Strengthen youth participation and empowerment).

In this Phase II, an integration of the CSA-IRA principles was achieved in the undergraduate, graduate and research programs in 15 universities, laying the foundations to strengthen youth participation and empowerment (P4) to educate future generations of professionals in these principles. The seminars brought together 317 university professors and 70 business leaders, who committed to promoting gender equality (P3) and youth empowerment (P4) in their respective sectors. The production of three master's theses (carried out at the Universidad Nacional de La Plata, Universidad Politécnica Salesiana and Universidad Politécnica de Madrid) and two doctoral dissertations based on research developed in previous years, allowed the generation of knowledge from research on the social aspects of responsible investment [19,20,23,28].

In relation to the economic dimension (prosperity), during Phase II, and at the suggestion of FAO, five case studies were prepared (in Argentina, Peru, Ecuador, Mexico and Colombia) from research aligned with the principles CSA-IRA P1 (Contribute to food security and nutrition), P2 (Contribute to sustainable and inclusive economic development) and P7 (Respect cultural heritage and traditional knowledge). These joint investigations [16,17,19,23,25] also enabled the building of trust and shared knowledge, meaningfully addressed both social (P3, P4) and economic (P1, P2, P7) principles, and paved the way for the "Shared Action" of Phase III.

3.3. Results of the Technical–Business Dimension (Phase III, 2022–2024): Shared Action

The third phase of this research–action process represented the culmination of a journey that led to the maturity necessary to establish a Metauniversity through joint projects. It was characterized by the transformation of shared knowledge into practical solutions for real challenges in sustainable development, encompassing teaching, research, societal engagement, administrative structuring, and strategic leadership as elements that laid the groundwork for the phase to be launched in September 2024: the creation of a Metauniversity for sustainable development [56].

3.3.1. Shared Actions in the Field of Postgraduate Studies

In this third phase, the above results allowed an innovative approach aimed at integral sustainability and institutional responsibility, carrying out various pioneering training actions in the field of higher education and executive training. These initiatives, endorsed by the Food and Agriculture Organization of the United Nations (FAO), constitute a milestone in the international articulation of postgraduate programs around the CSA-IRA Principles [57], with special impact in Latin America, the Caribbean and Spain. These actions in the field of postgraduate studies are summarized below.

(A) First international postgraduate program on CSA-IRA Principles, approved by FAO and aimed at faculty and researchers from universities in Latin America, the Caribbean, and Spain. It lasted two months, included 40 h of instruction, and had 65 participants in its first edition. The second and third editions, held in 2023 and 2024, respectively, were extended to four months and involved 135 participants from 23 universities across 9 countries.

(B) First international program for business executives, association leaders, and cooperatives on the CSA-IRA Principles, with a duration of two months and 40 h, and a total of 69 participants. The second and third editions maintained the same format, reaching 113 participants from 10 countries.

(C) Two executive training seminars on the IRA Principles and two teacher training seminars, organized in-person over two days in Lima (UCSS, PAD) and Santo Domingo (PUCMM), and held online for participants from other countries, with a total of 198 attendees.

(D) Certification programs for professors and business executives interested in becoming experts and future instructors in CSA-IRA seminars. Between 2023 and 2024, a total of 203 professors and business leaders from 45 university and business institutions across 12 countries were certified. These individuals went on to form the academic faculty of what would become the future Metauniversity.

These actions at the level of the postgraduate degree studies represent an educational innovation in the field of sustainability and institutional responsibility, by integrating academic, executive and certifying training around a common conceptual framework. The results achieved—more than 500 certified participants from more than 20 countries and the formation of an inter-university network—show the consolidation of a training ecosystem of international scope and the projection of the CSA-IRA Principles as a benchmark in education for sustainable development and good institutional governance.

3.3.2. Shared Actions in the Field of Societal Engagement

The actions developed in this phase constitute a novel contribution in the field of university-business cooperation and in the construction of international frameworks for research and training in sustainability.

(A) Second Meeting of the Network of Universities/Companies Committed to the CSA-IRA Principles: Held on 70December 2022, in Santo Domingo, hosted by the Pontificia Universidad Católica Madre y Maestra (PUCMM) in collaboration with UPM. Twenty universities from the network participated, along with three additional universities that joined during this meeting. University representatives formally endorsed the commitment of the 27 companies, associations, and cooperatives affiliated with the network.
(B) Three *Living Labs* as joint university–business initiatives were launched in Cayambe (Ecuador), Santiago Rodríguez (Dominican Republic), and Mantaro (Peru). These practical implementations of ongoing efforts were documented as case studies following FAO's format, under the term *living labs*, coined by FAO. Detailed information is available [47].
(C) As a result of a training plan in Mexico with researchers from the Colegio de Posgraduados (CP), the cooperative Sabores de Calpan, and entrepreneurs who founded the company Campo de Lima for marketing a new variety of corn, a new living lab was launched in June 2024, still in an early stage compared to the three previously described.
(D) Third Meeting in Santo Domingo, September 2024, with direct or delegated participation from nearly all network universities and companies, to discuss the creation of the Metauniversity: This new structure would replace the existing network with a higher-level research/teaching/societal engagement approach. The proposal was unanimously approved. The development of the Metauniversity is addressed later in this article.

The results obtained—the expansion of the network, the formalization of inter-institutional commitments, the creation of living laboratories, and the gestation of a Meta-University—demonstrate the capacity of the CSA-IRA Principles to generate academic, business, and social innovations. These actions not only consolidate an international community of practice but also constitute a pioneering platform for collective learning and institutional transformation in favor of integral sustainability.

3.3.3. Shared Actions in the Field of Research Serving Society

A central aspect in the consolidation of the CSA-IRA Principles has been the generation of research with academic and social impact. Between 2022 and 2025, the following significant advances were achieved that demonstrate the maturity of the scientific ecosystem linked to the network.

(A) Research articles published in impact journals (9): Peru (3), Ecuador (2), Colombia (1), Guatemala (1), Ethiopia (1), and Spain (1).
(B) Master's theses (4) completed in Colombia (2), Argentina (1), and Spain (1).

(C) Doctoral dissertation based on research conducted in Guatemala and Colombia, with related articles published in JCR Q1 journals. Four doctoral dissertations defended in the first seven months of 2025 by researchers from universities linked to the now-established Metauniversity (UPS, UCSS, and UNMSM), stemming from three prior research projects focused on the IRA Principles: Ecuador (UPS, UTPL) and Peru (UCSS, UNMSM).
(D) Creation of a repository, initiated by the FAO and officially presented at the Third Meeting, as a testament to nine years of work. Key materials were grouped as follows: Doctoral dissertations (3), Master's theses (12), JCR papers (16), Conference papers (12), Research project case studies (18), Teaching cases for business leaders (50).

The volume and diversity of scientific production achieved in this period shows the consolidation of research around the CSA-IRA Principles as an emerging field of international relevance. The combination of publications in impact journals, postgraduate theses, doctoral projects and a knowledge repository validated by the FAO demonstrates the model's ability to articulate academic research with advanced training and the generation of public knowledge goods at the service of society.

3.4. Global Impacts of the 2016–2024 Period Not Covered in Previous Sections

A particularly significant result of this cycle has been the strengthening of the links between academia and society. Over nine years, 25 workshops were organized that facilitated direct dialogue between academics, entrepreneurs and social actors, configuring spaces for co-creation and mutual learning that go beyond traditional models of unidirectional knowledge transfer. This interaction resulted in the following:

(A) Applied research projects. Implementation of 18 applied research projects addressing specific challenges of sustainable development in various regional contexts. These projects demonstrate the model's capacity to produce practical solutions to complex problems.
(B) Inclusive collaboration network. Another outstanding contribution was the creation of a multi-stakeholder network that is not limited to universities and companies but also incorporates 448 representatives of organized civil society. This breadth reinforces the participatory and inclusive nature of the model, which differentiates it from international experiences focused exclusively on academic institutions.
(C) Quantitative results. The scope of the program can be measured in the more than 11,299 participants directly involved: 441 business professionals, 764 university professors and directors, 151 representatives of the public administration, 8868 undergraduate students, 661 graduate students and 26 doctoral students. At the same time, academic production reached significant levels: 16 articles in high-impact journals, 12 master's theses and 7 doctoral dissertations (of which four were defended in the first months of 2025).

These indicators not only show the consolidation of a critical mass of research and advanced training, but also the capacity of the model to generate relevant and socially applicable scientific knowledge. Its novelty lies in the balanced integration of academia, business and civil society, which positions it as a benchmark in the construction of higher education models aimed at sustainable development and the strengthening of university-society interaction. Table 3 summarize the global impacts for each phase.

Table 3. Global impacts of the 2016–2024 period.

Category	Phase I (2016–2018)	Phase II (2019–2021)	Phase III (2022–2024)	TOTAL
Committed Entities	**27**	**75**	**98**	**98**
Universities	27	40	46	46
Companies and private associations	0	35	52	52
Countries	7	9	12	12
Societal Engagement	**13**	**15**	**2**	**30**
Workshops, Seminars	11	13	1	25
Publication Dissemination	2	2	1	5
Application of IRA Principles by Universities	**11**	**32**	**80**	**123**
Societal Engagement	1	5	11	17
Undergraduate	5	8	21	34
Postgraduate	4	13	33	50
Research	1	6	15	22
Teaching: Undergraduate and Postgraduate	**2**	**11**	**7**	**20**
Seminars	2	11	4	17
International Programs	0	0	3	3
Business Training	**8**	**1**	**57**	**66**
Seminars	8	1	4	13
International Programs	0	0	3	3
Cases	0	0	50	50
Research	**13**	**22**	**17**	**52**
Research Projects (Cases)	5	10	3	18
Impact Journals	4	3	9	16
Master's Theses	2	6	4	12
Doctoral Theses	2	0	1	3
Legislation Cases	0	3	0	3
Participants	**398**	**10,234**	**667**	**11,299**
Business Professionals	138	70	233	441
Faculty and University Executives	181	317	266	764
Public Administration	37	30	84	151
Undergraduate Students	0	8784	84	8868
Postgraduate Students	42	559	0	601
Research Students	0	26	0	26
Civil Society Engagement	0	448	0	448

Source: Own elaboration.

These results clearly demonstrate that the Metauniversity proposal presented in the following section is built upon foundational elements developed over time through the university–business network. This network holds strong potential to undertake new scientific, educational, and societal engagement ventures of a different order.

3.5. A Metauniversity for Sustainable Development

Such a proposal must be grounded in an established knowledge base, which is outlined below. It is evident that the concept of sustainable development does not permeate all areas of academic knowledge taught at universities, such as fields like the exact sciences, history, philology, or computer science. However, it is equally valid to argue that many other disciplines—particularly those related to economic, social, business, and environmental sciences, as well as certain engineering fields—are deeply affected by the overarching concept of sustainable development.

Thus, this transversal knowledge could greatly enhance the preparation of future professionals, enriching them with multidisciplinary and multicultural perspectives from countries such as those in Latin America, the Caribbean, and Spain. The results previously outlined enable the conceptualization and design of a Metauniversity proposal framed within sustainable development. It is oriented toward "working on policy and project proposals to strengthen governance for rural development and food and nutritional security", [31] with a more suitable structure as envisioned by the authors, based on the outcomes achieved.

This Metauniversity model, emerging from joint work since 2016 and rooted in the previously described conceptual framework, originates within the GESPLAN Research Group at UPM [31]. However, it comprises professors and senior academics from other Latin American and Caribbean universities, as well as their collaborations with businesses, associations, and cooperatives spaces, where experienced professionals collectively embark on this intellectual and practical journey.

Although this Metauniversity vision integrates technology and many aspects of the European educational model, it is based on the following key elements according to Cazorla & De los Ríos [56]:

- The role of faculty as guides, coordinators, and project leaders, responsible for educational quality, grounded in competencies and the CSA-IRA Principles aligned with applied methodologies.
- A strong ethical commitment in teaching and governance, centered on educational values derived from the CSA-IRA Principles.
- A shared conceptual foundation based on the fields of Project Studies and Planning, fostering project-based learning as a starting point for the integration of CSA-IRA Principles.
- A unified learning framework that incorporates diverse educational components from participating universities, contributing to the richness of the model.
- A degree of curricular homogeneity built around CSA-IRA Principles across interdisciplinary undergraduate and postgraduate programs, designed to address real-world problems and needs.
- "*Living labs*" in ongoing construction and evolution, built on shared experiences, facilitating research, education, and societal engagement.
- Technological tools to share knowledge and experience and to support distance learning (e-learning), combined with competency-based in-person training.

Figure 4 summarizes components and subcomponents of this Metauniversity, drawing from the previously explained research university model, and representing an integrated framework. The Metauniversity represents an innovative higher education paradigm, integrating research, teaching, societal engagement, and governance to address global sustainability challenges.

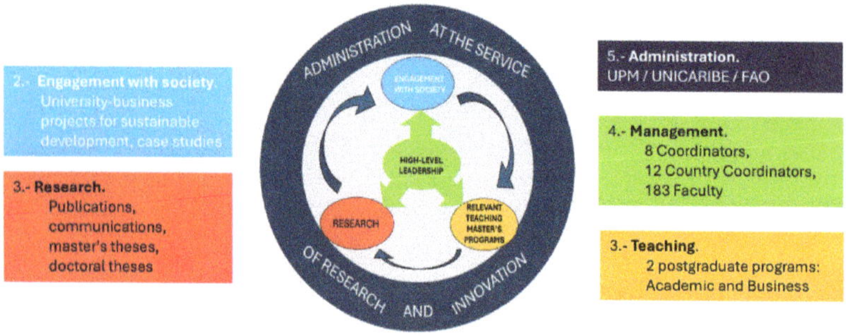

Figure 4. Characteristics of the Metauniversity. Source: Own elaboration.

Research: A central digital repository, the "crown jewel" of the network's digital identity [19], currently hosts 111 published documents and ongoing works. By centralizing outputs from 46 universities, the repository strengthens collaboration and evidence-based approaches critical for sustainable development.

Societal Engagement: Case studies developed by 117 certified faculty members provide diverse real-world scenarios from multiple ecosystems. This approach enhances academia–society interactions and supports applied learning aligned with sustainability goals.

Teaching: The Metauniversity has launched postgraduate programs integrating CSA-IRA Principles and Artificial Intelligence, targeting leaders of agrifood cooperatives across Latin America and the Caribbean. These programs cultivate skills for sustainable food systems and leadership in complex socio-environmental contexts.

High-Level Governance: A leadership team of 13 senior professionals from six countries, supported by 20 PhD coordinators across 12 nations, ensures strategic oversight. Administrative coordination by UPM, UNICARIBE University, and PAD enables efficient management of a complex transnational network.

Metauniversity Administration: UPM leads the administrative coordination, supported by UNICARIBE University and PAD (School of Government at the University of Piura) for the private agrifood sector. This was likely the first key success factor of the initiative, enabling a highly effective management model for a complex structure with first-rate online coordination.

4. Discussion

The results obtained over the nine uninterrupted years of development and structuring of the program validate the Metauniversity proposal and reveal significant patterns that deserve analysis: both in terms of achievements and challenges. This discussion is organized around the most relevant aspects, considering their theoretical and practical implications for the future of higher education.

4.1. Transformation of the University Model

The experience gained through the FAO–UPM program demonstrates that the transformation of the traditional university model into a more collaborative and flexible structure is not only possible but necessary in the face of global and complex challenges—especially those of a territorial nature—where traditional university collaboration proves limited due to a lack of institutional commitment.

The success of building a network of 46 universities and 52 companies suggests there is substantial willingness within the academic and business sectors to adopt new paradigms of collaboration. This finding aligns with Vest's [3] observations on the emergence of new forms of university organization in response to global challenges.

However, it is important to note that such transformation is not without complexity. Balancing rigorous academic standards while promoting innovation and flexibility requires a delicate balance. The project results suggest that the Metauniversity model offers an effective framework to manage this tension, allowing traditional academic excellence to coexist with more agile and collaborative modes of knowledge creation [56].

4.2. Social Capital and Collaboration Networks

A particularly revealing aspect of the results is the fundamental role of social capital in the model's success. Building trust-based relationships among institutions and across sectors proved to be a critical factor for the sustainability of the initiatives. Several years of inter-university work were needed to strengthen ties, engage the productive sector, and develop dynamic, practical training systems. This observation reinforces the arguments [1] regarding the importance of collaboration networks in contemporary higher education. The evolution of these networks over time suggests that trust-building is a long-term process requiring sustained commitment. The first two years of the program, focused on the political–contextual dimension, were essential for laying the groundwork of trust that enabled the later development of more ambitious initiatives.

4.3. Innovation in Educational Practice

The results related to innovation in educational practice deserve special attention. The creation of joint academic programs and the implementation of new teaching–learning methodologies have shown that the Metauniversity model can drive significant transformations in how higher education is delivered. The active participation of 9495 students in various programs suggests strong student acceptance of these new approaches.

The effective integration of digital technologies and participatory methodologies has been key to these achievements. However, it must be acknowledged that implementing these innovations required considerable effort in terms of faculty training and institutional adaptation. The Metauniversity's first two programs have already fostered innovation, especially in terms of didactic knowledge, focusing on two key areas: question-based learning methodologies and the incorporation of Artificial Intelligence in rural settings as an ambitious challenge in the LAC region.

4.4. Sustainability and Scalability of the Model

The long-term sustainability of the Metauniversity model emerges as a crucial topic for discussion. The results suggest that creating shared value between universities and companies can provide a solid foundation for both financial and operational sustainability. The active participation of 52 companies in the project indicates recognition of the value this kind of collaboration can generate for the productive sector.

The Metauniversity model represents a flexible and inclusive learning ecosystem that integrates emerging technologies and educational innovation approaches based on global collaboration and sustainable development. The conceptual framework of the Metauniversity is highly adaptable, which allows its effective extension to other fields and productive sectors, including industry, through a dual application: technological innovation and the structuring of collaborative networks. Multisectoral collaboration and the creation of shared value (WWP model) facilitates the integration of academia, business and civil society, allowing its possible application in any sector linked to sustainable development, as well as in areas such as health, education, technological innovation, energy and urban management, promoting comprehensive solutions with social, environmental and economic impact.

This scalability of the model to other sectors is based on its dual application from: (a) the transfer of immersive technologies for practical training and simulation of critical industrial processes; and (b) the use of a proven methodological framework (WWP) to build sustainable collaborative networks that generate shared economic value and practical solutions, essential for governance and innovation in the productive sector. However, the model's scalability presents significant challenges. The network's expansion from 27 to 46 universities during the project demonstrates that growth is possible, but it requires careful management of resources and institutional relationships. The results indicate that expansion must be step by step and accompanied by a proportional strengthening of coordination and support structures.

4.5. Implications for Sustainable Development

The contribution of a Metauniversity to sustainable development deserves special consideration. The results demonstrate that integrating the CSA-IRA Principles into academic and research activities can generate tangible impacts in the form of practical solutions to sustainability challenges. The production of 111 research documents and implementation of 33 applied projects illustrate the model's capacity to generate relevant and applicable knowledge.

The nine-year experience validates the use of the WWP model, structured in political-contextual, ethical-social and technical-business dimensions, as an effective framework to guide institutional transformation and the generation of knowledge oriented towards practical solutions. The development of applied projects and case studies shows the ability to produce concrete solutions to sustainability problems in real contexts.

A particularly revealing finding consistent with other research [49,50] is the critical role of social capital and trust for long-term success. The sustained construction of relationships of trust between institutions and productive sectors was a critical factor for the sustainability of the initiatives developed under the CSA-IRA principles.

However, it is important to recognize that measuring real impact in terms of sustainable development poses significant methodological challenges. The results from the project suggest the need to develop more robust frameworks to evaluate and document these impacts over the long term.

In summary, the Metauniversity, driven by the CSA-IRA principles from the WWP model, acts as an engine of sustainable development by fostering cross-sectoral collaboration and producing applicable solutions, although its long-term impact requires overcoming challenges in rigorous measurement and documentation.

5. Conclusions

This research has shown that the Metauniversity model represents a significant evolution in the conceptualization and operation of higher education, particularly in its ability to address sustainable development challenges. Through a detailed analysis of the FAO–UPM program (2016–2024) [57], this study provides substantial evidence of the viability and effectiveness of a more integrated and collaborative approach to higher education. Cazorla & De los Ríos [56] elaborate on this approach in a communication presented at a recent international congress.

The successful implementation of a network comprising 46 universities and 52 companies across 12 countries demonstrates that it is possible to overcome the traditional limitations of university structures and create more dynamic and adaptable systems. The WWP model, with its three fundamental dimensions—political–contextual, ethical–social, and technical–business—has provided an effective framework to guide this transformation.

A particularly significant finding has been the importance of the gradual construction of trust and social capital among participating institutions [57]. The program's progression through its three phases has shown that effective institutional transformation requires a careful balance between innovation and respect for academic traditions and values. The active participation of more than 11,299 individuals across various components of the program over nine years reflects the model's potential to generate meaningful impact within the academic community and beyond.

The integration of the CSA-IRA Principles [58] into the institutional fabric of participating universities represents a major step toward aligning higher education with the Sustainable Development Goals (SDGs).

The development of 18 applied research projects and 50 business-oriented case studies demonstrates that the model can produce not only theoretical knowledge but also practical solutions to real-world sustainability challenges. In conclusion, the Metauniversity exemplifies a scalable and replicable model that aligns higher education with the SDGs, particularly quality education (SDG 4), responsible production (SDG 12), and partnerships for the goals (SDG 17), offering a viable framework for institutional transformation aligned with sustainable development. Its success in building effective collaborative networks and generating tangible impacts suggests that this model could serve as a valuable response to the challenges facing higher education in the 21st century. The results documented in this study provide a solid foundation for future implementation and adaptation of the model in diverse global contexts.

Experience also shows that the Metauniversity faces significant limitations, starting with the challenge of assessing the long-term impact on a scarcely explored field of study such as the CSA-IRA Principles. At the operational level, the model also faces the lack of trained human resources in this field and who are the basis for the sustainability of the initiatives. The governance of the Metauniversidad network must also resolve outstanding issues of intellectual property of shared content and the transfer of international credits.

Future research directions focus on expanding the Metauniversity model and the framework of the CSA-IRA Principles in new global contexts. Over time, the metauniversity must develop more robust frameworks to assess and document the real impact and long-term sustainability of projects in new contexts. Another future direction is technological innovation, which includes research on the incorporation of Artificial Intelligence in training processes to improve the governance and operational efficiency of managers and professionals of institutions.

Author Contributions: Methodology, C.L.; Validation, A.C. (Adhemir Cáceres); Investigation, A.C. (Adolfo Cazorla); Writing—original draft, A.C. (Adolfo Cazorla). All authors have read and agreed to the published version of the manuscript.

Funding: This research received no external funding.

Institutional Review Board Statement: Not applicable.

Informed Consent Statement: Not applicable.

Data Availability Statement: The original contributions presented in this study are included in the article. Further inquiries can be directed to the corresponding author.

Acknowledgments: The authors gratefully acknowledge the support of the GESPLAN research group and the Metaniversity member´s, whose collaboration and contributions were essential to the development of this research.

Conflicts of Interest: The authors declare no conflicts of interest.

References

1. San Martín, F.; Cazorla, A. *Buen Gobierno en las Universidades de Investigación en un Entorno Digital Global*; Fondo Editorial de la UNMSM: Lima, Peru, 2021.
2. Costanza, R.; Kubiszewski, I.; Kompas, T.; Sutton, P.C. A global metauniversity to lead by design to a sustainable well-being future. *Front. Sustain.* **2021**, *2*, 653721. [CrossRef]
3. Vest, C.M. The Emerging Meta University. In *Part III Global Strategies for Emerging Universities*; 2008; Volume 217.
4. Couturier, L. *The American Research University from World War II to World Wide Web: Governments, the Private Sector, and the Emerging Meta-University*; The Review of Higher Education; University of California Press: Oakland, CA, USA, 2008; Volume 1, pp. 356–358. [CrossRef]
5. Manapbayeva, Z.; Daineko, Y. Metauniversity as a Tool for Achieving Multiple Sustainable Development Goals. In Proceedings of the 1st International Workshop on Application of Immersive Technology, Almaty, Kazakhstan, 5 March 2025.
6. Turzhanov, U.; Daineko, Y.; Tsoy, D.; Ipalakova, M. Metauniversity as a Tool for Inclusive Learning. *Procedia Comput. Sci.* **2024**, *251*, 550–554. [CrossRef]
7. Lozano, R.; Barreiro-Gen, M.; Lozano, F.J.; Sammalisto, K. Teaching sustainability in European higher education institutions: Assessing the connections between competences and pedagogical approaches. *Sustainability* **2019**, *11*, 1602. [CrossRef]
8. Wiek, A.; Withycombe, L.; Redman, C.L. Key competencies in sustainability: A reference framework for academic program development. *Sustain. Sci.* **2011**, *6*, 203–218. [CrossRef]
9. Brundiers, K.; Wiek, A.; Redman, C.L. Real-world learning opportunities in sustainability—Concept, competencies, and implementation. *Int. J. Sust. Higher Educ.* **2010**, *11*, 308–324. [CrossRef]
10. Requelme, N.; Afonso, A. The Principles for Responsible Investment in Agriculture (CFS-RAI) and SDG 2 and SDG 12 in Agricultural Policies: Case Study of Ecuador. *Sustainability* **2023**, *15*, 15985. [CrossRef]
11. Purvis, B.; Mao, Y.; Robinson, D. Three Pillars of Sustainability: In Search of Conceptual Origins. *Sustain. Sci.* **2019**, *14*, 681–695. [CrossRef]
12. Stojanović, I.; Ateljević, J.; Stević, R.S. Good governance as a tool of sustainable development. *Eur. J. Sustain. Dev.* **2016**, *5*, 558–573. [CrossRef]
13. De los Ríos-Carmenado, I.; Mereles, M.L.A.; Deza, X.N.D.L. Connecting Sustainable Rural Development Projects and Principles of Responsible Investment in Agriculture and Food Systems from WWP Model: Lessons from Case Studies in Seven Countries. *Preprints* **2025**, 2025090058. [CrossRef]
14. Fontana, A.; Velasquez-Fernandez, A.; Rodriguez-Vasquez, M.I.; Cuervo-Guerrero, G. Territorial Analysis Through the Integration of CFS-RAI Principles and the Working with People Model: An Application in the Andean Highlands of Peru. *Sustainability* **2025**, *17*, 1380. [CrossRef]
15. Acosta Mereles, M.L.; Mur Nuño, C.; Stratta Fernández, R.R.; Chenet, M.E. Good Practices of Food Banks in Spain: Contribution to Sustainable Development from the CFS-RAI Principles. *Sustainability* **2025**, *17*, 912. [CrossRef]
16. Cachipuendo, C.; Requelme, N.; Sandoval, C.; Afonso, A. Sustainable Rural Development Based on CFS-RAI Principles in the Production of Healthy Food: The Case of the Kayambi People (Ecuador). *Sustainability* **2025**, *17*, 2958. [CrossRef]
17. Regalado-López, J.; Maimone-Celorio, J.A.; Pérez-Ramírez, N. Strengthening of the Rural Community and Corn Food Chain Through the Application of the WWP Model and the Integration of CFS-RAI Principles in Puebla, México. *Sustainability* **2025**, *17*, 5442. [CrossRef]
18. Committee on World Food Security (CFS). *Principles for Responsible Investment in Agriculture and Food Systems*; FAO: Rome, Italy, 2014. Available online: https://www.fao.org/3/au866e/au866e.pdf (accessed on 26 August 2024).
19. Cazorla-Montero, A.; de los Rios-Carmenado, I.; Pasten, J.I. Sustainable development planning: Master's based on a project-based learning approach. *Sustainability* **2019**, *11*, 6384. [CrossRef]
20. Cazorla-Montero, A.; de los Ríos-Carmenado, I. From "Putting the Last First" to "Working with People" in Rural Development Planning: A Bibliometric Analysis of 50 Years of Research. *Sustainability* **2023**, *15*, 10117. [CrossRef]
21. Ávila Cerón, C.A.; de los Rios-Carmenado, I.; Martín Fernández, S. Illicit Crops Substitution and Rural Prosperity in Armed Conflict Areas: A Conceptual Proposal Based on the Working with People Model in Colombia. *Land Use Policy* **2018**, *72*, 201–214. [CrossRef]
22. Cazorla, A.; Negrillo, X.; Montalvo, V.; De Nicolas, V.L. Institutional Structuralism as a Process to Achieve Social Development: Aymara Women's Community Project Based on the Working with People Model in Peru. *J. Sociol. Soc. Welf.* **2018**, *45*, 55. [CrossRef]
23. Jiménez Aliaga, R.; de los Ríos-Carmenado, I.; San Martín Howard, F.; Calle Espinoza, S.; Huamán Cristóbal, A. Integration of the Principles of Responsible Investment in Agriculture and Food Systems CFS-RAI from the Local Action Groups: Towards a Model of Sustainable Rural Development in Jauja, Peru. *Sustainability* **2022**, *14*, 9663. [CrossRef]
24. De los Ríos, I.; Cadena, J.; Díaz, J.M. Creating local action groups for rural development in México: Methodological approach and lessons learned. *Agrociencia* **2011**, *45*, 815–829.

25. Jiménez Aliaga, R.; de los Ríos-Carmenado, I.; Huamán Cristóbal, A.E.; Aliaga Balbín, H.; Marroquín Heros, A.M. Competencies and capabilities for the management of sustainable rural development projects in the value chain: Perception from small and medium-sized business agents in Jauja, Peru. *Sustainability* **2023**, *15*, 15580. [CrossRef]
26. López, M.; Cazorla, A.; Panta, M.D.P. Rural Entrepreneurship Strategies: Empirical Experience in the Northern Sub-Plateau of Spain. *Sustainability* **2019**, *11*, 1243. [CrossRef]
27. Šūmane, S.; Kunda, I.; Knickel, K.; Strauss, A.; Tisenkopfs, T.; Rios, I.D.I.; Rivera, M.; Chebach, T.; Ashkenazy, A. Local and Farmers' Knowledge Matters! How Integrating Informal and Formal Knowledge Enhances Sustainable and Resilient Agriculture. *J. Rural Stud.* **2018**, *59*, 232–241. [CrossRef]
28. De los Ríos-Carmenado, I.; Sastre-Merino, S.; Díaz Lantada, A.; García-Martín, J.; Nole Correa, P.; Pérez-Martínez, J. Building world class universities through innovative teaching governance. *Stud. Educ. Eval.* **2021**, *70*, 1–13. [CrossRef]
29. Cazorla, A.; de los Ríos, I. *Jornadas de Diálogo Para la Inclusión de los Principios Para la Inversión Responsable en la Agricultura (IAR) y las Directrices Voluntarias de Gobernanza de la Tierra (DVGT) en el Ecosistema Universitario de Latinoamérica*; Conclusiones; Working paper 2018/1; Grupo GESPLAN, Universidad Politécnica de Madrid: Madrid, Spain, 2018. Available online: https://oa.upm.es/50898/ (accessed on 22 August 2025).
30. Cazorla, A.; de Los Ríos, I.; Salvo, M. Working with People (WWP) in Rural Development Projects: A Proposal from Social Learning. *Cuad. Desarro. Rural.* **2013**, *10*, 131–157.
31. UN General Assembly. *Transforming Our World: The 2030 Agenda for Sustainable Development*; A/RES/70/1, 21 October 2015; United Nations: New York, NY, USA, 2015. Available online: https://refworld.org/legal/resolution/unga/2015/en/111816 (accessed on 1 April 2024).
32. Kristjánsson, K. There is something about Aristotle: The pros and cons of Aristotelianism in contemporary moral education. *J. Educ.* **2014**, *48*, 48–69. [CrossRef]
33. De los Ríos, I.; Rodríguez, F.; Perez, C. Promoting professional Projects Management skills in Engineering Higher Education: Project-based learning (PBL) strategy. *Int. J. Eng. Educ.* **2015**, *31*, 184–198.
34. Castañeda, R.; Avila, C.A.; de los Ríos, I.; Domínguez, L.; Gomez, S. Implementing the voluntary guidelines on the responsible governance of tenure of land, fisheries and forests from the working with people model: Lessons from Colombia and Guate-mala. *J. Peasant. Stud.* **2022**, *50*, 1820–1851. [CrossRef]
35. Acosta, M.L.; de los Ríos, I.; Ávila, C.A.; Castañeda, R. Responsible investment in the processes of substitution of illicit crops from the "working with people" model: Case study Guaviare (Colombia). In Proceedings of the 27th International Congress on Project Management and Engineering, Donostia-San Sebastian, Spain, 10–13 July 2023. [CrossRef]
36. Rivera, M.; Knickel, K.; de los Rios, I.; Ashkenazy, A.; Pears, D.Q.; Chebach, T.; Šūmane, S. Rethinking the connections between agricultural change and rural prosperity: A discussion of insights derived from case studies in seven countries. *J. Rural. Stud.* **2018**, *59*, 242–251. [CrossRef]
37. De los Ríos-Carmenado, I.; Becerril-Hernandez, H.; Rivera, M. Ecological agriculture and its influence on rural prosperity: An agricultural company's vision (Murcia, Spain). *Agrociencia* **2016**, *50*, 375–389. Available online: https://oa.upm.es/44023/ (accessed on 22 August 2025).
38. Fontana, A.; de los Ríos Carmenado, I.; Villanueva-Penedo, J.; Ulloa-Salazar, J.; Santander-Peralta, D. Strategy for the Sustainability of a Food Production System for the Prosperity of Low-Income Populations in an Emerging Country: Twenty Years of Experience of the Peruvian Poultry Association. *Sustainability* **2018**, *10*, 4035. [CrossRef]
39. De Los Ríos, I.; Mur, C. Inclusive and transparent governance structures from IRA Principles: Case Fesbal "food bank". In Proceedings of the 27th International Congress on Project Management and Engineering, Donostia-San Sebastian, Spain, 10–13 July 2023.
40. Cazorla, A.; Fernández-Pello, C.; Fontana, A.; García, A.J.; De Giusti, M.R.; Gómez, F.; Herrán, J.; Mínguez, E.; Muñoz, J.P.; Nicolás, V.L.D.; et al. *La Universidad: Motor de Transformación de la Sociedad*; Editores Cazorla & Stratta: Madrid, Spain, 2017.
41. Altbach, P.; Salmi, J. *The Road to Academic Excellence: The Making of World-Class Research Universities*; The World Bank: Washington, DC, USA, 2011.
42. Blackmore, P.; Blackwell, R. Strategic leadership in academic development. *Stud. High. Educ.* **2006**, *31*, 373–387. [CrossRef]
43. Cazorla, A. *El papel de Los Profesores en la Universidades de Investigación: Relevancia en Los Ranking*; Ediciones UNICARIBE: Santo Domingo, Dominican Republic, 2023.
44. Aguillo, I.F.; Ortega, J.L.; Fernández, M. Webometric Ranking of World Universities: Introduction, Methodology, and Future Developments. *High. Educ. Europe* **2008**, *33*, 233–244. [CrossRef]
45. De Nicolas, V. Los rankings, un reflejo de la universidad en el mundo: Webometrics. In *La Universidad: Motor de Transformación de la Sociedad*; Grupo GESPLAN, Universidad Politécnica de Madrid: Madrid, Spain, 2017; pp. 32–51.
46. GESPLAN. Grupo de Investigation en Planification de Proyectos de Desarrollo y Gestion Sostenible. Available online: https://www.ruraldevelopment.es (accessed on 29 September 2025).
47. PrincipiosiraMetauniversidad. Available online: http://www.principiosirametauniversidad.com. (accessed on 10 October 2025).

48. Negrillo, X. Los Proyectos de Desarrollo Rural Como Proceso a Través de Sus Protagonistas: El Caso de la Coordinadora de Mujeres Aymaras (Puno, Perú). PhD. Thesis, Universidad Politécnica de Madrid, Madrid, Spain, 2018. [CrossRef]
49. Ferber, J.; Gutknecht, O. A meta-model for the analysis and design of organizations in multi-agent systems. Proceedings of International Conference on Multi Agent Systems (Cat. No. 98EX160), Paris, France, 3–7 July 1998; IEEE Computer Society: Los Alamitos, CA, USA; pp. 128–135. [CrossRef]
50. Farber, D.L.; Melton, D.; Alger, M. Building a Meta-University. In *The Bridge*; National Academy of Engineering: Washington, DC, USA, 2020; p. 16.
51. Cazorla, A. Metamodelo para la comercialización de un espacio. *Estud. Geográficos* **1988**, *49*, 57–58.
52. Didham, R.J.; Ofer-Manu, P. Social learning for sustainability: Advancing Community-Based inquiry and collaborative learning for sustainable lifestyles. In *Responsible Living: Conceps, Education and Future Perspectives*; Springer Nature: Berlin/Heidelberg, Germany, 2015; pp. 233–252. [CrossRef]
53. Mieg, H.A. *Inquiry-Based Learning-Undergraduate Research: The German Multidisciplinary Experience*; Springer Nature: Berlin/Heidelberg, Germany, 2019; p. 406.
54. De los Ríos, I.; Afonso, A.; Quintanero, S.; Ortega, R.; Nole, P.; Claudia, L.; Roncancio, M. FESBAL-UPM Food Bank Chair and the Service-Learning Projects from the Working with People perspective. In *International Conference on Economic Reseach-Theoretical, Empirical and Practical Approaches*; Springer Nature: Berlin/Heidelberg, Germany, 2021; pp. 79–90. [CrossRef]
55. Nole Correa, P.; Aliaga, G.; De Los Ríos, I. Service-Learning Projects for Responsible Consumption: 10 Years of Experience from the Food Bank Chair UPM. In Proceedings of the 27th International Congress on Project Management and Engineering, Donostia-San Sebastián, Spain, 10–13 July 2023.
56. Cazorla, A.; de los Ríos, I. Towards a Metauniversity for sustainable development: Responsible Investment in Agriculture and Food Systems. In Proceedings of the 28th International Congress on Project Management and Engineering, Jaén, Spain, 5–6 July 2024.
57. Afonso, A.; Cazorla, A.; Echevarria, E. *Avanzando en Los Principios de Inversión Responsable en Agricultura Y Sistemas Alimentarios. Seminarios Internacionales Para Capitalizar Experiencias de la Universidad Y La Empresa Agroalimentaria en Latinoamérica Y España. Working paper*; Gesplan Research Group: Madrid, Spain, 2021; pp. 43–44.
58. FAO. *La FAO y los Objetivos de Desarrollo Sostenible. Cumplir la Agenda 2030 Mediante el Empoderamiento de las Comunidades Locales.*; FAO: Rome, Italy, 2022. [CrossRef]

sustainability

MDPI

Article

Connecting Sustainable Rural Development Projects and the Principles of Responsible Investment in Agriculture and Food Systems from the WWP Model: Lessons from Case Studies Across Seven Countries

Ignacio de los Ríos-Carmenado *, María Leticia Acosta Mereles and Xavier Negrillo Deza

Planificación y Gestión Sostenible del Desarrollo Rural-Local (GESPLAN), Universidad Politécnica de Madrid, Avda Puerta de Hierro 2, 28040 Madrid, Spain; marialeticia.acosta@alumnos.upm.es (M.L.A.M.); xavier.negrillo@upm.es (X.N.D.)

* Correspondence: ignacio.delosrios@upm.es

Abstract

The international literature shows significant growth in relation to sustainable rural development in response to ongoing problems. The Principles for Responsible Investment in Agriculture and Food Systems (CFS-RAI) enable projects to be aligned with the Sustainable Development Goals (SDGs). In this article, we present an empirically grounded analysis of these RAI principles based on in-depth case studies in seven countries (Spain, Ecuador, Peru, Dominican Republic, Bolivia, Colombia, and Mexico). This experience comes from an international project coordinated by the GESPLAN Research Group at the Polytechnic University of Madrid. The Working with People model is incorporated into the methodological process to analyze rural actors' understanding of the CFS-RAI principles in different countries and in university–business relation contexts. The results show the effectiveness of the WWP model based on the integration of three dimensions—ethical–social, technical–business, and political–contextual—as an effective method for planning sustainable rural development projects in various contexts. The empirical evidence presented indicates that combining the WWP model with the principles of CFS-RAI in rural contexts allows progress toward sustainable development, balancing economic aspects with human, social, and environmental well-being.

Keywords: rural communities; working with people; project planning; sustainable development

1. Introduction

Sustainable development has been on the global agenda for decades and has experienced significant growth as a field of study, establishing itself as one of the most important areas of research. There are numerous challenges [1] to meeting the current needs without compromising future generations [2].

Sustainable development has evolved from economic development to a more holistic approach that seeks to balance social, economic, and environmental dimensions [3,4]. More recently, many have suggested integrating a fourth dimension into sustainable development: governance [5,6].

In this regard, for several years now, governance has become the basic paradigm for rural development programs, projects, policies, and strategies [7].

Rural areas play an important role in sustainable development [8] and, over time, have been exposed to multiple global crises [9], including the COVID-19 pandemic [10,11] which has exacerbated pre-existing problems such as poverty and food insecurity [12], as well as other phenomena such as international migration [13], the population shift from rural to urban areas [14,15], and land grabbing [16–19]. In the face of these challenges, achieving sustainability in rural regions takes on a key importance [9].

For this reason, international organizations have shown a growing interest in sustainable rural development, integrating human rights and recognizing the promotion of equitable and responsible progress [20,21]. The Committee on World Food Security (CFS), an intergovernmental body of the United Nations coordinated by the FAO, approved the Principles for Responsible Investment in Agriculture and Food Systems in 2014 [22]. Subsequently, the Sustainable Development Goals (SDGs) were approved by the United Nations General Assembly on 25 September 2015, as part of the 2030 Agenda for Sustainable Development [23]. Both instruments were created to guide and promote sustainable and responsible interventions that preserve the availability of resources for future generations [24,25].

Approved a year before the adoption of the SDGs, the CFS-RAI principles anticipate and reinforce many of the goals contained in the 2030 Agenda. However, little research has been conducted on how the CFS-RAI principles function as a practical tool for implementing the SDGs in the agricultural and food sectors and for guiding governments, businesses, and institutions in making responsible and sustainable decisions.

In 2016, the Polytechnic University of Madrid (UPM) and the Food and Agriculture Organization of the United Nations (FAO) entered into a collaboration agreement to implement CFS-RAI principles in specific territories through university–business relationships in the fields of agriculture, food, the environment, and climate change.

Within the framework of this agreement, coordinated by the GESPLAN Research Group at the UPM (https://www.ruraldevelopment.es), an RU-RAI Network has been created in recent years, currently comprising 49 universities and 50 companies from 13 countries in Latin America, the Caribbean, and Spain [26], with the aim of promoting the implementation of the CFS-RAI principles in teaching, research, and development cooperation, as well as through their application in collaborative sustainable and responsible investment initiatives [27].

Following this period of collaboration, this research analyzes the conceptual framework developed by the GESPLAN Research Group, which is recognized for its multidisciplinary approach and extensive experience in the planning and management of sustainable rural development projects [27].

1.1. *Rural Development Projects Based on the Working with People Model*

Sustainable rural development projects have evolved significantly over the years [28]. In the 1960s and 1970s, the traditional model prevailed, which viewed rural regions as incapable of developing on their own, with a strong emphasis on modernization and economic growth [29]. Given the failure of these models, starting in the 1980s and 1990s, a more people-centered approach emerged, which emphasized local participation, recognizing the importance of endogenous development and the creation of local organizational structures [30–32].

An important milestone in this evolution was the LEADER (Liaisons entre activités de Développement de l'Économie Rurale) approach in the European Union in the 1990s. This new perspective emerged in the context of transition, giving way to a postmodern vision of rural development based on a territorial approach, the creation of new participatory local government structures, and decentralized management [32].

https://doi.org/10.3390/su18031402

This approach also emphasizes social innovation, multilevel governance, and cooperation between local actors, enabling the promotion of sustainable projects that are more adapted to local contexts and the specific needs of each region, with significant impacts on social cohesion and territorial development [33].

Sustainable rural development recognizes people as the central axis of change processes, promoting their active participation in building resilient and equitable territories [34–36]. As an integral process, it involves the incorporation of social, cultural, and environmental values that strengthen community cohesion and contribute to improving the population's quality of life [37].

In this context, rural development projects have incorporated various strategies aimed at strengthening territorial sustainability. Among them are the value chain approaches [38], which promote economic sustainability through productive articulation, and perspectives of territorial resilience [39], which enhance the adaptive capacity of local systems to social and environmental changes.

The "Working with People" model [28,40] is presented as a model that transcends the traditional technical vision of rural development by emphasizing local participation, capacity building, and strengthened governance through alliances between actors in the territory.

The WWP model is primarily being applied within the framework of the European LEADER initiative in collaboration with Local Action Groups in response to a new experimental approach to rural development in the EU [31]. In addition to the specific features of the LEADER initiative, the WWP model incorporates elements of planning as social learning [41,42], the logic of participation [43–45], methodologies for formulating and evaluating rural development projects [31,46], and international skills for project management [32,47,48].

Numerous studies have shown that integrating project management skills based on the International Project Management Association (IPMA) standard is a significant factor in the sustainability of rural development projects in contexts other than those of the European Union, enabling the formation of Local Action Groups that can act effectively in diverse rural areas [49–51]. Over the years, the WWP model has been successfully implemented in rural development processes in countries across Europe [52–54] and Latin America, such as Mexico [32], Peru [55,56], Ecuador [57], Argentina [50], Colombia [58], and Chile [59], where it has contributed to improving rural prosperity. In its latest evolution, it has aligned with the CFS-RAI principles, reinforcing responsible investment practices and their contribution to the Sustainable Development Goals [49,60–63].

Projects planned and executed using the social learning approach based on the WWP model have therefore been found to be suitable for promoting ethical values of social responsibility and for developing the capacities of the affected population. By placing people at the center of development, they empower them, turning them into leaders of their own development, which contributes to strengthening the social structure and sustainability of rural development projects [64].

Rural development projects based on the WWP model, like rural development itself, are influenced by various factors [65] and integrate a holistic approach from its three dimensions. One of the critical factors is ethical governance, which establishes the necessary framework for planning, executing, and evaluating projects effectively, ensuring teamwork and trust in sustainability [66,67]. Similarly, the development of skills within work teams, and especially among project managers, is a decisive factor for sustainability and is key to the success of projects and rural development [49,50].

https://doi.org/10.3390/su18031402

1.2. Principles for Responsible Investment in Agriculture and Food Systems (CFS-RAI) of the Committee on World Food Security

The Principles for Responsible Investment in Agriculture and Food Systems (CFS-RAI) are presented as a guide to guide projects toward the fulfillment of the Sustainable Development Goals (SDGs). They have been demonstrated to be a strategic framework for orienting investments toward the achievement of the SDGs in rural areas [62,63].

This link between the SDGs and the CFS-RAI is based on five interrelated pillars, known as the 5 Ps: People, Planet, Prosperity, Peace, and Partnerships [23,61]. This structure facilitates the organization and analysis of experiences based on the different dimensions that make up sustainability.

Achieving the SDGs requires concrete actions and responsible investments, particularly in rural areas affected by complex challenges such as climate change, energy, food security, migration, and poverty [68].

For this reason, the FAO led a global participatory process involving various stakeholders from five continents, with the aim of establishing a consensus-based set of principles to enable the achievement of sustainable development. As a result of this initiative, in October 2014, the ten principles of the CFS-RAI were approved [22], designed as voluntary guidelines, focusing on rural areas and promoting sustainable and responsible investment in agriculture and food systems [49,62,63,69].

The multi-stakeholder nature of these principles allows for the active participation of a wide range of actors, including governments, businesses, civil society organizations, universities, and producers, creating a collaborative environment that promotes consistency and effectiveness in interventions. The CFS-RAI principles establish a framework that guides projects toward comprehensive development, ensuring that they contribute effectively to the achievement of the SDGs.

In this process of disseminating, raising awareness of, and implementing the CFS-RAI Principles, since the agreement was signed between the UPM and the FAO in 2016, strategic actions have been coordinated to promote sustainable projects that integrate these principles under the leadership of GESPLAN. The FAO selected the GESPLAN Research Group at the Polytechnic University of Madrid (UPM) for its experience and planning of rural development and sustainable management projects based on the WWP model to promote teaching, research, and links with societies in Latin America, the Caribbean, and Spain based on these principles. Its extensive network of doctors specializing in Projects and Planning in Sustainable Rural Development, many of them graduates of the UPM, has enabled it to create an innovative approach to the development of rural areas.

In this context, various training initiatives have been developed and promoted by the RU-RAI Network, such as seminars and workshops, which have facilitated the training of various university professors, professionals in the agri-food sector, representatives of public and private organizations, and members of civil society. These activities have contributed to knowledge transfer, capacity building, and the creation of multi-stakeholder dialog spaces for the implementation of the CFS-RAI principles, creating a community committed to sustainable rural development based on these principles.

In this study, an empirical analysis of the implementation of CFS-RAI principles in 11 case studies from seven countries (Spain, Ecuador, Peru, Dominican Republic, Bolivia, Colombia, and Mexico) was conducted to present collaborative experiences of sustainable and responsible investments that are being continuously developed. The case studies and the criteria by which they were selected are presented in Section 2 (Table 1).

To apply and integrate the CFS-RAI principles into rural development projects, this study proposes a metamodel based on the WWP model. Eleven case studies conducted in different countries and contexts formed the basis for the comparative analysis that supports

https://doi.org/10.3390/su18031402

the proposal, providing a diverse set of empirical data and territorial contexts for its formulation. The metamodel integrates the principles as a guiding framework and links them with participatory governance processes, territorial planning, and capacity building, contributing to sustainability and improving decision-making in rural development projects.

Table 1. Case study selection.

Case Study	Selection Criteria [1]				Included
	C1	C2	C3	C4	
FESBAL-Food Banks (Spain)	✓	✓	✓	✓	Yes
Governance Model with CFS-RAI Principles in Social Organizations: Case Study of the Tejemujeres Women Artisans' Cooperative	✓	×	×	✓	No
Yesera Educational Basin (Bolivia)	✓	✓	✓	✓	Yes
Risaralda Specialty Coffee Cluster (Colombia)	✓	✓	✓	✓	Yes
Healthy Food: Cayambe (Ecuador)	✓	✓	✓	✓	Yes
Corn Agri-Food Chain, Puebla (Mexico)	✓	✓	✓	✓	Yes
ACRICUCEN Mantaro Valley (Peru)	✓	✓	✓	✓	Yes
CMA Aymara Women Artisans (Peru)	✓	✓	✓	✓	Yes
Territorial Development in Mountain Areas (Peru)	✓	✓	✓	✓	Yes
Bioregeneration of Tepetate Soils for Productive Restoration	✓	×	×	✓	No
Integration of CFS-RAI Principles in Postgraduate Programs (Peru)	✓	✓	✓	✓	Yes
Villa Poppy—Family Farming (República Dominicana)	✓	✓	✓	✓	Yes
University–Business Alliance UPM-UTESA–BANELINO–ISM (República Dominicana)	✓	✓	✓	✓	Yes
Cultivating the Future: Responsible Investment Strategies and Integrated Pest Management for Sustainable Agriculture	✓	×	×	✓	No
Healthy Nutrition through Hydroponic Cultivation	✓	×	×	✓	No

[1] Selection criteria: C1: Common conceptual framework (WWP model integrated with CFS-RAI principles); C2: Relevance and sufficiency of available information; C3: Maturity of the process to document implementation; C4: Generalizable learning contributing to other contexts and SDGs.

1.3. Research Questions

To guide the analysis of the application of the Principles for Responsible Investment in Agriculture and Food Systems (CFS-RAI) in sustainable rural development projects, the following research questions were posed:

- How do the CFS-RAI principles contribute to sustainable rural development in different international contexts?
- How can the CFS-RAI principles be integrated into the Working with People (WWP) model to contribute to sustainable rural development?
- How do university–business relationships influence rural actors' understanding and adoption of the CFS-RAI principles?

2. Materials and Methods

The methodological approach used is the WWP model [28,40], which has been applied to implement the CFS-RAI principles in various contexts, in collaboration with universities, businesses, and civil society.

https://doi.org/10.3390/su18031402

2.1. Working with People Model

The WWP model prioritizes the value of people in the planning of sustainable rural development projects and the connection between knowledge and action from its three dimensions: political–contextual, ethical–social, and technical–entrepreneurial [40].

This WWP model integrates the specific features of LEADER [31] and the development of project management skills [47,48], contributing significantly to the sustainable development of various rural areas [28,70].

The specific features of the LEADER approach comprise seven elements [31,71]: (a) a bottom-up approach; (b) a territorial approach; (c) a comprehensive and multisectoral approach; (d) innovation; (e) the creation of Local Action Groups (LAGs); (f) proximity management and financing; (g) learning in collaborative networks [31,72].

On the other hand, IPMA (International Project Management Association) competencies are structured in three dimensions: (a) personal competencies (People), related to behavior, social skills, and the ability to lead and work with others; (b) technical competencies (Practice) necessary for managing projects and programs; and (c) contextual competencies (Perspective), related to understanding the project environment and its alignment with the organization's strategic objectives [47].

The WWP model seeks to connect knowledge and action through its three dimensions or components as follows.

The political–contextual dimension encompasses the capacity of different actors to build relationships with political organizations and international, regional, and local public administrations in the context in which they operate. This dimension considers the territorial approach and proximity management of the LEADER approach for sustainable development [56] and the IPMA's perspective on competency. The implementation of CFS-RAI principles begins with this dimension, providing the essential framework for the political support, resources, and collaborative environment necessary for the successful implementation and long-term sustainability of projects.

The ethical–social dimension includes the behaviors, attitudes, and values of the people involved in actions at the personal and collective levels. It constitutes a system of social relations based on people's moral attitudes and behaviors that enable effective teamwork and cooperation among actors through commitment, trust, and personal freedom. It is related to the personal competencies area of the IPMA model and to the bottom-up approach and partnership building of the LEADER approach. The processes of awareness raising, participation, and empowerment of local communities for the integration of the CFS-RAI principles correspond to this dimension. The creation of the LAG partnership promotes coordination and strategic planning for the efficient management of development projects [62].

The technical–entrepreneurial dimension allows the different actors involved to be the main source of innovation and to create products and services for society in accordance with quality standards based on their technical skills. It focuses on promoting business strengthening and integrates the practical skills of the IPMA model for execution. It connects with the multisectoral and innovative approach of the LEADER approach to generate synergies and added value for the project. This technical–entrepreneurial dimension is also considered crucial for success and sustainability based on the CFS-RAI principles.

Finally, the social learning component provides the WWP model with an integrative component to ensure the space and processes for learning among the different subsystems, which leads to learning change agents.

Figure 1 summarizes these three dimensions of the WWP model with the specific features of LEADER [28,40] for the process of integrating CFS-RAI principles based on the

https://doi.org/10.3390/su18031402

WWP model and learning among multiple stakeholders from different areas of politics, public administration, business, universities, and civil society.

Figure 1. Synergies of the WWP model and the LEADER approach. Source: Adapted from [28].

Each actor, through their own behavior, attitudes, and values, interacts and contributes to the process of integration and learning, contributing, and receiving knowledge from technical, organizational, and political practices and changes [40].

2.2. Case Studies

This research is based on 11 case studies within the framework of the international FAO-UPM project coordinated by the GESPLAN Research Group at the UPM, in which the CFS-RAI principles from the WWP model were integrated. The cases were implemented in various contexts over the eight years of research by the international network. The criteria used to select the 11 case studies were as follows:

C1: Common conceptual framework: all case studies incorporated the WWP model to implement the CFS-RAI principles by promoting university–business collaboration.

C2: Relevance and richness of available information: each case study provided important data that enabled us to answer the central research questions related to the application of the CFS-RAI principles in various contexts.

C3: Maturity: the selected cases were sufficiently developed and advanced to offer solid information with accessible, sufficient, and representative data on the implementation of the principles and model in different environments.

C4: Learning: the knowledge gained had to be generalizable, allowing for its application in other contexts and contributing to progress toward the SDGs.

These criteria ensured that the selected cases were representative and added value to the international study.

A total of 15 case studies were identified through the RU–RAI international network. The selection was carried out by the research team applying the criteria described above. After the evaluation, four case studies were excluded for not meeting Criterion 2 (relevance

https://doi.org/10.3390/su18031402

and sufficiency of information) and Criterion 3 (process maturity). Consequently, 11 case studies were included in the final analysis.

Table 1 presents the evaluation results for each case study, showing the compliance with the criteria and the final decision of inclusion or exclusion.

Table 2 provides an overview of the selected case studies (Figure 2).

Figure 2. The body of the figure identifies the territorial location of the 11 case studies on the FAO map of Latin America and the Caribbean Member States UN Geospatial information. Source: Prepared by the authors, based on the FAO map [73].

Table 2. Overview of the 11 case studies.

Country	Case Study/Key Topics	Source(s)
Spain	FESBAL-Food Banks: best practices in reducing food waste and responsible consumption.	[61]
Bolivia (Tarija, Cercado)	Yesera Educational Basin: Sustainable water resource management by local communities, linking research and local knowledge.	[74]
Colombia (Risaralda)	Risaralda Specialty Coffee Cluster: Optimizing the cluster's logistics, strengthening business capabilities.	[75]

https://doi.org/10.3390/su18031402

Table 2. *Cont.*

Country	Case Study/Key Topics	Source(s)
Ecuador (Cayambe, Pichincha)	FCCC Healthy Food Cayambe: Cooperation and microcredit for rural women for sustainable food production and marketing.	[76]
Mexico (Puebla)	Corn Agri-Food Chain: Comprehensive improvement of the corn agri-food chain in rural communities, strengthening partnerships for the application of CFS-RAI principles.	[77]
Peru (Jauja, Junín)	ACRICUCEN Mantaro Valley: Formation of the GAL partnership for sustainable practices, innovation, and improvement of the guinea pig value chain.	[49,62]
Peru (Puno)	CMA—Aymara Women Artisans: skills development and production and marketing of sustainable artisanal textiles.	[64,78]
Peru (Paucar del Sara Sara, Ayacucho)	Territorial development in mountain areas: Territorial model for optimizing resource allocation for sustainable development in mountain areas.	[60]
Peru (three ecosystems)	Integration of CFS-RAI Principles into Postgraduate Programs in three ecosystems in Peru: the coast, the Andes, and the Amazon rainforest.	[74]
Dominican Republic (Constanza, La Vega)	Villa Poppy–Family Farming: A Family Farming System in accordance with CFS-RAI principles: sustainable practices and partnerships with tourism.	[69]
Dominican Republic (Santiago Rodríguez, Valverde, Dajabón)	University–Business Alliance UTESA, BANELINO, ISM: for sustainable territorial development through CFS-RAI principles: family social responsibility program with entrepreneurship.	[79]

2.3. Data Collection: Instruments and Processes

Mixed methods, combining quantitative and qualitative data, has been used to provide a comprehensive view of the analyzed cases [80]. To this end, primary data obtained through surveys, conferences, and workshops conducted by project teams, local coordinators of the RU-RAI Network, and FAO agents have been used. Secondary sources, such as scientific articles and project monitoring reports, have also been used.

2.3.1. Participatory Process: Conferences, Workshops, and Seminars

During the eight years that the RU-RAI Network has been in operation, conferences and workshops have been held with local actors, academics, organizations, and communities linked to the projects. These participatory spaces have made it possible to capture the perceptions, experiences, and proposals of participants regarding the implementation of the CFS-RAI principles and their impact on sustainable rural development processes. These have reached more than 11,257 direct participants, including 441 entrepreneurs, 1205 teachers and business executives, 218 representatives of public administration bodies, 8868 undergraduate students, and 559 graduate students [26]. Table 3 shows a summary of these key events for promoting the CFS-RAI principles within the framework of the UPM-FAO agreement.

https://doi.org/10.3390/su18031402

Table 3. Conferences and workshops to promote the CFS-RAI principles.

Year	Event	Location
2016	Good Business Practices from the CFS-RAI.	Universidad Politécnica de Madrid, Madrid
2017	The CFS-RAI Principles as a Driver for Sustainable Economic Development: New University–Business Partnerships.	Escuela de Alta Dirección de la Universidad de Piura, Peru
2018	I CFS-RAI Network Meeting for the dissemination of CFS-RAI.	Universidad Nacional Mayor de San Marcos, Peru
2019	Consolidation of CFS-RAI commitments in university curriculum.	Universidad Politécnica de Madrid, Madrid
2020	Towards consistent business conduct based on the CFS-RAI Principles.	Virtual; 11 countries in the network
2021	I International academic/business program: "The principles of responsible investment in agriculture and food systems."	Pontificia Universidad Católica Madre and Maestra, Dominican Republic
2022	II Meeting of the Network of Universities Committed to the Dissemination of the CFS-RAI Principles, RU-RAI.	Pontificia Universidad Católica Madre and Maestra, Dominican Republic
2023	Seminar of the RU-RAI Network of Universities and Companies.	Universidad Nacional de la Plata (UNLP) La Plata, Argentina
2024	III Meeting Towards a Meta-University for Sustainable Development: Responsible Investment in Agriculture and Food Systems.	UNICARIBE—Santo Domingo, República Dominicana
2024	III International Business Program on CFS-RAI Principles in collaboration with FAO.	Virtual
2025	Meeting: Keys to Good Governance in Agri-Food Cooperatives in Latin America and the Caribbean: Principles of Responsible Investment in Agriculture and Food Systems.	UNICARIBE—Santo Domingo, República Dominicana

2.3.2. Survey of Project Leaders

A survey was designed and implemented in 2024 to collect qualitative and quantitative data on the progress of the projects. The Survey123-ESRI (Environmental Systems Research Institute) tool in ArcGIS Online software (version 3.20.69) was used. This geocoded platform facilitated the collection, analysis, and generation of data reports [81].

The survey was organized into four thematic blocks:

1. Project characterization: identification of stakeholders, collaborating entities, and beneficiaries.
2. Implemented activities: description of actions carried out to apply the CFS-RAI principles.
3. Results and difficulties: main achievements and barriers encountered during implementation.
4. Assessment of actions and contribution to sustainable development: a set of items using a five-point Likert scale 1 (very low) to 5 (very high), that measured the level of implementation of the CFS-RAI principles based on actions grouped into the three dimensions of the WWP model (political–contextual, ethical–social, and technical–entrepreneurial), as well as their perceived contribution to achieving the Sustainable Development Goals in the territory.

Participants were selected through purposive sampling, inviting three profiles for each of the 11 case studies: the academic coordinator, the project manager or operational

https://doi.org/10.3390/su18031402

lead, and an additional key informant (e.g., community leader or representative of an associated organization). This approach ensured triangulation of perspectives and sectoral and territorial representation of the projects analyzed. In total, 33 invitations were sent, and 26 complete responses were obtained, corresponding to a response rate of 78.8%.

The survey was conducted between January and May 2024. The data collected were complemented with scientific publications, technical reports, and records from seminars and workshops in which the project teams presented and discussed their progress, thereby strengthening the internal validity of the study through source triangulation.

Prior to participation, all respondents were informed about the objectives of the research and provided their informed consent. Data collection complied with the ethical guidelines of the Universidad Politécnica de Madrid (UPM) and the FAO.

2.4. Data Analysis

Data analysis integrated quantitative and qualitative information from the results [82–85], providing a solid analytical framework for testing the validity of the findings. The survey data was downloaded from ArcGIS Survey123 in shapefile format and analyzed using Microsoft Excel for Microsoft 365. The information from the conferences and workshops was analyzed using the common methodological framework of the WWP model, grouping the results according to the three dimensions (political–contextual, ethical–social, and technical–entrepreneurial) using the CFS-RAI principles [40,62], as shown in Table 4.

The analysis was based on the 11 case studies. From the conceptual framework of the WWP model, the actions were defined, classified into the three mentioned dimensions, and analyzed in relation to the degree of implementation of the ten CFS-RAI principles using a 1-to-5 scale [86]. From these evaluations, proportional values were calculated, expressed as the relative weight of each dimension across the set of cases and the weight of each action within its dimension, always using the global sum of all valued actions (as the denominator. In this way, the tables present the relative importance of dimensions and actions in the implementation of the CFS-RAI principles.

Table 4. Relationship between the CFS-RAI principles and the dimensions of the WWP model.

CFS-RAI Principles	WWP Model Dimension [1]		
	P-C	E-S	T-E
P1. Contribute to food security and nutrition		X	
P2. Contribute to economic development and poverty eradication			X
P3. Promote gender equality and women's empowerment		X	
P4. Enhance the participation and empowerment of young people		X	
P5. Respect the tenure of land, fisheries, forests, and access to water	X		
P6. Conserve and sustainably manage natural resources, increase resilience, and reduce disaster risks	X		
P7. Respect cultural heritage and traditional knowledge, and support diversity and innovation		X	
P8. Promote safe and healthy agricultural and food systems			X
P9. Incorporate inclusive and transparent governance structures, processes, and grievance mechanisms	X		
P10. Evaluate and address impacts and promote accountability	X		

[1] WWP Model Dimensions: P-C: political–contextual, E-S: ethical–social, T-E: technical–entrepreneurial. Source: [60].

https://doi.org/10.3390/su18031402

3. Results and Discussion

The results of 11 case studies across seven countries (Spain, Bolivia, Colombia, Ecuador, Mexico, Peru, and the Dominican Republic) are presented here, where the subsections correspond to the main research questions presented in the introduction.

3.1. Contribution of the CFS-RAI Principles to Sustainable Development in the Territories

The implementation of the CFS-RAI principles in different territories suggests a contribution to rural development and to progress toward the SDGs [49,60–63,76].

Analysis of the case studies shows how application of these principles based on the WWP model has enabled the development of strategies and actions and the consolidation of processes to promote responsible investments adapted to local contexts, ranging from the development of stakeholders' skills and the creation of university–business partnerships to the transformation and innovation of agri-food systems.

Table 5 shows how the CFS-RAI principles were implemented in each case study according to the initial agreements in each of the projects, and Table 6 provides the overall results according to the three dimensions (political–contextual, ethical–social, and technical–entrepreneurial) of the actions undertaken associated with the CFS-RAI principles.

Table 5. Integration of the CFS-RAI principles into case studies.

Case Study	CFS-RAI Principles [1]									
	P1	P2	P3	P4	P5	P6	P7	P8	P9	P10
FESBAL-Food Banks (Spain)	X	X	X	X		X	X		X	X
Yesera Educational Basin (Bolivia)	X	X	X	X	X	X	X	X		
Risaralda Specialty Coffee Cluster (Colombia)	X	X		X				X		
Healthy Food: Cayambe (Ecuador)	X	X	X	X	X	X	X	X	X	X
Corn Agri-Food Chain, Puebla (Mexico)	X	X	X	X		X	X	X	X	
ACRICUCEN Mantaro Valley (Peru)	X	X	X			X		X	X	X
CMA Aymara Women Artisans (Peru)	X	X	X			X	X		X	X
Territorial development in mountain areas (Peru)	X	X	X	X	X	X	X	X	X	X
Integration of CFS-RAI Principles in Postgraduate Programs (Peru)	X	X	X	X	X		X	X		
Villa Poppy—Family Farming (República Dominicana)	X	X	X	X	X	X	X	X	X	X
University–Business Alliance UPM-UTESA–BANELINO–ISM (República Dominicana)	X	X	X	X		X	X	X	X	X
General total	11	11	10	9	5	9	9	9	8	7

[1] CFS-RAI principles: P1. Contribute to food security and nutrition. P2. Contribute to sustainable economic development and poverty eradication. P3. Promote gender equality and the empowerment of women. P4. Promote participation and empower youth. P5. Respect the tenure of land, fisheries, forests, and access to water. P6. Conserve and sustainably manage natural resources, build resilience, and reduce disaster risks. P7. Respect cultural heritage and traditional knowledge and support diversity and innovation. P8. Promote safe and healthy food and agricultural systems. P9. Incorporate inclusive and transparent governance structures, processes, and grievance mechanisms, P10. assess and address impacts and promote accountability.

The results show a heterogeneous adoption of the principles in the projects developed, with a predominant trend towards P1, P2, and P3 (Table 5), indicating that the projects prioritize food security, sustainable economic development, and women's empowerment as key determinants. This orientation was chosen in response to research highlighting the role of women in good governance [55] and the critical structural conditions of rural areas in Latin America, marked by high levels of poverty, food insecurity, and inequality [87–89].

https://doi.org/10.3390/su18031402

The focus on these principles is relevant and consistent with the most urgent needs of the territory, in line with what has been pointed out by Woodhill et al. [90], who assert that the future well-being of people in rural areas depends on the transformation of food systems towards more equitable, sustainable, and resilient models.

Table 6. Integration of the CFS-RAI principles into case studies in relation to the dimensions of the WWP model.

WWP Model Dimension	CFS-RAI	Importance of Actions
Political–Contextual	P5, P6, P9, and P10	34.72%
Ethical–Social	P1, P3, P4, and P7	32.66%
Technical–Entrepreneurial	P2 and P8	32.62%
Total		100.00%

Similarly, the incorporation of principles P4, P6, P7, and P8 shows a shift toward comprehensive approaches to rural sustainability. Strengthening the role of young people reflects progress in building human capital and the challenge of generational renewal in the context of high socioeconomic vulnerability [91]. At the same time, there has been a gradual adoption of sustainable natural resource management practices, and the valorization of traditional knowledge has lent sociocultural legitimacy to development strategies, integrating local knowledge into technical and organizational processes [92]. In addition, the promotion of healthy and safe food systems is contributing to improved food security and community health in rural areas [93].

Principles P9 and P10, relating to governance, impact assessment, and accountability, are being implemented to a lesser extent, but some projects are beginning to institutionalize monitoring and transparency mechanisms. This process is essential for building trust, legitimacy, and sustainability in rural interventions.

Principle P5 (respect for land tenure, fisheries, forests, and water), while the least developed, is highly relevant in five cases. This low frequency can be explained by two factors: in some contexts, tenure rights are already formalized and therefore do not require direct intervention, while in others, the legal and political complexity of these issues limits community projects' capacity for action. Previous studies have pointed out that the effective implementation of principle 5 requires profound institutional reform, recognition of customary rights, and inclusive mechanisms of territorial governance [94]. These results should be considered in the context of global challenges such as population growth, increased demands for food, and pressure on natural resources, which require more sustainable and resilient strategies.

At the global level, the political–contextual dimension (P5, P6, P9, and P10) stands out (34.72%), followed by the ethical–social dimension (P1, P3, P4, and P7) (32.66%) and the technical–entrepreneurial dimension (P2 and P8) (32.62%), highlighting the need for the CFS-RAI principles to be implemented with a holistic and balanced approach regarding the three dimensions of the WWP model, as shown in Table 6.

All case studies illustrate processes that contribute to the sustainable development of territories, reflecting not only their technical aspects but also the fact that they serve as instruments for delivering multisectoral initiatives in economic, social, environmental, and governance areas [49,60–62,76].

Figure 3 shows the results of local actors and leaders' assessments of the contribution of these principles to sustainable development, with a significant majority rating their contribution to implementation as high (level 5).

https://doi.org/10.3390/su18031402

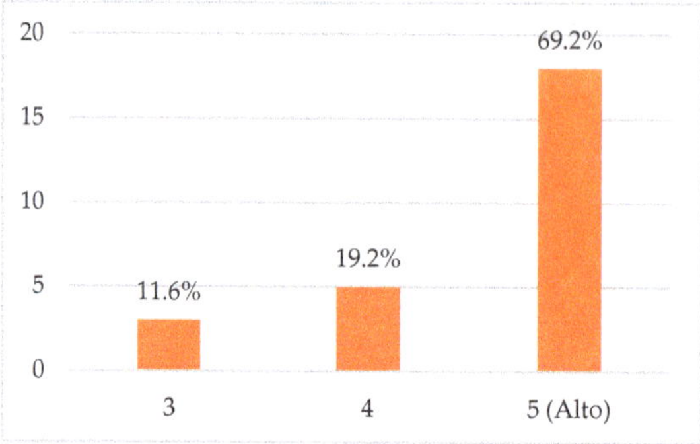

Figure 3. Perception of the contribution of the CFS-RAI principles to sustainable development.

3.2. Actions from the Political–Contextual Dimension: P5, P6, P9, and P10

Actions from the political–contextual dimension accounted for 34.72% of all those implemented, making it the most prominent dimension among the three dimensions of the WWP model. This dimension emphasizes the importance of understanding and managing the political, regulatory, and contextual structures in which rural development projects are carried out. This dimension also incorporates the perspective competencies defined by the IPMA standard. Actions in this dimension are therefore essential for interventions to be adapted to local realities, with participatory governance structures, greater coordination between actors, and the capacity to respond to the global challenges of sustainable rural development (social, environmental, economic, and political).

Table 7 shows the actions carried out from this political–contextual dimension during rural development projects as a central axis for initiating processes and achieving the sustainability of rural development [28]. Its relevance is evident in the need to form organizational structures to establish solid relationships both internally—among community actors—and externally through partnerships with public, private, academic, and international institutions [95,96].

Table 7. Main actions from the political–contextual dimension.

Political–Contextual Actions	34.72%
University–business agreements	6.42%
Partnerships for cooperation between agents	6.25%
Creation of new local–regional governance structures	6.04%
Legal formalization of associations and cooperatives	5.80%
Participation in national and international networks	5.18%
Monitoring, evaluation, and accountability system	5.03%

The results highlight the central role that governance and institutional coordination processes have come to play in rural areas. Among the most important activities are university–business agreements (6.42%), partnerships for cooperation between agents (6.25%), and the creation of new local–regional governance structures (6.04%), which reflects a sustained focus on building collaborative frameworks between various actors, as shown in Table 8 below.

University–business agreements (6.42%) are present in all case studies, playing a key role in linking technical and academic knowledge with the socioeconomic dynamics of the territory, promoting networks with management, innovation, and sustainability capabilities. This link has been consolidated through the exchange of expert and experiential knowledge, as a social learning process characteristic of the WWP model [28,40]. Various researchers

https://doi.org/10.3390/su18031402

agree that universities act as facilitators of sustainable development in rural contexts by actively engaging with local actors [97], promoting co-creation and applied research processes that address real needs [98,99]. In the context of the FAO-UPM project and the CFS-RAI Network [26], university–business agreements are a key strategic component for the implementation of CFS-RAI principles through various engagement processes.

Table 8. Main actors in the case studies.

Case Study	Business/Association	University
FESBAL-Food Banks (Spain).	Spanish Federation of Food Banks (FESBAL), 54 Food Banks.	Universidad Politécnica de Madrid, UPM.
Yesera Educational Basin (Bolivia).	Associations of women producers of strawberries, peaches, and fish farming.	Universidad Autónoma Juan Misael Saracho.
Risaralda Specialty Coffee Cluster (Colombia).	Cluster of 46 companies, 13 farmers' associations.	Universidad Católica de Pereira, Universidad Tecnológica de Pereira.
Healthy Food Cayambe People (Ecuador).	Cayambe Peasant House Foundation Association of Agroecological Producers. International Cooperation for Development (ICD).	Universidad Politécnica Salesiana, UPS.
Corn Agri-Food Chain Puebla (Mexico).	Guardians of Calpan Knowledge and Flavors Cooperative, Local Action Group.	Colegio de Postgraduados, COLPOS.
ACRICUCEN Mantaro Valley (Peru).	Local Action Group. Association of producers, agricultural suppliers, and collectors.	Universidad Nacional Mayor de San Marcos, UNMSM.
CMA Aymara Women Artisans (Peru).	Coordinator of Aymara Women, NGO Design for Development.	Universidad Politécnica de Madrid, UPM.
Territorial development in mountain areas (Peru).	Producers' associations, mining companies in Ayacucho and Paucar del Sara Sara.	Universidad de Piura.
Integration of CFS-RAI Principles in Postgraduate Programs (Peru).	APECAM, Association of Organic Coffee Producers of Alto Mayo. Inter-American Development Bank.	Universidad Católica Sedes Sapientiae, UCSS.
Villa Poppy—Family Farming (Dominican Republic).	COOPROVIPO, Neighborhood Council, FAO, ASONAHORES (National Association of Hotels and Restaurants), Grupo Raya, SuperFrech.	Universidad Politécnica de Madrid, UPM.
University–Business Alliance UTESA–BANELINO–ISM (Dominican Republic).	Industrias San Miguel (ISM) Northwest Organic Banana Association (BANELINO).	Universidad Tecnológica de Santiago, Universidad Politécnica de Madrid.

Partnerships and cooperation with other actors (6.25%) are also a key activity in all implemented projects, playing a crucial role in achieving sustainable development by enabling joint learning and synergies between endogenous and exogenous knowledge and in optimizing resources. Collaboration with governments, business entities, non-governmental organizations (NGOs), and community groups fosters innovation and the alignment of common goals, which, in turn, leads to more comprehensive and resilient solutions [100]. In line with this perspective, the SDGs emphasize the importance of establishing effective public–private partnerships, maximizing the acquisition of experience and the use of resources [23].

These collaborations have also made it possible to create and consolidate local governance structures (6.04%), enabling the creation of innovative interventions consistent with territorial priorities. These types of structures are facilitating the generation of trust and

https://doi.org/10.3390/su18031402

the empowerment of actors, especially women, in line with the notion of "institutional structuralism" proposed by Midgley [101], where multiple actors actively participate in planning and implementing local developments [49,60,62,74,76,77].

As other research shows [64,102], the WWP model reinforces and highlights the need for institutional structuralism, with solid organizations for local governance that guarantee sustainable development processes, especially in contexts where the state does not play an active role. In this sense, universities have played a key role as articulators of technical, scientific, and social knowledge in the different case studies, promoting this institutional structuralism and local governance for participatory planning processes, applied research, local capacity building, and technical support for producers, women, youth, and community associations. Through international cooperation networks and university–business links, they have promoted the transformation of agri-food systems towards more sustainable models, consolidating themselves as key agents in rural development [49,60,74,76,77].

Along with the above actions, associations and cooperatives (5.80%) have been legally formalized to strengthen the institutional legitimacy of producer organizations [62,64,77] as a determining factor for their entry into more demanding markets.

Finally, within this political–contextual dimension, the implementation of impact monitoring and evaluation mechanisms (5.03%) and participation in national and international networks (5.18%) represent relevant aspects for the integration of the CFS-RAI principles and project sustainability. The evaluation of results and accountability is closely linked to building trust among actors, which is a fundamental aspect for the establishment of good governance [103].

In this dimension, the project of the Catholic University Sedes Sapientiae (Integration of CFS-RAI Principles in Postgraduate Programs in Peru) stands out. Through training programs, it has incorporated an innovative territorial approach in relation to principle 5 (respect land, fishing, and forest tenure, and access to water) and principle 6 (conserve and sustainably manage natural resources and increase resilience and reduce disaster risks) through rural development actions and by contributing to strengthening the professional capacities aimed at responsible rural development management in three natural regions of Peru (coast, highlands, and jungle). Another case that highlights the high relevance of principles 5 and 6 in the Peruvian context is territorial development in mountain areas [60], which reflects principle 5 by prioritizing land and water tenure and access as key elements of territorial planning, an essential condition for reducing inequalities in rural areas. At the same time, the sustainable and long-term use of these resources in local development plans is encouraged, territorial management is linked with the social and economic stability of communities, and the importance of principles 5 and 6 as a basis for responsible and sustainable investment in highly vulnerable territories is reinforced.

Other relevant principles in this dimension are P9 (incorporate inclusive and transparent governance structures, processes, and grievance mechanisms) and P10 (evaluate and address impacts and promote accountability), which highlight the experiences of FESBAL-Food Banks (Spain), Corn agri-food chain Puebla (Mexico), and ACRICUCEN Mantaro Valley (Peru), projects that promote transparent and inclusive governance structures.

In the ACRICUCEN Mantaro Valley project (Peru), the implementation of principles 9 and 10 is evident in the creation of an LAG (Local Action Group), a governance space that brings together actors from different levels and sectors through participatory processes. This mechanism has not only improved inclusion in decision-making but also generated trust and social cohesion among producers, institutions, and local governments. Openness to participation and the establishment of strategic alliances have strengthened organizational capacity and commercial innovation, factors that contribute to increased incomes and a better quality of life in rural areas [49,62]. Similarly, Corn agri-food chain

https://doi.org/10.3390/su18031402

Puebla (Mexico), through its LAG, brings together a variety of actors, including producers, public institutions, and academics, who interact through participatory and transparent mechanisms [77]. This formalization of organizations, in addition to strengthening the management of local structures, allows access to new resources, generates greater transparency and legal stability, and promotes participation in production chains [104].

Another innovative case is that of FESBAL-Food Banks (Spain), which stands out for its ability to coordinate with different sectors focused on reducing waste and ensuring food security. In this context, the implementation of principles 9 and 10 of the CFS-RAI principles is reflected in the consolidation of solid, transparent, and inclusive governance structures, positioning them as benchmarks for governance in the third sector [61].

3.3. Actions from the Ethical–Social Dimension: P1, P3, P4, and P7

Actions in the ethical–social dimension represent 32.66% of all actions implemented, making it the second most important of the three dimensions of the WWP. This dimension focuses on building values, attitudes, and behaviors that guide people's participation in development processes and enable the construction of relationships based on trust, cooperation, and commitment. The ethical–social dimension, through its various actions, is vital because it provides the basis for trust, shared values, and the collaborative relationships necessary to effectively implement projects aligned with CFS-RAI principles, contributing to sustainable development from a human perspective [28].

This dimension relates to the personal competencies of the IPMA standard—leadership, communication, reliability, conflict management, and results orientation—which are key to facilitating participatory processes and empowering the population. The bottom-up approach of the WWP model, aligned with the LEADER approach, reinforces this ethical–social framework by placing communities at the forefront when identifying needs, making decisions, and in rural development planning. This participatory logic has made it possible to establish spaces for dialog between diverse actors—producers, universities, local governments, and businesses—creating environments of mutual trust.

The total number of actions implemented in the case studies reflects their relevance to the rural development processes analyzed, recognizing the need for training to strengthen the social fabric, ethical foundations, and active involvement of communities as necessary conditions for advancing toward the integration of CFS-RAI principles and sustainable development [105,106].

The main strategies promoted in this area are detailed in Table 9, highlighting those aimed at developing skills and capacities, awareness-raising, and training processes as key axes for promoting transformative and sustainable processes in the territories.

Table 9. Actions from the ethical–social dimension.

Ethical–Social Actions	32.66%
Skills and capacity building	7.05%
Awareness-raising and training processes	6.90%
Mechanisms for active and inclusive participation	6.53%
Promotion of cultural identity and traditional knowledge	6.35%
Partnership building/LAGs	5.83%

Skill development, awareness-raising, and training actions are carried out in the 11 case studies. However, these are not merely support activities but are understood as fundamental and necessary components for the effective implementation of the CFS-RAI principles and the achievement of sustainable development, especially through the empowerment of communities and the construction of ethical and collaborative systems.

https://doi.org/10.3390/su18031402

Both actions are essential for building bridges between "knowledge and action" and are deeply intertwined.

Skills and abilities (7.05%) have been developed in all projects through three international training programs coordinated by the UPM and supervised by the FAO. These programs were based on IPMA standards, focusing on the development of skills in the following different dimensions: practice (project management elements), perspective (contextual understanding), and person (personal and interpersonal skills, including ethical, emotional, and social aspects). These personal competencies are considered essential for success in managing responsible investment projects, in organizational strengthening, and in community leadership [49,62,64,74,76].

Awareness-raising and training processes (6.90%) include educational campaigns, training workshops, and public awareness initiatives. The FESBAL–UPM Food Bank Chair project (Spain) stands out as a far-reaching experience in raising awareness and promoting ethical values aimed at reducing food waste and strengthening solidarity towards vulnerable populations. This experience includes volunteer network and service-learning projects from the university, aligned with the CFS-RAI principles. P1 contributes to food security and nutrition [61], for which the Healthy Food: Cayambe People project stands out, where P1 constitutes the ethical basis, establishing the right to food and food security as central pillars of community management [76].

Mechanisms for active and effective participation (6.53%) stood out for their practices linked to principle 3 (promote gender equality and women's empowerment), such as those in the Yesera Educational Basin project in Bolivia, which promotes the inclusion of women through their active participation in water management committees and technical training that combines local knowledge and research and their involvement in dialog spaces for the development of a Local Watershed Management Plan. These mechanisms not only recognize their strategic role in water conservation but also strengthen their leadership [74].

Other cases stand out for their practices related to principle 4 (enhance the participation and empowerment of young people), such as FESBAL-Food Banks, where youth volunteering has become a central mechanism for inclusion and learning. Through their active participation in the collection, sorting, and distribution of food, young people have not only contributed to reducing waste but also acquired key skills in leadership, organizational management, and teamwork. This process reinforced values of solidarity and social commitment [61,107].

The promotion of cultural identity and traditional knowledge (6.35%) is visible in the Healthy Food: Cayambe People, CMA Aymaras, and Yesera Educational Basin (Bolivia) projects, where traditional agroecological and textile practices were recovered, revaluing ancestral knowledge as an essential part of their territorial development, as also recognized in principle 7 of the CFS-RAI principles. The CMA Aymara Women Artisans project, for example, focused on strengthening female leadership, with a comprehensive vision that included economic autonomy, the valuation of traditional knowledge, and the construction of networks of trust among participants, allowing for the recovery of weaving as ancestral knowledge and a means of cultural expression. In addition to articulating this knowledge with market colors and trends, new products have been developed for the global market using local production techniques [64].

Finally, the creation of partnerships and Local Action Groups (5.83%) was key in the Corn agri-food chain Puebla (Mexico), ACRICUCEN Mantaro Valley (Peru), Villa Poppy—Family Farming (Dominican Republic) projects, where multi-stakeholder collaboration structures were created to facilitate joint work between producers, institutions, universities, and local governments. These partnerships made it possible to articulate

https://doi.org/10.3390/su18031402

long-term strategies and generate synergies for the inclusive and sustainable development of the territory.

From a theoretical perspective, this approach coincides with that proposed by Friedman [108], who emphasizes that planning processes must be based on personal relationships and shared values that reflect the lived reality of communities. Likewise, various authors highlight that active participation and a sense of belonging strengthen the legitimacy and sustainability of decisions collectively made [109–111].

Empowerment in rural development is a fundamental process that enables rural communities to actively participate, make decisions, negotiate, influence, and control key aspects of their environment [112].

All the case studies studied in this dimension strengthened social capital and community cohesion, and social empowerment and active participation were promoted. Women's entrepreneurship in rural areas is key to sustainable development and social inclusion. Through leading business initiatives, women not only empower themselves but also contribute to reducing poverty, improving family income, and decreasing inequalities, thereby transforming rural communities [113,114].

3.4. *Actions from the Technical–Entrepreneurial Dimension: P2 and P8*

The main objective of sustainable rural development is to improve the quality of life of rural populations [37]. The technical–entrepreneurial dimension, which accounts for 32.62% of the total actions evaluated in the case studies, comprises a set of activities aimed at strengthening productive capacities, incorporating appropriate technologies, improving the quality of products and processes, and the development of technical–entrepreneurial strategies in line with the conditions of rural areas. It is also linked to IPMA practical skills that enable efficient project implementation. Table 10 shows the main strategies and actions developed in the case studies: transformation and innovation to generate value for products; marketing and access to differentiated markets; logistics management and optimization of production systems; standardized production systems and product quality; product certification; and digitization and improvement of production processes.

Table 10. Actions from the technical–entrepreneurial dimension.

Technical–Entrepreneurial Actions	32.62%
Transformation and innovation to generate value for products	6.58%
Marketing and access to differentiated markets	6.51%
Logistics management and optimization of production systems	6.46%
Standardized production systems, product quality, and certification	6.54%
Digitization and improvement of production processes	6.53%

The most highly valued action was transformation and innovation to generate value for products (6.58%), which suggests the momentum behind strategies aimed at developing new products, diversifying supply, and increasing added value in agri-food chains. Secondly, marketing and access to differentiated markets (6.51%) is a priority in these projects, promoting the competitive positioning of local products in higher value segments and favoring the insertion of small producers into sustainable marketing circuits.

Logistics management and production system optimization (6.46%) highlight the interest in improving operational efficiency and coordination at different stages of the value chain, which helps reduce losses, costs, and distribution times. In turn, the implementation of standardized production systems and product certification (6.54%) shows progress toward compliance with quality, sustainability, and traceability standards, which are increasingly demanded by national and international markets. Finally, the digitization

https://doi.org/10.3390/su18031402

and improvement of production processes (6.53%) mark the beginning of a technological transition that incorporates digital tools for production, commercial, and organizational management, increasing the responsiveness and adaptability of rural actors to the current challenges of the agri-food system.

Based on the analysis of the case studies, projects linked to the implementation of the CFS-RAI principles are being developed in a wide variety of sectors, particularly in agriculture, agri-food, textiles, and academia. This sectoral diversity reflects the versatility of the CFS-RAI principles in different contexts, strengthening both productive innovation and learning and knowledge transfer processes in these territories.

The link between these technical processes and the CFS-RAI principles is clearly expressed. Principle 2, which promotes economic development and poverty eradication, is reflected in the creation of local capacities to sustain autonomous, market-oriented production processes. Principle 8, which advocates for safe and healthy agri-food systems, is observed in the adoption of quality, traceability, food safety, and responsible marketing practices, applied at multiple stages of the technical–productive processes of the projects implemented.

Noteworthy projects in this area include the Risaralda Specialty Coffee Cluster (Colombia), Villa Poppy—Family Farming (Dominican Republic), and the University–Business Alliance UPM-UTESA–BANELINO–ISM (Dominican Republic).

In the case of the University–Business Alliance UPM-UTESA–BANELINO–ISM, the company BANELINO has obtained international certifications such as certification from the Fairtrade Labeling Organization (FLO International), among others, which endorse the social and environmental sustainability of its agricultural production. These achievements have been made possible thanks to continuous technical assistance processes in traceability, agronomic management, and compliance with quality standards. The adoption of these systems requires technical records, control protocols, and the training of producers in management and international marketing [79].

For its part, the Villa Poppy—Family Farming project (Dominican Republic) applied precision agricultural technologies such as weather stations, which have contributed to efficient crop planning. Improved agronomic practices and management tools such as inventory control, sales planning, and production systematization were introduced and integrated into a joint marketing strategy through a local cooperative [69].

In the Risaralda's Specialty coffee sector project, local capacity building strengthened coffee growers' autonomy and consolidated market-oriented production processes, contributing to income generation and rural poverty reduction. Likewise, the adoption of quality, traceability, and food safety practices ensured the production of differentiated and competitive coffee, while strengthening safe and sustainable agri-food systems for communities in the region [75].

To highlight the scope and processes developed by the projects analyzed in the technical–business dimension, Table 11 below shows the number of direct beneficiaries in each of the case studies. This information is key to understanding the scale of the initiatives, as well as their contribution to strengthening the productive fabric, boosting the economy, and improving living conditions in the rural areas involved.

Actions within the technical–entrepreneurial dimension were carried out within the framework of collaborative innovation processes, where local actors actively participate in the adaptation of technical solutions. This approach is in line with that proposed by Pandey et al. [115], who highlight the importance of cooperation for innovation and knowledge appropriation. In turn, the results coincide with the work of Knickel et al. [116] by considering rural innovation as a social process based on networks and collective learning. In addition, the projects featured both extension services and technical training,

https://doi.org/10.3390/su18031402

an aspect that Mapiye et al. [117] identify as key to facilitating the efficient management of productive systems and their sustainability in rural contexts.

Table 11. Number of direct beneficiaries per case study.

Case Study	Number of Beneficiaries
FESBAL-Food Banks (Spain)	1,187,976
Yesera Educational Basin (Bolivia)	589
Risaralda Specialty Coffee Cluster (Colombia)	19,163
Healthy Food Cayambe People (Ecuador)	129
Corn Agri-Food Chain Puebla (Mexico)	800
ACRICUCEN Mantaro Valley (Peru)	350
CMA Aymara Women Artisans (Peru)	327
Territorial development in mountain areas (Peru)	9609
CFS-RAI in Postgraduate Programs (Peru)	1406
Villa Poppy—Family Farming (Dominican Republic)	52
University–Business Alliance UTESA–BANELINO–ISM (Dominican Republic)	279

Note: Direct beneficiaries reported by the project managers.

3.5. Social Learning Through University–Business Relationships

The results obtained from applying the WWP model to rural development projects suggest its applicability and methodological soundness for addressing rural development project planning. It constitutes a novel methodological approach that allows the complexity of rural development to be addressed from a participatory perspective.

The WWP model seeks to promote a balance between its three dimensions: political–contextual, ethical–social, and technical–entrepreneurial. Figure 4 reflects the degree of balance achieved in the cases analyzed. The results show that, although there is variability between cases, in general, the three dimensions are balanced through processes of social learning, active participation, and institutional cooperation, which are key for the implementation of the CFS-RAI principles and the promotion of sustainable development.

The WWP model promotes sustainable rural development by integrating training in IPMA standard competencies, which enables the professionalization of project management in its key dimensions: technical, personal, and contextual. This capacity building contributes to consolidating local leadership, improving resource management efficiency, and creating conditions for organizational and productive autonomy. For its part, the LEADER approach provides a territorial, multisectoral, and participatory framework, notable for its bottom-up approach, which places local actors at the forefront of identifying problems and solutions. The complementarity between the WWP model, the IPMA competency approach, and the LEADER approach reinforces their use in the planning and management of projects aimed at sustainable rural development.

In addition, the Principles for Responsible Investment in Agriculture and Food Systems (CFS-RAI) were adopted as a guide to orient actions towards sustainable rural development. Principle 1, linked to food security, was operationalized through strategies that ensured stable and sufficient access to high-quality food. Principle 2, encompassing economic development and poverty eradication, was related to the strengthening of enterprises and value chains. Principles 3 and 4, on gender equality and youth empowerment, were integrated through mechanisms for participation and inclusion, while principle 5, considering respect for land tenure, fisheries, forests, and access to water, together with principle 6—which promotes the conservation and sustainable management of natural resources, increased resilience, and risk reduction—were addressed through the implementation of agroecological practices, efficient technologies, and local knowledge to conserve soils, manage water, and strengthen environmental resilience. Principle 7, relating to re-

https://doi.org/10.3390/su18031402

spect for culture and traditional knowledge, was incorporated through the valorization of local knowledge and territorial practices, and finally, principles 9 and 10, considering responsible governance and accountability, were integrated through participatory organizational structures, inter-institutional coordination, and transparent monitoring and evaluation mechanisms.

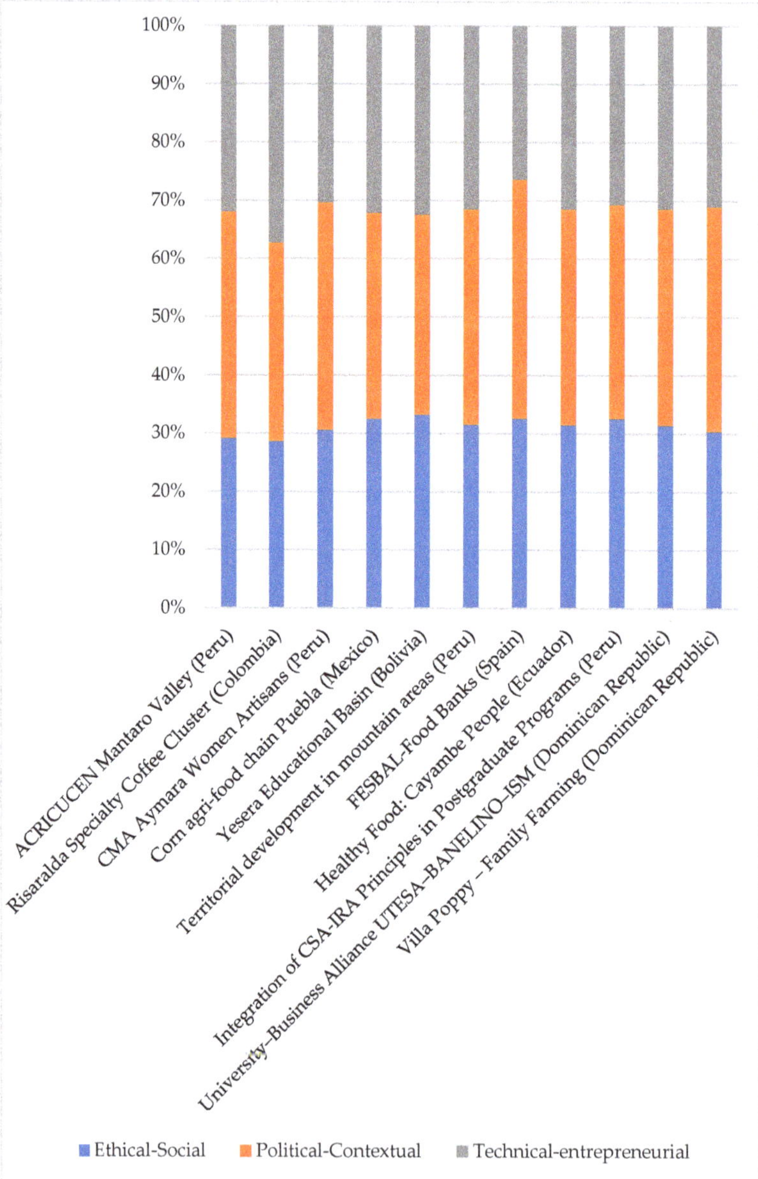

Figure 4. Balance between the three dimensions (Political–Contextual, Ethical–Social and Technical–Entrepreneurial) of the WWP model in case studies. The results show the total percentage of actions taken in each of the case studies, according to their integration into each of the three dimensions of the WWP model.

https://doi.org/10.3390/su18031402

Applying the WWP model, the LEADER approach, IPMA competencies, and the CFS-RAI principles together provides a novel framework for implementing rural development processes aimed at inclusion, sustainability, and capacity building in local territories.

3.6. Limitations

This study presents some limitations that should be acknowledged. First, there is a potential selection bias, as the cases analyzed are part of the network and depend on the availability of information provided by those responsible. Second, there is an advocacy risk, since the network carries out both implementation and evaluation functions, which could influence the interpretation of the results. Future research should consider comparative designs with contexts outside the network and complementary methodologies to help reduce these biases and strengthen the validity of the findings.

4. Conclusions

The findings obtained from the analysis of the experiences of the case studies presented suggest that combining the Principles of Responsible Investment in Agriculture and Food Systems (CFS-RAI) with the Working with People (WWP) model constitutes a solid, adaptable, and innovative methodological proposal. This integration responds to the need to rethink intervention approaches in rural areas, incorporating not only technical criteria but also ethical and social criteria from local contexts.

The WWP model, structured around three dimensions—ethical–social, technical–entrepreneurial, and political–contextual—offers a framework that transcends the instrumental logic of traditional planning. Its application allows development processes to be understood as collective learning experiences, where rural actors are not only recipients of projects but also active protagonists in the formulation, execution, and evaluation of development processes.

The implementation of the CFS-RAI principles has provided a coherent regulatory framework to guide projects toward responsible and sustainable investments. In territories where these principles were incorporated from the project design stage, processes such as community organization, the inclusion of women and young people, and inter-institutional coordination have been strengthened, highlighting their potential to guide local governance processes and promote the creation of Local Action Groups (LAGs) as innovative territorial management structures. The presented case studies show that project progress does not depend solely on the capital invested, but also on the quality of inter-institutional relations, the recognition of local knowledge, and the ability to generate participatory processes.

One of the most significant contributors identified is university–business partnerships, which act as coordinating agents, facilitators of knowledge, and promoters of the sharing of both scientific and local knowledge, thereby facilitating training and innovation.

In short, not only does this methodological combination illustrate innovations in project management, but it also promotes ethical development based on respect for human dignity and social equity.

Author Contributions: Conceptualization: I.d.l.R.-C. and M.L.A.M.; methodology: I.d.l.R.-C. and M.L.A.M.; validation: I.d.l.R.-C., M.L.A.M. and X.N.D.; formal analysis: I.d.l.R.-C. and M.L.A.M.; investigation: I.d.l.R.-C., M.L.A.M. and X.N.D.; resources: I.d.l.R.-C. and M.L.A.M.; data curation: I.d.l.R.-C. and M.L.A.M.; writing—original draft preparation: I.d.l.R.-C. and M.L.A.M.; writing—review and editing: I.d.l.R.-C., M.L.A.M. and X.N.D.; visualization: I.d.l.R.-C., M.L.A.M. and X.N.D.; supervision: I.d.l.R.-C. and M.L.A.M.; project administration: I.d.l.R.-C.; funding acquisition: I.d.l.R.-C. All authors have read and agreed to the published version of the manuscript.

https://doi.org/10.3390/su18031402

Funding: This research was funded through the collaboration agreement between the Food and Agriculture Organization of the United Nations (FAO) and the Universidad Politécnica de Madrid (UPM) within the framework of the project TCP/RLC/4018 for the implementation of the CFS-RAI Principles in Latin America, the Caribbean, and Spain.

Institutional Review Board Statement: This study is waived for ethical review by the Polytechnic University of Madrid, as no personal information from the people interviewed and surveyed has been used.

Informed Consent Statement: Informed consent was obtained from all subjects involved in the study.

Data Availability Statement: The raw data supporting the conclusions of this article will be made available by the authors on request.

Acknowledgments: The authors gratefully acknowledge the support of the GESPLAN research group and the RU-RAI university network, whose collaboration and contributions were essential to the development of this study.

Conflicts of Interest: The authors declare no conflicts of interest.

References

1. Farrukh, M.; Meng, F.; Raza, A.; Tahir, M.S. Twenty-Seven Years of Sustainable Development Journal: A Bibliometric Analysis. *Sustain. Dev.* **2020**, *28*, 1725–1737. [CrossRef]
2. World Commission on Environment and Development (WCED). *Our Common Future*; Oxford University Press: Oxford, UK, 1987.
3. Hajian, M.; Kashani, S.J. Evolution of the Concept of Sustainability. From Brundtland Report to Sustainable Development Goals. In *Sustainable Resource Management: Modern Approaches and Contexts*; Elsevier: Amsterdam, The Netherlands, 2021; pp. 1–24. [CrossRef]
4. Purvis, B.; Mao, Y.; Robinson, D. Three Pillars of Sustainability: In Search of Conceptual Origins. *Sustain. Sci.* **2019**, *14*, 681–695. [CrossRef]
5. Ozili, P.K. Sustainability and Sustainable Development Research around the World. *Manag. Glob. Transit.* **2022**, *20*. [CrossRef]
6. Stojanović, I.; Ateljević, J.; Stević, R.S. Good governance as a tool of sustainable development. *Eur. J. Sustain. Dev.* **2016**, *5*, 558–573. [CrossRef]
7. Adamowicz, M.; Zwolinska-Ligaj, M. The "Smart Village" as Away to Achieve Sustainable Development in Rural Areas of Poland. *Sustainability* **2020**, *12*, 6503. [CrossRef]
8. Yu, Y.; Appiah, D.; Zulu, B.; Adu-Poku, K.A. Integrating Rural Development, Education, and Management: Challenges and Strategies. *Sustainability* **2024**, *16*, 6474. [CrossRef]
9. Liu, L.; Cao, C.; Song, W. Bibliometric Analysis in the Field of Rural Revitalization: Current Status, Progress, and Prospects. *Int. J. Environ. Res. Public Health* **2023**, *20*, 823. [CrossRef]
10. Kumareswaran, K.; Jayasinghe, G.Y. Systematic Review on Ensuring the Global Food Security and COVID-19 Pandemic Resilient Food Systems: Towards Accomplishing Sustainable Development Goals Targets. *Discov. Sustain.* **2022**, *3*, 29. [CrossRef]
11. Mueller, J.T.; McConnell, K.; Berne Burow, P.; Pofahl, K.; Merdjanoff, A.A.; Farrell, J. Impacts of the COVID-19 Pandemic on Rural America. *Proc. Natl. Acad. Sci. USA* **2021**, *118*, 2019378118. [CrossRef]
12. Liu, Y.; Zhou, Y. Reflections on China's Food Security and Land Use Policy under Rapid Urbanization. *Land Use Policy* **2021**, *109*, 105699. [CrossRef]
13. Obi, C.; Bartolini, F.; Brunori, G.; D'Haese, M. How Does International Migration Impact on Rural Areas in Developing Countries? A Systematic Review. *J. Rural Stud.* **2020**, *80*, 273–290. [CrossRef]
14. Busso, M.; Chauvin, J.P.; Herrera, L.N. Rural-Urban Migration at High Urbanization Levels. *Reg. Sci. Urban Econ.* **2021**, *91*, 103658. [CrossRef]
15. Ge, D.; Long, H.; Qiao, W.; Wang, Z.; Sun, D.; Yang, R. Effects of Rural–Urban Migration on Agricultural Transformation: A Case of Yucheng City, China. *J. Rural Stud.* **2020**, *76*, 85–95. [CrossRef]
16. Wolford, W.W.; White, B.; Scoones, I.; Hall, R.; Edelman, M.; Borras, S.M. Global Land Deals: What Has Been Done, What Has Changed, and What's Next? *J. Peasant Stud.* **2024**, *7*, 1409–1446. [CrossRef]
17. Mansilla-Quiñones, P.; Uribe-Sierra, S.E. Rural Shrinkage: Depopulation and Land Grabbing in Chilean Patagonia. *Land* **2024**, *13*, 11. [CrossRef]
18. Yang, B.; He, J. Global Land Grabbing: A Critical Review of Case Studies across the World. *Land* **2021**, *10*, 324. [CrossRef]
19. Burja, V.; Tamas-Szora, A.; Dobra, I.B. Land Concentration, Land Grabbing and Sustainable Development of Agriculture in Romania. *Sustainability* **2020**, *12*, 2137. [CrossRef]

https://doi.org/10.3390/su18031402

20. Yeshanew, S. Rights in the Collaboration between the World Bank and the United Nations in the Areas of Investment in Agriculture, Rural Development and Food Systems. *Int. J. Hum. Rights* **2023**, *27*, 1086–1106. [CrossRef]
21. Mingst, K.A.; Karns, M.P.; Lyon, A.J. *The United Nations in the 21st Century*; Routledge: London, UK, 2022.
22. Committee on World Food Security (CFS). *Principles for Responsible Investment in Agriculture and Food Systems*; FAO: Rome, Italy, 2014. Available online: https://www.fao.org/3/au866e/au866e.pdf (accessed on 26 August 2024).
23. United Nations. *Transforming Our World: The 2030 Agenda for Sustainable Development*; UN: New York, NY, USA, 2015. Available online: https://sdgs.un.org/2030agenda (accessed on 26 August 2024).
24. Leal Filho, W.; Tripathi, S.K.; Andrade Guerra, J.B.S.O.D.; Giné-Garriga, R.; Orlovic Lovren, V.; Willats, J. Using the *Sustainable* Development Goals towards a Better Understanding of Sustainability Challenges. *Int. J. Sustain. Dev. World Ecol.* **2019**, *26*, 179–190. [CrossRef]
25. Bolis, I.; Morioka, S.N.; Sznelwar, L.I. Are We Making Decisions in a Sustainable Way? A Comprehensive Literature Review about Rationalities for Sustainable Development. *J. Clean. Prod.* **2017**, *145*, 310–322. [CrossRef]
26. Cazorla, A.; De los Ríos, I. Towards a Meta-University for Sustainable Development: Responsible Investment in Agriculture and Food Systems. In Proceedings of the XXVIII International Congress on Project Management and Engineering, Jaén, Spain, 3–4 July 2024; pp. 1959–1973.
27. GESPLAN. Principios CSA-IRA: Inversión Responsable En Agricultura y Sistemas Alimentarios En La Universidad y La Empresa. Available online: https://ruraldevelopment.es/index.php/es/docencia-lms/principios-iar-universidad (accessed on 1 January 2025).
28. Cazorla-Montero, A.; De los Ríos-Carmenado, I. From "Putting the Last First" to "Working with People" in Rural Development Planning: A Bibliometric Analysis of 50 Years of Research. *Sustainability* **2023**, *15*, 10117. [CrossRef]
29. Friedmann, J. *Planning as Social Learning*; IURD Working Paper Series; Institute of Urban and Regional Development, University of California: Berkeley, CA, USA, 1981.
30. Pieterse, J.N. My Paradigm or Yours? Alternative Development, Post-Development, Reflexive Development. *Dev. Change* **1998**, *29*, 343–373. [CrossRef]
31. Cazorla-Montero, A.; De los Ríos-Carmenado, I.; Díaz-Puente, J.M. The LEADER Community Initiative as Rural Development Model: Application in the Capital Region of Spain. *Agrociencia* **2005**, *39*, 697–708.
32. Ottomano Palmisano, G.; Govindan, K.; Boggia, A.; Loisi, R.V.; De Boni, A.; Roma, R. Local Action Groups and Rural Sustainable Development: A Spatial Multiple Criteria Approach for Efficient Territorial Planning. *Land Use Policy* **2016**, *59*, 12–26. [CrossRef]
33. Dax, T.; Oedl-Wieser, T. Rural Innovation Activities as a Means for Changing Development Perspectives—An Assessment of More than Two Decades of Promoting LEADER Initiatives across the European Union. *Stud. Agric. Econ.* **2016**, *118*, 30–37. [CrossRef]
34. Peeters, W.; Dirix, J.; Sterckx, S. Putting Sustainability into Sustainable Human Development. *J. Human Dev. Capab.* **2013**, *14*, 58–76. [CrossRef]
35. Pascual, U.; Balvanera, P.; Díaz, S.; Pataki, G.; Roth, E.; Stenseke, M.; Watson, R.T.; Başak Dessane, E.; Islar, M.; Kelemen, E.; et al. Valuing Nature's Contributions to People: The IPBES Approach. *Curr. Opin. Environ. Sustain.* **2017**, *26–27*, 7–16. [CrossRef]
36. Liu, J.; Hull, V.; Godfray, H.C.J.; Tilman, D.; Gleick, P.; Hoff, H.; Pahl-Wostl, C.; Xu, Z.; Chung, M.G.; Sun, J.; et al. Nexus Approaches to Global Sustainable Development. *Nat. Sustain.* **2018**, *1*, 466–476. [CrossRef]
37. Suárez Roldan, C.; Méndez Giraldo, G.A.; López Santana, E. Sustainable Development in Rural Territories within the Last Decade: A Review of the State of the Art. *Heliyon* **2023**, *9*, e17555. [CrossRef]
38. Moretti, M.; Belliggiano, A.; Grando, S.; Felici, F.; Scotti, I.; Ievoli, C.; Blackstock, K.; Delgado-Serrano, M.M.; Brunori, G. Characterizing Value Chains' Contribution to Resilient and Sustainable Development in European Mountain Areas. *J. Rural Stud.* **2023**, *100*, 103022. [CrossRef]
39. Brunetta, G.; Ceravolo, R.; Barbieri, C.A.; Borghini, A.; de Carlo, F.; Mela, A.; Beltramo, S.; Longhi, A.; De Lucia, G.; Ferraris, S.; et al. Territorial Resilience: Toward a Proactive Meaning for Spatial Planning. *Sustainability* **2019**, *11*, 2286. [CrossRef]
40. Cazorla, A.; De Los Ríos, I.; Salvo, M. Working with People (WWP) in Rural Development Projects: A Proposal from Social Learning. *Cuad. Desarro. Rural* **2013**, *10*, 131–157.
41. Cazorla, A.; Friedmann, J. *Planificación e Ingeniería: Nuevas Tendencias*; Taller de Ideas; Universidad Politecnica de Madrid: Madrid, Spain, 1995.
42. Holden, M. Social Learning in Planning: Seattle's Sustainable Development Codebooks. *Prog. Plann.* **2008**, *69*, 1–40. [CrossRef]
43. Uphoff, N. Paraprojects as New Modes of International Development Assistance. *World Dev.* **1990**, *18*, 1401–1411. [CrossRef]
44. Cernea, M.M. *Putting People First: Sociological Variables in Rural Development*, 2nd ed.; Oxford University Press: New York, NY, USA, 1991.
45. Chambers, R. Participatory Rural Appraisal (PRA): Challenges, Potentials and Paradigm. *World Dev.* **1994**, *22*, 1437–1454. [CrossRef]
46. Martínez-Falero, E.; Trueba, I.; Cazorla, A.; Alier, J.L. Optimization of Spatial Allocation of Agricultural Activities. *J. Agric. Eng. Res.* **1998**, *69*, 1–13. [CrossRef]

https://doi.org/10.3390/su18031402

47. IPMA. *Individual Competence Baseline for Project, Programme, and Portfolio Management, version 4.0*; International Project Management Association: Zurich, Switzerland, 2015. Available online: https://www.ipma-greece.gr/wp-content/uploads/2022/09/IPMA_ICB_4_0_WEB.pdf (accessed on 25 January 2025).
48. El-Sabaa, S. The Skills and Career Path of an Effective Project Manager. *Int. J. Project Manag.* **2001**, *19*, 1–7. [CrossRef]
49. Jiménez Aliaga, R.; De los Ríos-Carmenado, I.; Huamán Cristóbal, A.E.; Aliaga Balbín, H.; Marroquín Heros, A.M. Competencies and Capabilities for the Management of Sustainable Rural Development Projects in the Value Chain: Perception from Small and Medium-Sized Business Agents in Jauja, Peru. *Sustainability* **2023**, *15*, 15580. [CrossRef]
50. Stratta Fernández, R.; De los Ríos Carmenado, I.; López González, M. Developing Competencies for Rural Development Project Management through Local Action Groups: The Punta Indio (Argentina) Experience. In *International Development*; Appiah-Opoku, S., Ed.; IntechOpen: Rijeka, Croatia, 2017. [CrossRef]
51. Martínez Arroyo, F.; Sacristán López, H.; Yagüe Blanco, J.L. Are Local Action Groups, under LEADER Approach, a Good Way to Support Resilience in Rural Areas? *Ager* **2015**, *18*, 39–63.
52. López, M.; Cazorla, A.; Panta, M.d.P. Rural Entrepreneurship Strategies: Empirical Experience in the Northern Sub-Plateau of Spain. *Sustainability* **2019**, *11*, 1243. [CrossRef]
53. Herrera-Reyes, A.T.; Carmenado, I.D.L.R.; Martínez-Almela, J. Project-Based Governance Framework for an Agri-Food Cooperative. *Sustainability* **2018**, *10*, 1881. [CrossRef]
54. Šūmane, S.; Kunda, I.; Knickel, K.; Strauss, A.; Tisenkopfs, T.; Rios, I.D.I.; Rivera, M.; Chebach, T.; Ashkenazy, A. Local and Farmers' Knowledge Matters! How Integrating Informal and Formal Knowledge Enhances Sustainable and Resilient Agriculture. *J. Rural Stud.* **2018**, *59*, 232–241. [CrossRef]
55. Cazorla, A.; Negrillo, X.; Montalvo, V.; De Nicolas, V.L. Institutional Structuralism as a Process to Achieve Social Development: Aymara Women's Community Project Based on the Working with People Model in Peru. *J. Sociol. Soc. Welf.* **2018**, *45*, 55. [CrossRef]
56. Fontana, A.; Sastre-Merino, S.; Baca, M. The Territorial Dimension: The Component of Business Strategy That Prevents the Generation of Social Conflicts. *J. Bus. Ethics* **2017**, *141*, 367–380. [CrossRef]
57. Salgado, J.P.; De los Ríos, I.; López, M. Management of Innovative and Entrepreneurship Projects from Project-Based Learning: A Case Study on the UPS University in Ecuador. In *Case Study of Innovative Projects. Successful Real Cases*; Llamas, B., Mazadiego, F., Storch, L., Eds.; IntechOpen: London, UK, 2017. [CrossRef]
58. Ávila Cerón, C.A.; De los Ríos-Carmenado, I.; Martín Fernández, S. Illicit Crops Substitution and Rural Prosperity in Armed Conflict Areas: A Conceptual Proposal Based on the Working with People Model in Colombia. *Land Use Policy* **2018**, *72*, 201–214. [CrossRef]
59. Bugueño, F.; De, I.; Ríos, L.; Castañeda, R. Responsible Land Governance and Project Management Competences for Sustainable Social Development. The Chilean-Mapuche Conflict. *Int. J. Econ. Financ. Issues* **2017**, *7*, 202–211.
60. Fontana, A.; Velasquez-Fernandez, A.; Rodriguez-Vasquez, M.I.; Cuervo-Guerrero, G. Territorial Analysis Through the Integration of CFS-RAI Principles and the Working with People Model: An Application in the Andean Highlands of Peru. *Sustainability* **2025**, *17*, 1380. [CrossRef]
61. Acosta Mereles, M.L.; Mur Nuño, C.; Stratta Fernández, R.R.; Chenet, M.E. Good Practices of Food Banks in Spain: Contribution to Sustainable Development from the CFS-RAI Principles. *Sustainability* **2025**, *17*, 912. [CrossRef]
62. Jiménez Aliaga, R.; De los Ríos-Carmenado, I.; San Martín Howard, F.; Calle Espinoza, S.; Huamán Cristóbal, A. Integration of the Principles of Responsible Investment in Agriculture and Food Systems CFS-RAI from the Local Action Groups: Towards a Model of Sustainable Rural Development in Jauja, Peru. *Sustainability* **2022**, *14*, 9663. [CrossRef]
63. Requelme, N.; Afonso, A. The Principles for Responsible Investment in Agriculture (CFS-RAI) and SDG 2 and SDG 12 in Agricultural Policies: Case Study of Ecuador. *Sustainability* **2023**, *15*, 15985. [CrossRef]
64. Negrillo Deza, X. Los Proyectos de Desarrollo Rural Como Proceso a Través de Sus Protagonistas: El Caso de La Coordinadora de Mujeres Aymaras (Puno, Perú). Ph.D. Thesis, Universidad Politécnica de Madrid, Madrid, Spain, 2018. Available online: https://doi.org/10.20868/UPM.thesis.53126 (accessed on 26 August 2024).
65. Kutsenko, M. Application of the Project Approach as Part of the Territory Development Strategy. In Proceedings of the 2021 IEEE 16th International Conference on Computer Sciences and Information Technologies (CSIT), Zakharkiv, Ukraine, 22–25 September 2021; Volume 2, pp. 375–378. [CrossRef]
66. Musawir, A.U.; Abd-Karim, S.B.; Mohd-Danuri, M.S. Project Governance and Its Role in Enabling Organizational Strategy Implementation: A Systematic Literature Review. *Int. J. Proj. Manag.* **2020**, *38*, 1–16. [CrossRef]
67. Young, R.; Chen, W.; Quazi, A.; Parry, W.; Wong, A.; Poon, S.K. The Relationship between Project Governance Mechanisms and Project Success: An International Data Set. *Int. J. Manag. Proj. Bus.* **2020**, *13*, 1496–1521. [CrossRef]
68. Horlings, L.G.; Kanemasu, Y. Sustainable Development and Policies in Rural Regions; Insights from the Shetland Islands. *Land Use Policy* **2015**, *49*, 310–321. [CrossRef]

https://doi.org/10.3390/su18031402

69. Ávila Cerón, C.A.; Castañeda, R.; De Los Ríos, I. Development of a Family Agriculture Model Aligned with the Principles of Responsible Investment: Integration Tourism Chain in the Dominican Republic. In Proceedings of the XXVIII International Congress on Project Management and Engineering (AEIPRO 2024), Jaén, Spain, 3–4 July 2024; pp. 1974–1987.
70. De Los Ríos-Carmenado, I.; Díaz-Puente, J.M.; Cadena-Iñiguez, J. La iniciativa LEADER como modelo de desarrollo rural: Aplicación a algunos territorios de México. *Agrociencia* **2011**, *45*, 609–624.
71. Olvera Hernández, J.I.; Cazorla Montero, A.; Ramírez Valverde, B. La Política de Desarrollo Rural En La Unión Europea y La Iniciativa LEADER, Una Experiencia de Éxito. *Reg. Soc.* **2009**, *21*, 3–25. [CrossRef]
72. Navarro Valverde, F.A.; Cejudo García, E.; Cañete Pérez, J.A. A long-run analysis of neoendogenous rural development policies: The persistence of businesses created with the support of LEADER and PRODER in three districts of Andalusia (Spain) in the 1990s. *AGER Rev. De Estud. Sobre Despoblación Y Desarro. Rural.* **2018**, *25*, 189–219.
73. FAO. Latin American and Caribbean Member States: UN Geospatial Information. 2024. Available online: https://www.fao.org/geospatial/en (accessed on 26 August 2024).
74. Cazorla, A.; De los Ríos-Carmenado, I.; Afonso, A. *Principles for a Responsible Investment in Agriculture and Voluntary Guidelines on the Responsible Governance of Tenure of Land: Links with University*; Working Paper; GESPLAN Group, Universidad Politécnica de Madrid: Madrid, Spain, 2018. Available online: https://oa.upm.es/50898/ (accessed on 1 January 2025).
75. Morris, L.; Salazar, O.; Garmendia, J. Characterization of the Technical Operations to Strengthen the Logistics Processes of the Cluster of Specialty Coffees of Risaralda. In Proceedings of the XXVII International Congress on Project Management and Engineering, San Sebastián, Spain, 10–13 July 2023; pp. 2382–2392.
76. Cachipuendo, C.; Requelme, N.; Sandoval, C.; Afonso, A. Sustainable Rural Development Based on CFS-RAI Principles in the Production of Healthy Food: The Case of the Kayambi People (Ecuador). *Sustainability* **2025**, *17*, 2958. [CrossRef]
77. Regalado-López, J.; Maimone-Celorio, J.A.; Pérez-Ramírez, N. Strengthening of the Rural Community and Corn Food Chain Through the Application of the WWP Model and the Integration of CFS-RAI Principles in Puebla, México. *Sustainability* **2025**, *17*, 5442. [CrossRef]
78. Sastre-Merino, S.; Negrillo, X.; Hernández-Castellano, D. Sustainability of Rural Development Projects within the Working with People Model: Application to Aymara Women Communities in the Puno Region, Peru. *Cuad. Desarro. Rural.* **2013**, *10*, 219–244.
79. De los Ríos-Carmenado, I.; Cazorla, A.; Gandini, G.; Eberth, E. *Los Principios CSA-IRA para una Integración Universidad-Empresa en las Provincias de Santiago Rodríguez, Valverde (Mao), Dajabón y en República Dominicana*; Working Paper; GESPLAN Group, Universidad Politécnica de Madrid: Madrid, Spain, 2023. Available online: https://principiosirametauniversidad.com/wp-content/uploads/2024/10/caso-pira-utesa_ism_banelino_upm_ok.pdf (accessed on 21 January 2025).
80. Creswell, J.W.; Plano Clark, V.L. *Designing and Conducting Mixed Methods Research*; Sage Publications: Thousand Oaks, CA, USA, 2017.
81. Teucher, M.; Thürkow, D.; Alb, P.; Conrad, C. Digital In Situ Data Collection in Earth Observation, Monitoring and Agriculture—Progress towards Digital Agriculture. *Remote Sens.* **2022**, *14*, 393. [CrossRef]
82. Bender, D.E.; Ewbank, D. The Focus Group as a Tool for Health Research: Issues in Design and Analysis. *Health Transit. Rev.* **1994**, *4*, 63–80.
83. Rabiee, F. Focus-Group Interview and Data Analysis. *Proc. Nutr. Soc.* **2004**, *63*, 655–660. [CrossRef]
84. Timmermans, S.; Tavory, I. Theory Construction in Qualitative Research: From Grounded Theory to Abductive Analysis. *Sociol. Theory* **2012**, *30*, 167–186. [CrossRef]
85. South, L.; Saffo, D.; Vitek, O.; Dunne, C.; Borkin, M.A. Effective Use of Likert Scales in Visualization Evaluations: A Systematic Review. *Comput. Graph. Forum* **2022**, *41*, 43–55. [CrossRef]
86. Likert, R. A Technique for the Measurement of Attitudes. *Arch. Psychol.* **1932**, *22*, 1–55.
87. Hernández-Vásquez, A.; Visconti-Lopez, F.J.; Vargas-Fernández, R. Factors Associated with Food Insecurity in Latin America and the Caribbean Countries: A Cross-Sectional Analysis of 13 Countries. *Nutrients* **2022**, *14*, 3190. [CrossRef]
88. Hossain, M.; Mendiratta, V.; Savastano, S. Agricultural and Rural Development Interventions and Poverty Reduction: Global Evidence from 16 Impact Assessment Studies. *Glob. Food Sec.* **2024**, *43*, 100806. [CrossRef]
89. Peng, B.; Shuai, C.; Yin, C.; Qi, H.; Chen, X. Progress toward SDG-2: Assessing Food Security in 93 Countries with a Multidimensional Indicator System. *Sustain. Dev.* **2024**, *32*, 815–862. [CrossRef]
90. Woodhill, J.; Kishore, A.; Njuki, J.; Jones, K.; Hasnain, S. Food Systems and Rural Wellbeing: Challenges and Opportunities. *Food Secur.* **2022**, *14*, 1099–1121. [CrossRef]
91. Trivelli, C.; Morel, J. Rural Youth Inclusion, Empowerment, and Participation. *J. Dev. Stud.* **2021**, *57*, 635–649. [CrossRef]
92. Leakey, R.R.B.; Tientcheu Avana, M.-L.; Awazi, N.P.; Assogbadjo, A.E.; Mabhaudhi, T.; Hendre, P.S.; Degrande, A.; Hlahla, S.; Manda, L. The Future of Food: Domestication and Commercialization of Indigenous Food Crops in Africa over the Third Decade (2012–2021). *Sustainability* **2022**, *14*, 2355. [CrossRef]

https://doi.org/10.3390/su18031402

93. Mabhaudhi, T.; Hlahla, S.; Chimonyo, V.G.P.; Henriksson, R.; Chibarabada, T.P.; Murugani, V.G.; Groner, V.P.; Tadele, Z.; Sobratee, N.; Slotow, R.; et al. Diversity and Diversification: Ecosystem Services Derived from Underutilized Crops and Their Co-Benefits for Sustainable Agricultural Landscapes and Resilient Food Systems in Africa. *Front. Agron.* **2022**, *4*, 859223. [CrossRef] [PubMed]
94. Castañeda, R.; Cerón, C.A.; Ríos-Carmenado, I.D.L.; Domínguez, L.; Gomez, S. Implementing the Voluntary Guidelines on the Responsible Governance of Tenure of Land, Fisheries and Forests from the Working with People Model: Lessons from Colombia and Guatemala. *J. Peasant Stud.* **2023**, *50*, 1820–1851. [CrossRef]
95. Gargano, G. The Bottom-Up Development Model as a Governance Instrument for the Rural Areas. The Cases of Four Local Action Groups (LAGs) in the United Kingdom and in Italy. *Sustainability* **2021**, *13*, 9123. [CrossRef]
96. Cronin, E.; Fosselle, S.; Rogge, E.; Home, R. An Analytical Framework to Study Multi-Actor Partnerships Engaged in Interactive Innovation Processes in the Agriculture, Forestry, and Rural Development Sector. *Sustainability* **2021**, *13*, 6428. [CrossRef]
97. Sabrina, T.; Alessio, C.; Chiara, A.; Gigliola, P.; Concetta, F.; Federica, B.; Paolo, P. Civic Universities and Bottom-up Approaches to Boost Local Development of Rural Areas: The Case of the University of Macerata. *Agric. Food Econ.* **2021**, *9*, 15. [CrossRef] [PubMed]
98. Trencher, G.P.; Yarime, M.; Kharrazi, A. Co-Creating Sustainability: Cross-Sector University Collaborations for Driving Sustainable Urban Transformations. *J. Clean Prod.* **2013**, *50*, 40–55. [CrossRef]
99. Atterton, J.; Thompson, N. University Engagement in Rural Development: A Case Study of the Northern Rural Network. *J. Rural Community Dev.* **2010**, *5*, 123–132. Available online: https://journals.brandonu.ca/jrcd/article/view/338/78 (accessed on 15 July 2025).
100. Almeida, F. The Role of Partnerships in Municipal Sustainable Development in Portugal. *Int. J. Urban Sustain. Dev.* **2024**, *16*, 231–244. [CrossRef]
101. Midgley, J.O. *Social Development: Theory and Practice*; SAGE Publications Ltd.: London, UK, 2013; p. 296.
102. Ortega, R.; Cazorla, A. The International Cooperation for Development Based on Easterly and Sachs: The Institutional Structuralism as the Pathway to the Future. In Proceedings of the XXIV International Congress on Project Management and Engineering, Alcoi, Spain, 7–10 July 2020; pp. 1076–1693.
103. De los Ríos, I.; García, C.; Herrera, A.T.; Rivera, M. *Innovation and Social Learning in Organic Vegetable Production in the Region of Murcia, Camposeven, Spain*; RETHINK Case Study Report; GESPLAN, Universidad Politécnica de Madrid: Madrid, Spain, 2015.
104. López-García, D.; Carrascosa-García, M. Agroecology-Oriented Farmers' Groups. A Missing Level in the Construction of Agroecology-Based Local Agri-Food Systems? *Agroecol. Sustain. Food Syst.* **2023**, *47*, 996–1022. [CrossRef]
105. Nordberg, K.; Mariussen, Å.; Virkkala, S. Community-Driven Social Innovation and Quadruple Helix Coordination in Rural Development. Case Study on LEADER Group Aktion Österbotten. *J. Rural Stud.* **2020**, *79*, 157–168. [CrossRef]
106. Quiroz-Niño, C.; Murga-Menoyo, M.Á. Social and Solidarity Economy, Sustainable Development Goals, and Community Development: The Mission of Adult Education & Training. *Sustainability* **2017**, *9*, 2164. [CrossRef]
107. De los Ríos-Carmenado, I.; Acosta Mereles, M.L.; Farías Estrada, I.J.; Nole Correa, P.; Zuluaga Zuluaga, C.L.; Mur Nuño, C.; Peral, C.; García Gutiérrez, J. Los Principios de Inversión Responsable en la agricultura y los sistemas alimentarios desde el Proyecto la Gran Recogida de la Federación Española de Bancos de Alimentos. In *Experiencias de Aprendizaje-Servicio en la UPM: 2024*; Universidad Politécnica de Madrid, Oficina de Aprendizaje-Servicio: Madrid, Spain, 2025. Available online: https://oa.upm.es/88689/ (accessed on 1 August 2025).
108. Friedmann, J. *Insurgencies: Essays in Planning Theory*; Routledge: London, UK, 2011.
109. Reed, M.S. Stakeholder Participation for Environmental Management: A Literature Review. *Biol. Conserv.* **2008**, *141*, 2417–2431. [CrossRef]
110. Forester, J. *The Deliberative Practitioner: Encouraging Participatory Planning Processes*; Mit Press: Cambridge, MA, USA, 1999.
111. Leeuwis, C.; Pyburn, R.; Röling, N.G. (Eds.) *Wheelbarrows Full of Frogs: Social Learning in Rural Resource Management—International Research and Reflections*; Van Gorcum: Assen, The Netherlands, 2002.
112. del Arco, I.; Ramos-Pla, A.; Zsembinszki, G.; de Gracia, A.; Cabeza, L.F. Implementing SDGs to a Sustainable Rural Village Development from Community Empowerment: Linking Energy, Education, Innovation, and Research. *Sustainability* **2021**, *13*, 12946. [CrossRef]
113. Sutter, C.; Bruton, G.D.; Chen, J. Entrepreneurship as a Solution to Extreme Poverty: A Review and Future Research Directions. *J. Bus. Ventur.* **2019**, *34*, 197–214. [CrossRef]
114. Ge, T.; Abbas, J.; Ullah, R.; Abbas, A.; Sadiq, I.; Zhang, R. Women's Entrepreneurial Contribution to Family Income: Innovative Technologies Promote Females' Entrepreneurship Amid COVID-19 Crisis. *Front. Psychol.* **2022**, *13*, 828040. [CrossRef]
115. Pandey, N.; de Coninck, H.; Sagar, A.D. Beyond Technology Transfer: Innovation Cooperation to Advance Sustainable Development in Developing Countries. *Wiley Interdiscip. Rev. Energy Environ.* **2022**, *11*, e422. [CrossRef]

https://doi.org/10.3390/su18031402

116. Knickel, K.; Brunori, G.; Rand, S.; Proost, J. Towards a Better Conceptual Framework for Innovation Processes in Agriculture and Rural Development: From Linear Models to Systemic Approaches. *J. Agric. Educ. Ext.* **2009**, *15*, 131–146. [CrossRef]
117. Mapiye, O.; Makombe, G.; Molotsi, A.; Dzama, K.; Mapiye, C. Towards a Revolutionized Agricultural Extension System for the Sustainability of Smallholder Livestock Production in Developing Countries: The Potential Role of ICTs. *Sustainability* **2021**, *13*, 5868. [CrossRef]

https://doi.org/10.3390/su18031402

sustainability

MDPI

Article

Service Learning Projects and CFS-IRA Principles: Application to the Food Bank Chair from the Working with People Model

Priscila Nole Correa [1,2,*], Irely Joelia Farías Estrada [1,3], Guillermo Aliaga [1] and Claudia Zuluaga [1]

1 Planificación y Gestión Sostenible del Desarrollo Rural-Local (GESPLAN), Cátedra Banco de Alimentos, Universidad Politécnica de Madrid, Avda. Puerta de Hierro 2, 28040 Madrid, Spain; ij.farias@alumnos.upm.es or ij.farias@ce.pucmm.edu.do (I.J.F.E.); gf.aliaga@alumnos.upm.es (G.A.); catedrafesbal.etsiaab@upm.es (C.Z.)

2 Escuela de Ciencias Agrícolas y Ambientales, Pontificia Universidad Católica del Ecuador-Ibarra, Av. Jorge Guzmán Rueda y Av. Padre Aurelio Espinosa Polit, Ibarra 100112, Ecuador

3 Facultad de Ciencias Sociales, Humanidades y Arte, Pontificia Universidad Católica Madre y Maestra, Autopista Duarte Km 1 ½, Santiago 51000, Dominican Republic

* Correspondence: pl.nole@alumnos.upm.es or pnole@pucesi.edu.ec

Abstract

This research study addresses the critical contradiction within global food systems: unsustainable consumption patterns and persistent food insecurity coexist and are exacerbated by food waste, which deepens socioeconomic inequalities and generates negative environmental externalities. In this scenario, higher education plays a central role in adopting comprehensive strategic frameworks to develop specialized human capital and influence society. This study analyzes a Service Learning model that integrates the CFS-IRA Principles to promote the SDGs and ensure responsible consumption. Based on a case study of the Food Bank Chair spanning 10 years and 212 projects, the implementation of this model was evaluated using the Working with People (WWP) method, which combines the development of postgraduate students' skills with community service to address social problems. The results demonstrated the effectiveness of the SL-WWP model in strengthening students' technical, social, and ethical competencies while reducing food waste. The evaluation showed strong alignment with key SDGs, with outstanding performance in governance, although the need to strengthen environmental and social criteria was identified. The originality lies in integrating the CFS-IRA Principles into an SL model that encourages innovative cooperation among universities, civil society, and public–private sectors, offering a replicable proposal for higher education institutions to establish themselves as agents of change towards sustainability.

Keywords: Service Learning project; SL model; responsible consumption; food waste; Sustainable Development Goals (SDGs); Food Banks

1. Introduction

The promotion of sustainability in three dimensions—economic, social, and environmental—including the reduction in food waste, is high on the agenda of higher education institutions [1]. Their role in the formation of responsible leaders and their ability to influence sustainability policies and practices position them as key actors in the fulfillment of the United Nations Sustainable Development Goals (SDGs) [2].

The problem addressed in this research study is the need to frame these efforts within a global reality characterized by the unsustainable use of resources. Currently, humanity consumes the equivalent of 1.7 planets to maintain its lifestyle, and it is estimated that if this trend continues, 3 planets would be needed to meet future demand [3]. This situation contrasts with persistent levels of poverty and hunger; in 2023, more than 733 million persons faced food insecurity [4]. This paradox is accentuated when considering that, while millions of people suffer from extreme hunger, more than 1000 million tons of food are wasted annually globally, equivalent to 19% of the products available for consumption [5].

In Spain, in 2024, food waste accounted for 3.7% of the total food and beverages purchased, which is equivalent to more than 1125 million kilos/liters [6]. This phenomenon not only aggravates social inequalities but also leads to severe environmental consequences, such as soil erosion, deforestation, water and air pollution, and increased greenhouse gas emissions [7–9].

Against this backdrop, the motivation behind this research study, i.e., to highlight the need for higher education to adopt strategic frameworks such as the Sustainable Development Goals (SDGs) and the Responsible Investment in Agriculture and Food Systems (CFS-IRA) Principles, which provide concrete guidance for addressing critical social, environmental, economic, and governance challenges, becomes a priority [10].

Many studies have proposed specific actions to mitigate the problem of food waste, with emphasis on raising consumer awareness regarding the responsible use of food [11,12]. These initiatives have been developed at different levels—international, national, and local—and include campaigns aimed at diverse audiences, especially young people [13–15].

This study presents a model of Service Learning projects that integrates the CFS-IRA Principles and the SDGs as implemented by the so-called Chair of Food Banks (FBC) of the Polytechnic University of Madrid (UPM), as an innovative structure in the educational field. This FBC partnership, formed in 2013 between FESBAL and the GESPLAN Research Group, has made it possible to carry out 212 projects over a decade, aimed at raising awareness about food waste and promoting responsible consumption, using the Working with People methodological approach [16].

This research study aims to analyze the SL model as a methodological proposal that enables a strengthening of the competencies of students at the postgraduate level while generating impacts on society in collaboration with third-sector organizations and Food Banks which work for sustainable development [17–20]. It also highlights the strategic role of the university as a platform for the design of sustainable interventions, aligned with the principles of social justice, environmental sustainability, and participatory governance [21].

The findings indicate that the integration of the SL model with the CFS-IRA Principles within the FBC framework results in consistently high ratings across all sustainability dimensions (social, environmental, economic, and governance). Furthermore, it is important to emphasize that the distinctive contribution of this research study lies in the introduction of an innovative evaluation framework for applying the Principles to the SL model [3,16–18,22]. The empirical evidence demonstrates that this methodology serves as a valid instrument for the systematic analysis and optimization of sustainability-oriented projects, enabling the identification of specific improvement areas across all dimensions.

Literature Review: SL Projects, CFS-IRA Principles, and FBC

Service Learning projects (SL model) have established themselves as fundamental tools for linking higher education with contemporary social challenges [22,23]. This methodological approach makes it possible to address complex problems, such as food waste, environmental sustainability, and social inclusion, by implementing learning projects with community impact [24]. Various authors have highlighted the benefits of the SL model, which, for example, fosters key competencies, promotes experiential learning, and strengthens student civic engagement, especially in the university environment [25].

These SL projects combine academic theory with community practice, encouraging students' participation in their own learning while addressing real problems in their communities [26,27]. This methodology not only contributes to the development of academic skills but also promotes values such as social responsibility and teamwork, preparing students to face the challenges of today's world [28]. SL projects involve "serving while learning" or "learning while serving" [29–31] and are thus focused on improving students' understanding of theory by enabling them to provide community service and reflecting on the experience [32].

Although it is often considered a recent proposal, the SL model has its roots in the nineteenth century in the United States, with projects linking academic learning and community engagement [33,34]. In 1916, John Dewey, who was among the first to lay the foundations of what we know today as the SL model, argued that students would learn more effectively and become better citizens if they committed themselves to community service and integrated this experience into their academic curriculum [35,36].

In 1969, the term as we know it today was consolidated at the first Service Learning Conference in Atlanta [33]. According to educators, this term implies that service is an important value in integral formation, since it represents a link among authentic community service, intentional academic learning, and reflection [37].

In recent decades, multiple definitions have emerged to describe Service Learning, considering it a methodology, experience, project, pedagogy, and even educational philosophy [38]. Despite this conceptual diversity, there is consensus that the SL model is articulated around organized activities that seek to satisfy community needs while enriching academic learning through reflection [39,40].

Various studies underscore its potential to connect theory and practice, strengthen civic engagement, and broaden citizen participation [41–43]. From an applied perspective, the SL model has also been adopted as an instrument of community intervention, with the aim of transforming the social, educational, and environmental spheres and promoting the development of personal, social, and professional skills in young people [44]. In this sense, a well-designed Service Learning project allows students to assume meaningful responsibilities within their educational and social communities, contributing to the strengthening of the social fabric [45–47].

This approach is particularly relevant in the context of sustainable development. Through SL projects, students not only apply acquired knowledge but also become active agents of change by promoting responsible consumption practices and environmental sustainability. These initiatives align closely with the Sustainable Development Goals (SDGs) and the Responsible Investment in Food and Agriculture (CFS-IRA) Principles, providing a robust operational framework for the design of educational interventions with transformative impact. In this way, Service Learning transcends academic training to become a strategic tool for social transformation.

The SL model is tightly linked to Project-Based Learning (PBL), with which it shares the methodological foundation: active, meaningful, and contextualized learning [48]. The SL model can be considered to be PBL with a social purpose, with projects not only becoming vehicles for learning and developing skills but also opportunities for sustainable, ethical development and social responsibility [49].

Among the various human activities, consumption is among those with the greatest impact on the environment [50–52]. George Fisk defined responsible consumption as a rational and efficient use of resources considering the needs of the global population, a vision underscoring the importance of adopting ethical and conscious consumption habits. Numerous studies have shown that many environmental problems can be mitigated by modifying the behaviors at their root [53].

Responsible consumption has traditionally been understood as a rational practice in which informed citizens make decisions that consider the social and environmental impact of their actions [54,55]. In this context, the growing academic and social interest in behaviors related to food waste has driven research aimed at identifying its causes and developing intervention strategies [56–58].

Since the 1980s, the study of consumption habits among children, adolescents, and young people has intensified, as these groups have been highly exposed to commercial stimuli and have played an increasingly active role as consumers [59,60]. This evolution has been influenced by a culture of leisure and consumption that often relegates other fundamental values [61]. Consequently, the transformation of individual ethical codes has placed consumption as the structuring axis of social identity [62,63]. It is crucial to encourage critical thinking in education, promoting individual and collective responsibility as a basis for sustainable development [64,65].

In this sense, the SDGs and the CFS-IRA Principles offer a strategic framework that seeks to balance economic growth, social equity, and environmental conservation. The SDGs, adopted by the United Nations General Assembly in 2015, set out 17 global targets to eradicate poverty, protect the planet, and ensure prosperity for all. These are articulated around five fundamental pillars: People, Planet, Prosperity, Peace, and Partnership [66]. The CFS-IRA Principles are the result of a global consultation process led by the FAO between 2012 and 2014 and played a key role in the formulation of the SDGs, as they promote sustainable agricultural and food investments through a comprehensive vision that considers human rights, gender equity, sustainable resource management, and inclusive participation [67–69].

Both frameworks align with the four key dimensions of sustainability: social, environmental, economic, and governance [22,70,71].

Partnerships between universities and third-sector entities are essential to translating these principles into concrete actions. An example is the Chair of Food Banks (FBC), resulting from the collaboration between the Spanish Federation of Food Banks (FESBAL) and the GESPLAN Research Group. All FBC projects seek to contribute to solving the most pressing problems of society, such as hunger and food waste.

The Spanish Federation of Food Banks (FESBAL) is a non-political and non-denominational entity founded in 1995 which coordinates and represents the Food Banks of Spain in their fight against hunger, poverty, and food waste. Its mission includes the recovery and redistribution of food to people in need while promoting environmental sustainability [72].

FESBAL is made up of 54 Food Banks throughout Spain, which, in 2023, distributed food to 6919 charities, benefiting more than 1.2 million people. In addition, it is a member of the European Federation of Food Banks (FEBA) and the Global FoodBanking Network (GFN) and was awarded the Prince of Asturias Award for Concord in 2012 [73].

In 2013, the Food Bank Chair (FBC) was created with the purpose of strengthening, supporting, and raising awareness about rational and responsible food consumption; it promotes training in values of solidarity and sustainability and contributes to the reduction in food waste.

The Chair develops educational programs aimed at different levels, from primary and secondary education in schools and institutes to undergraduate and postgraduate degrees in higher education institutions [74], as well as specific training activities for professionals. Within this framework, conferences, seminars, and round tables are organized, and scholarships and support for the completion of doctoral theses and final degree projects are offered. The training provided is aimed to develop key competencies, such as leadership, sustainability, social responsibility, and community cohesion.

In 2014, the FBC launched the Rational Food Consumption (CORAL) Program, an educational initiative aligned with the SDGs that seeks to raise awareness about responsible and ethical food consumption. Aimed at primary, secondary, and baccalaureate students, the CORAL Program promotes awareness about the importance of avoiding food waste and reflection on its link with problems such as poverty and hunger. This program, in line with the mission of Food Banks, promotes social ethics from the earliest educational levels, providing students with practical tools to adopt more responsible consumption habits [75].

FBC and FESBAL projects are aligned with the SDGs and the CFS-IRA Principles as they promote food security, the reduction in food waste, and sustainability in the distribution chain [76,77]. These actions range from community awareness to the implementation of strategies for the collection, conservation, and redistribution of surplus food. Likewise, research is being carried out to optimize logistic processes, and strategic partnerships are consolidated among the public, private, and academic sectors, generating a comprehensive and sustainable intervention model [69,78].

Among the most outstanding projects is the implementation of good practices through the use of Geographic Information Systems (GISs), such as the GIS-FESBAL tool. This platform enables the management and updating of information on the activities and impact of Food Banks in Spain, improving their operational capacity and social reach, in alignment with the SDGs and the CFS-IRA Principles [79–81].

The various projects developed in this context allow students to be trained in fundamental values and competencies to perform in the workplace in an ethical and responsible way [82]. These values and skills are essential to promoting sensitivity to the most complex social realities. In terms of technical skills, priority is given to the acquisition of skills that allow projects to be carried out with impact, following the international standards of the International Project Management Association (IPMA). The competencies defined by these frameworks include technical, behavioral, and contextual aspects, which are essential to addressing the challenges of sectors such as construction, industry, the environment, energy, or technology [83–85].

Based on the conceptual foundation, the SL model developed by the FBC is applied to foster training in relevant values with the acquisition of skills, a reduction in food waste, and the establishment of sustainable food systems. Figure 1 summarizes this SL-WWP model for implementing Service Learning projects in three main areas: (1) training and educational cooperation, (2) dissemination and transfer of knowledge, and (3) promotion of research, development, and innovation.

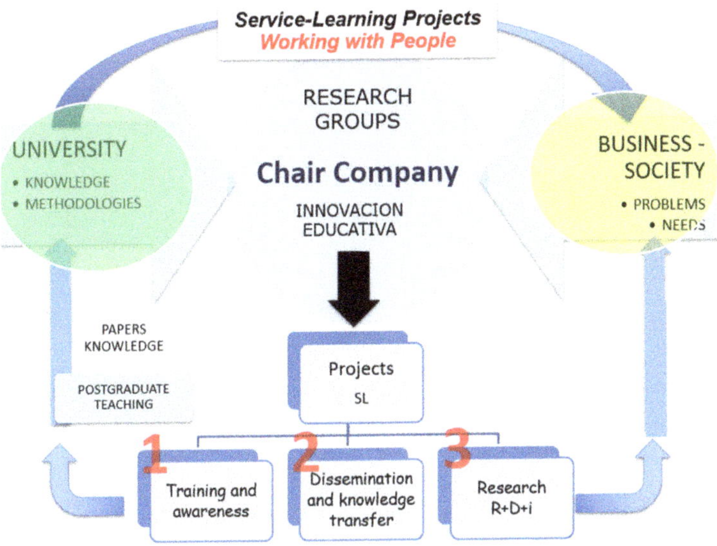

Figure 1. SL-WWP work model of FBC-UPM. Reprinted from [49] with permission.

2. Materials and Methods

We analyze the accumulated experience of ten years of work of the FBC, in collaboration with FESBAL and its 54 associated banks, as a case study. The SL-WWP [49] work model is used to evaluate the Service Learning projects developed by the FBC (Figure 1), assessing each one with a specific measurement tool.

Figure 2 shows the FBC's collaborative ecosystem that enables interaction among various public and private actors, the university, and related civil society organizations.

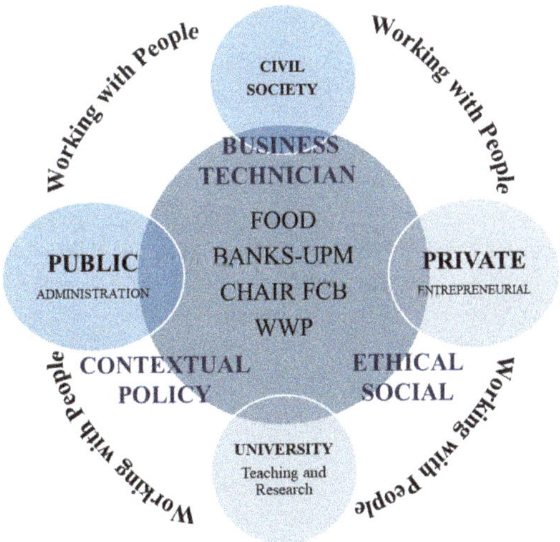

Figure 2. Working with People model.

In the development of this research, artificial intelligence (AI) tools were employed to assist in the literature search and review phases. Specifically, the platforms Consensus and SciSpace were used to locate relevant academic literature, synthesize key findings, and refine the accuracy of citation references. The purpose of using these tools was to streamline the document review process and ensure comprehensive bibliographic coverage. It should be noted that all ideas, analyses, and interpretations presented in the manuscript are solely those of the authors, and any content generated or suggested by AI was critically evaluated and verified before being incorporated into the study.

Implementation of SL Projects and CFS-IRA Principles

By applying the case study approach, we conducted a comprehensive analysis of a population of 212 SL model projects that integrate the Responsible Investment in Agriculture and Food Systems (CFS-IRA) Principles defined by the FAO [86], where the Principles constitute a comprehensive framework that addresses social, environmental, economic, and governance aspects to promote sustainable investments in agriculture.

Each project was evaluated according to 10 fundamental Principles, using a Likert scale from 1 to 5, where 1 represents a very low impact and 5 a very high impact in relation to each Principle, enabling standardized and comparative assessment. Furthermore, based on an internationally recognized sustainability framework, the Principles were grouped into four key dimensions to facilitate analysis:

Social Dimension: Food security and nutrition (P1), gender equality (P3), and youth participation (P4).

Environmental Dimension: Sustainable tenure and management of natural resources (P5 and P6) and sustainability of agricultural and food systems (P8).

Economic Dimension: Inclusive economic development (P2) and promotion of cultural heritage, diversity, and innovation (P7).

Governance Dimension: Responsible and inclusive governance (P9) and accountability (P10).

The final value of each dimension was calculated as the simple average of the grouped Principles, ensuring transparency, comparative assessment, and reproducibility of result synthesis. Equal weight was assigned to each Principle within its dimension, enhancing evaluation credibility based on the strong content validity of the CFS-IRA framework for measuring sustainability. The overall project rating was obtained as the average of the four dimensions.

This methodology facilitated the acquisition of normalized comparative measurements. Consequently, the dataset derived from the analysis is available in the Results Section, which compiles the calculated metrics. The instrument's validity is demonstrated through its capacity to identify improvement opportunities in each dimension, as shown in Results, establishing foundations for feedback and continuous improvement processes.

The evaluation was conducted by the interinstitutional research team that coordinates the FBC, which includes the authors, with support from experts of the GESPLAN Research Group. Therefore, project assessment was carried out according to a collaborative and rigorous process. The participation of experts ensured that the application of the evaluation instrument benefited from their empirical experience, as they are actively involved in project implementation, and validated the results.

However, it is significant to note that although the implementation of the CFS-IRA Principles, their application across the four sustainability dimensions, and SDG promotion served as the evaluative basis, this study did not explicitly document formal training processes to standardize evaluation criteria among raters, nor were inter-rater reliability measures calculated (such as the Kappa coefficient or correlation), which should be considered a methodological limitation.

The formal analysis of the 212 projects was supported by a rigorous data curation process, collaboratively executed by the four authors. This procedure involved two sequential phases: first, the verification of the effective implementation of the CFS-IRA Principles as a fundamental inclusion criterion; second, the removal of records with incomplete or inconsistent data, as can be seen in Table 1. Only after this comprehensive validation were the simple averages for the sustainability dimensions calculated, thereby ensuring result reliability.

Table 1. Assessment rubric used in the analysis.

Principle—Description (Dimension)	Observable Indicator	Data Source/Criterion
P1—Contribute to food security and nutrition (economic)	Degree to which projects promote food security and nutrition.	CBA project database (scoring system P1–P10, 0–5 scale).
P2—Contribute to sustainable and inclusive economic development and poverty eradication (economic)	Extent to which projects foster local economic development, employment, or sustainable income generation.	Project files and institutional CBA-UPM documentation (scale 0–5).
P3—Promote gender equality and women's empowerment (social)	Degree of integration of gender perspective or female participation in the projects.	Participation records and activity descriptions (scale 0–5).
P4—Enhance participation and youth empowerment (social)	Level of youth involvement in the planning or implementation or as main beneficiaries of initiatives.	Activity reports and participation records (scale 0–5).
P5—Respect tenure rights to land, fisheries, and forests and access to water (environmental)	Extent to which projects promote equitable access to and sustainable management of natural resources.	Technical project sheets and result reports (scale 0–5).
P6—Sustainably manage natural resources, build resilience, and reduce disaster risk (environmental)	Degree of incorporation of sustainable environmental practices or resilience-strengthening measures.	Project documentation and intermediate evaluations (scale 0–5).
P7—Respect cultural heritage and traditional knowledge to foster diversity and innovation (economic)	Degree of recognition or integration of traditional knowledge, local culture, or social innovation.	Activity descriptions and technical reports (scale 0–5).
P8—Promote safe and healthy agri-food systems (environmental)	Extent to which projects integrate sustainable agricultural practices or initiatives for safe food systems.	Project files and implementation reports (scale 0–5).
P9—Embed inclusive and transparent governance structures and grievance mechanisms (governance)	Degree of openness, transparency, and participation in project management and decision making.	Governance documents and institutional reports (scale 0–5).
P10—Assess and address impacts; ensure accountability (governance)	Existence and quality of evaluation, monitoring, and accountability mechanisms.	Final reports and internal/external evaluations (scale 0–5).

Likewise, SL projects are integrated in the curriculum of the master's degree and doctorate in Rural Development Project Planning and Sustainable Management at UPM. In 2009, this program was accredited within the IPMA Registry of Skills Development Programs, being a pioneer in the application of these competencies to sustainable rural development in Spain. In addition, the International Postgraduate Program for Sustainable Rural Development—Erasmus Mundus—has been developed with the participation of five universities of the European Union, which formed the Agris Mundus Alliance for Sustainable Development [87]. The development and assessment of competencies in this program are supported by the framework of the International Project Management Association (IPMA) based on the IPMA Competence Baseline (ICB) standard, and since 2016, the program has integrated the Responsible Investment in Agriculture and Food Systems (CFS-IRA)

Principles in its learning curriculum. The program is integrated into the meta-university and has been consolidated in an international network of 46 universities and 52 business organizations in 12 countries in Latin America, the Caribbean, and Spain committed to the implementation of these principles (https://principiosirametauniversidad.com, accessed on 29 September 2025) aligned with the Sustainable Development Goals (SDGs).

Among these CFS-IRA Principles, No. 4, focused on youth empowerment, stands out as it is directly aligned with target 4.7 of SDG 4, which promotes education oriented towards sustainable development.

Below, Table 2 summarizes the curricular contents of the master's degree, identifying the learning objectives, the IPMA competencies applied, and the service actions linked to the SL projects.

Table 2. Curricular contents of master's degree linked to SL model and IPMA competencies.

Subject	SDG/CFS-IRA Learnings	Competencies and Skills According to IPMA	Service Actions	Related SDGs	Related CFS-IRA Principles
Planning in the public sphere: from knowledge to action	Identification of community needs, participatory diagnoses, and design of interventions.	Teamwork, planning, and control in real contexts.	Management of local projects and promotion of community cooperation.	SDGs 1, 11, and 17	P1, P4, and P9
Project management for rural/local development	Execution of real projects and critical reflection on their impact.	Self-reflection, personal communication, and leadership.	Facilitation of forums and presentation of results to communities and stakeholders.	SDGs 8, 10, and 16	P2, P4, and P10
Basis for competency in project management	Evaluation of opportunities for intervention and execution of projects with social impact.	Strategy, stakeholders, planning, and control.	Development of initiatives applied in the community with a focus on sustainability and results.	SDGs 1, 11, and 13	P2, P5, and P6
Technologies for the study of the territory: remote sensing and GIS	Solving territorial problems using GIS tools and remote sensing.	Procurement, design, and personal communication.	Creation of participatory maps and management strategies for Food Banks.	SDGs 11, 13, and 15	P5, P6, and P8
Human Development—development magazine	Group dynamics, leadership, and project management in community and intercultural contexts.	Relationships and engagement, leadership, and planning.	Design of sustainable, student-led community projects.	SDGs 5, 10, and 16	P3, P4, and P9

Source: Authors' elaboration based on the curriculum of the master's degree in Rural Development Project Planning and Sustainable Management (UPM).

3. Results and Discussion

The research results obtained following the integration and analysis of the information collected during 2014–2024 are presented below. The SL projects carried out by the FBC were categorized into three lines of action to facilitate their analysis: training and awareness raising, dissemination and transfer of knowledge, and research. Table 3 summarizes the 212 projects, providing the average and total scores per dimension (social, environmental, economic, and governance). Additionally, Figure 3 compares the percentages of the dimensions across these categories.

Table 3. Projects by line of action.

Line of Action	Project Type	N°	Social	Environmental	Economic	Governance	Overall Rating
Dissemination and transfer	Scientific article	9	1.00	1.00	2.00	3.00	1.7
	Outreach article	2	2.00	1.67	2.67	4.00	2.5
	BLOG FBC-UPM	5	1.00	1.00	2.00	3.00	1.7
	Bulletin	2	1.00	1.00	1.33	3.00	1.5
	International congress	17	3.00	3.33	4.00	4.00	3.6
	National congress	2	2.00	2.00	2.33	4.00	2.5
	eBook	6	1.50	1.67	3.00	3.50	2.4
	Interview	2	1.00	1.00	1.33	3.00	1.5
	International conference	6	3.00	3.00	4.00	4.50	3.6
	National Day	11	2.00	2.00	2.67	4.00	2.6
	Social media	11	2.00	2.00	3.00	4.00	2.7
Total for line of action		73	1.77	1.79	2.58	3.64	2.39
Training and awareness	SL model: TFG-TFM	26	4.00	2.33	3.00	5.00	3.4
	SL model: volunteering	15	4.00	2.67	3.33	4.50	3.5
	Educational cooperation	46	4.50	3.00	4.00	5.00	4
	Other activities	4	2.00	1.67	1.67	3.00	2
	Internship	3	3.00	2.33	3.33	4.00	3.1
	International seminar	8	3.00	3.00	4.00	4.50	3.6
	National seminar	1	2.00	2.00	2.67	4.00	2.6
	GIS-FESBAL	5	3.00	3.00	3.67	4.50	3.5
Total for line of action		108	3.19	2.50	3.21	4.31	3.21
Research	Research report	25	2.00	4.00	4.33	4.50	3.8
	Research project	6	3.50	4.67	4.67	5.00	4.5
Total for line of action		31	2.75	4.33	4.50	4.75	4.15

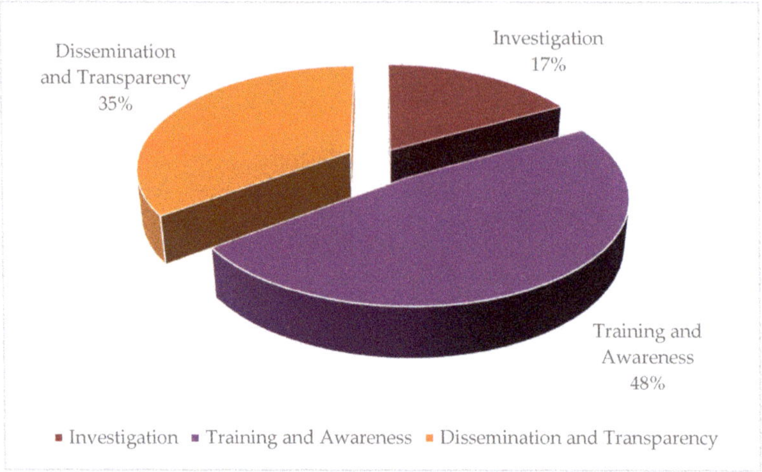

Figure 3. Project percentages by line of action.

3.1. Global Scope of Service Learning (SL) Projects

The SL-WWP model has generated a relevant impact on the training of students and teachers, as well as on the beneficiary communities, when implemented in the Service Learning (SL) projects developed along three lines of action: training and awareness (49%), research (17%), and dissemination and transfer of knowledge (34%).

The scope of the impact is evaluated based on the projects' contribution to society, particularly children and young people, the main recipients of the various initiatives [88]. It should be noted that this impact is also evident in the sustainability of many of these projects, such as the CORAL Program, which has been implemented over time since the model was established. This sustained growth has favored cooperation among different organizations, mainly Food Banks, promoting positive experiences with the most vulnerable sectors of society. The projects are coordinated by the FBC, which makes joint decisions with FESBAL and the UPM team, generating effective solutions for various social problems throughout the years. Table 4 shows the extent of these impacts.

Table 4. Impact of Service Learning (SL) projects.

SL Reach Indicators	2013/ 2014	2014/ 2015	2015/ 2016	2016/ 2017	2017/ 2018	2018/ 2019	2019/ 2020	2020/ 2021	2021/ 2022	2022/ 2023	2023/ 2024
SL projects carried out	16	19	16	24	12	12	16	15	24	37	21
Total number of entities involved	21	439	701	395	382	385	219	237	378	732	1099
Number of direct participants in SL	93	4459	3921	7969	8031	10,288	24,342	7549	8887	48,603	42,906
Number of participating schools (CORAL)		12	55	81	84	124	34	65	82	110	133
Number of participating BANs (CORAL)		4	9	10	13	16	6	9	11	19	19
Number of participating children (CORAL)		1580	2704	3809	4000	5124	758	3725	4284	4786	4869

Over the years, some projects, such as the CORAL Program (in the area of awareness), have been consolidated, reflecting growing acceptance of this methodology among Food Banks and educational entities. The cumulative impact of SL projects during the period analyzed reflects that this model is a key pillar in academic training, providing transformative and meaningful educational experiences [89–91].

Entities Involved According to Project Areas

The analysis of the entities involved according to project area reveals significant differences in the levels of participation, depending on the type of entity and the thematic axis addressed. Firstly, it is observed that public entities have outstanding participation rates in all areas, being especially prominent in training and awareness, with 2031 entities (46.70%), and in dissemination and transfer, with 191 entities (42.26%). This finding coincides with previous studies that highlight the central role of public institutions in promoting sustainable development policies and environmental education [92,93]. These entities mainly include regional and local governments, as well as public educational entities that actively collaborate in awareness and training programs.

On the other hand, private entities, mainly companies, show balanced participation in the different areas, with 1139 entities (26.19%) in training and awareness, 151 entities (33.41%) in dissemination and transfer, and 10 entities (5.35%) in research. Among these, the company ESRI Spain stands out, as it contributes to the GIS-FESBAL project by offering use of its technology, an innovative Geographic Information System (GIS) aimed at monitoring FBC projects [94]. This trend could reflect greater orientation towards technological innovation and product development rather than towards social awareness, as suggested by the literature on corporate social responsibility [95].

Civil society entities show outstanding participation rates in all SL project areas, standing out in research, with 161 entities involved (86.10%); the records show 1179 (27.11%) civil society entities in training and awareness and 110 (24.34%) in dissemination and transfer, as can be seen in Table 5. This diversity indicates a great adaptability of third-sector organizations, which usually act as a bridge between local communities and public policies, also collaborating in the generation and application of knowledge [96].

Table 5. Entities involved according to project area.

Area	Total, Entities	Public Entities	% Public	Civil Entities	% Civil Society	Private Entities	% Private
Dissemination and transfer	452	191	42.26	110	24.34	151	33.41
Training and awareness	4349	2031	46.7	1179	27.11	1139	26.19
Research	187	16	8.56	161	86.1	10	5.35

In general terms, it can be concluded that public entities lead in the number of participants in SL projects, followed by civil society entities and private entities. These results support the idea that the public sector assumes a key role in the implementation of development projects, while the private sector and civil society act as complementary collaborators with more specialized approaches.

3.2. *Service Learning (SL) Projects Aimed at Training and Raising Awareness*

In the category of training and awareness, 103 SL projects have been implemented, representing 49% of the total. These projects are the basis for concretely contributing to addressing the challenges associated with food production and consumption patterns while promoting responsible investments in the agricultural sector, in line with the principles of the CFS-IRA Principles and the Sustainable Development Goals (SDGs) [69,97].

One of the most outstanding projects is the CORAL Program, aimed at raising awareness about responsible food consumption. This initiative, coordinated by the FBC and FESBAL, involves educational centers throughout Spain. Each of the ten annual editions is an SL project mainly aimed at primary school students and organized into four phases: preliminary, regional, national, and awards ceremony. In the preliminary phase, Food Banks, with the support of the Food Bank Chair, carry out educational and awareness-raising activities in the participating schools. In the regional phase, local juries select the best drawings sent by the centers in their jurisdiction. The winning works (three from CET and one from CEE) are registered on the ArcGisSurvey123 platform or sent by email. The national phase involves the evaluation of these works by the jury of the Food Bank-UPM Chair, which proposes the winners and honorable mentions to FESBAL. Finally, at the awards ceremony, the four winning drawings at the national level receive a diploma and a prize during the annual "Golden Spike Awards" ceremony organized by FESBAL at CaixaForum—Madrid [98].

As shown in Table 6, the results of the ten editions of the CORAL Program's drawing competition show a progressive expansion both in territorial coverage and in the number of participants, consolidating this initiative as an effective strategy for education and awareness among children and adolescents about food security and sustainability. The participation of 780 schools and more than 38,000 students reflects the project's ability to generate meaningful learning through artistic and participatory methodologies, in line with transformative pedagogy approaches [99,100]. At the same time, the sustained increase in the number of Food Banks involved—from 4 in the first edition to 19 in the most recent ones, with a cumulative total of 116—demonstrates the strengthening of territorial cooperation networks, a key aspect in the framework of the Sustainable Development Goals, particularly SDG 2 (Zero Hunger) and SDG 17 (Partnerships to Achieve the Goals), based on the promotion of cooperation among social, educational, and third-sector actors [101,102]. The results of the drawing contest are illustrated in Figure 4, which shows one of the award-winning works from the latest edition.

Table 6. Results of the SL project of the Drawings in Schools contest carried out by the FBC.

	2014–2015	2015–2016	2016–2017	2017–2018	2018–2019	2019–2020	2020–2021	2021–2022	2022–2023	2023–2024	
Editions	**1ED**	**2ED**	**3ED**	**4ED**	**5ED**	**6ED**	**7ED**	**8ED**	**9ED**	**10ED**	**Total**
Number of Schools	12	55	81	84	124	34	65	82	110	133	780
Number of students/drawings	1580	2704	3809	4000	5124	758	3725	4284	4786	7596	38,366
Number of Food Banks involved	4	9	10	13	16	6	9	11	19	19	116

The drop observed in the sixth edition (2019–2020) in school participation and number of drawings is directly linked to the effects of the COVID-19 pandemic on educational processes and extracurricular activities [103]. However, the immediate recovery in subsequent editions shows both the resilience of the program and the sustained commitment of the educational community. This capacity for adaptation confirms the potential of non-formal education as a space for containment, motivation, and pedagogical continuity in crisis contexts [104]. The use of drawing as a means of expression has been shown to be effective in encouraging students' critical reflection on problems such as food waste, inequality, and solidarity. Previous research has underlined the value of creative practices for the development of socio-emotional and citizenship competencies in childhood [105,106], reinforcing the value of the contest as a comprehensive educational practice. In short, the contest has evolved not only quantitatively but also in its qualitative dimension, constituting a transformative educational experience with the potential for replicability and integration in public policies of education for sustainability.

Figure 4. Third place in the drawing contest, 3rd A Coruña Paula Botana Lozano Colegio San Francisco Javier.

This area also includes the Final Master's Project (FMP), as well as curricular internships and volunteer experiences that are undertaken as SL projects with the international master's degree in Rural Development Project Planning and Sustainable Management. These projects have proven to be an effective way to integrate academic training with social commitment, allowing students to apply their knowledge in real contexts while developing technical and social competencies that are fundamental for comprehensive learning [107–109]. The application of the SL model in the FBP not only responds to a growing trend in the university environment but also constitutes an opportunity to strengthen the link between university and community, generating a transformative impact on both the student and the environment in which they intervene [110,111]. Within the framework of the Food Bank Chair (FBC-UPM/FESBAL), a diverse and transversal academic production has been consolidated in the form of Final Master's Projects (FMPs) and Final Bachelor's Projects (FBPs), which comprehensively address the challenges of the contemporary food system. The more than 25 recent research projects show remarkable thematic diversity, ranging from food security and sovereignty to surplus management, food waste, rural planning, and food public policy analysis, both in local and international contexts.

Among the topics analyzed are alternative food webs, organic certification, Food Banks, the evaluation of market management models, innovative sources of food (such as edible insects), and legislative aspects of food waste based on participatory approaches such as the Working with People model. Likewise, works focused on international cooperation can be identified, with case studies in Ecuador, Cameroon, Nigeria, and Peru, which reinforces the global nature of the food problem and the way in which it is addressed in applied research.

This diversity responds to the strategic objectives of the FBC, which include the fight against food waste, the promotion of food security, the promotion of university social responsibility, environmental sustainability, and development cooperation. In particular, several works are directly aligned with the Sustainable Development Goals (SDGs), highlighting SDG 2 (Zero Hunger), SDG 12 (Responsible Consumption and Production) and SDG 17 (Partnerships to Achieve the Goals).

The academic research performed constitutes a key input for innovation in food policies, the design of sustainable intervention methodologies, and the raising of awareness among public and private actors about the importance of fair, resilient, and sustainable food systems.

Other SL projects are international and national seminars, which have a predefined scope, involve stakeholders, and evaluate learning outcomes, as can be seen in Table 7.

Table 7. Training and awareness projects.

Training and Awareness	Quantity	Percentage
National seminar	1	0.97%
Awareness contests	4	3.88%
Curricular internships	4	3.88%
International seminars	8	7.77%
Cooperative learning day workshops	11	10.68%
Service Learning volunteering	15	14.56%
Master's Degree Final Project	26	25.24%
CORAL Program	34	33.01%
Total	103	100.00%

3.3. Service Learning Projects in the Research Area

A total of 35 research projects have been conducted, representing 17% of the total. These are characterized by their applied orientation, their interdisciplinary approach, and the use of innovative technological tools aimed at solving real problems. The results of these projects are published by FESBAL in the form of technical reports, prepared according to a rigorous scientific methodology.

The topics addressed are diverse and include sustainable production, food waste, product analysis, good practices in the management of Food Banks, the study of poverty in Spain and the role of Food Banks (FBs), the analysis of public policies, research on food waste in school canteens, and the design of strategic plans to align Food Banks' activity with SDGs and the CFS-IRA framework [1,72].

One of the most relevant projects has been the design and implementation of the Geographic Information System (GIS) Geoportal GIS-FESBAL, which provides a technical approach to territorial analysis and contributes to more sustainable planning (Observatory of Good Practices in Food Banks). This project aligns with previous studies that highlight the value of GISs in socio-environmental research [112].

Below are some representative examples of Service Learning (SL) projects in the field of applied research:

- The project "GIS as a Tool for Calculating the Carbon Footprint of Food Banks" [4] involved students in the collection and analysis of data to estimate the carbon footprint generated by the logistic operations of FBs by applying GIS tools and environmental sustainability methodologies [81].
- In the study "Towards Project-Based Governance from the SDGs and ESG Criteria: The Case of Food Banks", the participation of students in the development of governance frameworks based on the SDGs and ESG criteria was promoted, strengthening the institutional and strategic dimension of FBs [113].
- The project "Learning from the Madrid Food Bank Experience: 30 Years of Work on Volunteer Projects to Help Vulnerable People" [114] made it possible to document the good practices accumulated over three decades of volunteer work by using qualitative methods, with the active collaboration of students in the analysis and systematization of data [115].
- In the research study "Exploratory Analysis of Plate Food Waste in School Cafeterias in Spain" [78], students participated in direct observation in schools and in the evaluation

of eating habits, contributing to the design of strategies aimed at the prevention of waste.

- Finally, during the COVID-19 pandemic, the study "Eating Behavior Changes Related to Individual and Household Factors during the COVID-19 Lockdown in Spain" [116], in which students actively participated in the design and application of surveys, as well as in the statistical analysis of the data obtained [117], was carried out.

These projects demonstrate the potential of Service Learning as a methodology that integrates academic training with the generation of useful knowledge for society, strengthening the research dimension of educational processes.

In the studied context, the work dynamics of the SL model are always collaborative, involving the interdisciplinary research team of the FBC and FESBAL. This enables the establishment of a knowledge foundation to address the current challenges of Food Banks and FESBAL in relation to sustainability, which, as pointed out in [118], requires integrating multiple dimensions to achieve significant impacts on public policies and local communities. In addition, the orientation towards practical application is inherent in action research approaches that seek to transform realities through the participation of various actors, as proposed in [119].

The inclusion of the social component in knowledge management and the participation of researchers from other universities and interest groups are key factors in the success of these activities. The authors of [120] similarly underline the relevance of stakeholder engagement in the context of sustainability-oriented projects.

Within the framework of the research initiatives analyzed, three main types of outputs were identified: Research reports represent the largest proportion, with 68.57% of the total, which shows a strong focus on the systematization of results and the generation of applied knowledge. Research projects constitute 17.14%, reflecting the promotion of more structured and far-reaching processes. Finally, the development of the GIS-FESBAL Geoportal represents 14.29%, highlighting the incorporation of technological tools for the visualization and spatial analysis of information related to food safety and the activity of Food Banks.

3.4. *Service Learning (SL) Projects for the Dissemination and Transfer of Knowledge*

Finally, in the category of dissemination and transfer of knowledge, 73 projects were carried out, representing 35% of the total.

These projects are aimed at disseminating results that reinforce academic training with dissemination and transfer actions while helping students develop oral and written communication skills. The students participating in these projects are mainly from the doctoral program in planning rural development projects and sustainable management, coordinated by the research professors collaborating with the FBC.

Various actions aimed at the dissemination and transfer of knowledge have been implemented, covering both general discussion spaces and specialized forums with wider audiences, e.g., international and national conferences presenting communications that enabled the exchange of research advances with the global and national scientific community. The transfer of knowledge is an essential function of universities, as part of their mission to contribute to social and economic development [121]. Through this function, higher education institutions not only generate knowledge but also apply it to solve community challenges, thus promoting collective well-being. This commitment is manifested in the so-called "third mission" of universities, which encompasses activities such as innovation, entrepreneurship, and collaboration with various sectors of society [122]. In this context, education emerges as a transformative tool, capable of empowering communities and fostering a more equitable and sustainable society [123].

The direct impact on beneficiary communities ensures the implementation of evidence-based strategies and promotes sustainable development through innovative and collaborative solutions.

As for the dissemination and transfer of knowledge, 22.97% of projects involved the presentation of communications at international conferences, constituting the most recurrent format. These are followed by national conferences and workshop- or seminar-type events (14.86%), which indicate active participation in training and discussion spaces at the local level. Studies published in research journals accounted for 12.16% of the total, while international conferences and other similar contributions accounted for 9.46% each. On the other hand, studies in popular journals and communications at national congresses showed lower percentages, 2.70% and 1.35%, respectively, evidencing a lower preference for formats involving the general public and discussion spaces of national scope.

3.5. Relationship of Projects with IRA Principles and the SDGs in the Dimensions of Sustainability

A total of 21 projects were evaluated by applying the Committee on World Food Security's Responsible Investment in Agriculture and Food Systems (CFS-IRA) Principles [88] and organized according to the four pillars of sustainability: social, environmental, economic, and governance. This assessment was complemented by the United Nations Sustainable Development Goals (SDGs) framework [124].

The results show that the governance dimension had the highest score, with an average score of 4.0, followed by the economic dimension (3.0), the social dimension (2.4), and the environmental dimension (2.3), as can be seen in Table 8. This suggests a strong institutional commitment to transparency, participation, and organizational management, although significant challenges remain in integrating social and environmental criteria, which are crucial to comprehensive sustainability [125].

Table 8. SL projects carried out by the FBC and their relationship with the CFS-IRA Principles and the pillars of sustainability.

Area	Project Type	No.	P1	P2	P7	P3	P4	P5	P6	P8	P9	P10	Social	Environmental	Economic	Governance	Overall Rating
Dissemination and transfer	Scientific article	9	3	2	1	1	1	0	1	2	3	3	1.00	1.00	2.00	3.00	1.7
Dissemination and transfer	Outreach article	2	4	3	1	2	2	0	2	3	4	4	2.00	1.67	2.67	4.00	2.5
Dissemination and transfer	BLOG FBC-UPM	5	2	3	1	1	1	0	1	2	3	3	1.00	1.00	2.00	3.00	1.7
Dissemination and transfer	Bulletin	2	2	2	0	1	1	0	1	2	3	3	1.00	1.00	1.33	3.00	1.5
Dissemination and transfer	International congress	17	5	4	3	3	3	2	3	5	4	4	3.00	3.33	4.00	4.00	3.6
Dissemination and transfer	National congress	2	3	2	2	2	2	1	2	3	4	4	2.00	2.00	2.33	4.00	2.5
Dissemination and transfer	eBook	6	4	3	2	2	1	0	2	3	3	4	1.50	1.67	3.00	3.50	2.4
Dissemination and transfer	Interview	2	2	2	0	1	1	0	1	2	3	3	1.00	1.00	1.33	3.00	1.5
Dissemination and transfer	International conference	6	5	4	3	3	3	2	3	4	5	4	3.00	3.00	4.00	4.50	3.6
Dissemination and transfer	National Day	11	3	3	2	2	2	1	2	3	4	4	2.00	2.00	2.67	4.00	2.6
Dissemination and transfer	Social media	11	4	3	2	2	2	1	2	3	4	4	2.00	2.00	3.00	4.00	2.7
Training and awareness	SL: FBP-FMP	26	2	4	3	3	5	3	2	2	5	5	4.00	2.33	3.00	5.00	3.4
Training and awareness	SL: volunteering	15	3	5	2	4	4	2	3	3	4	5	4.00	2.67	3.33	4.50	3.5

Table 8. *Cont.*

Area	Project Type	No.	P1	P2	P7	P3	P4	P5	P6	P8	P9	P10	Social	Environmental	Economic	Governance	Overall Rating
Training and awareness	Educational cooperation	46	4	5	3	4	5	3	3	3	5	5	4.50	3.00	4.00	5.00	4
Training and awareness	Other activities	4	2	2	1	2	2	1	2	2	3	3	2.00	1.67	1.67	3.00	2
Training and awareness	Internship	3	3	4	3	3	3	2	2	3	4	4	3.00	2.33	3.33	4.00	3.1
Training and awareness	International seminar	8	5	4	3	3	3	2	3	4	5	4	3.00	3.00	4.00	4.50	3.6
Training and awareness	National seminar	1	3	3	2	2	2	1	2	3	4	4	2.00	2.00	2.67	4.00	2.6
Training and awareness	SIG-FESBAL	5	4	4	3	3	3	2	3	4	5	4	3.00	3.00	3.67	4.50	3.5
Research	Research report	25	5	5	3	2	2	3	4	5	5	4	2.00	4.00	4.33	4.50	3.8
Research	Research project	6	5	5	4	3	4	4	5	5	5	5	3.50	4.67	4.67	5.00	4.5
													50.5	48.33	63.00	84.00	
													2.40	2.30	3.00	4.00	

Therefore, in the considered Service Learning projects, these dimensions show low integration of the related CFS-IRA Principles, thus of environmental and social indicators. This implies that, while there may be an intent to address socio-environmental issues, strengthening the systematic application and measurement of impact across all initiatives should be a priority; it could be accomplished by addressing the identified gaps, including incorporating food waste metrics as Key Performance Indicator (KPIs) in new projects and continuing to explore the inclusion of tools for estimating the carbon footprint.

Among the types of projects evaluated, research project stood out as the most balanced and robust, with a total score of 4.5, standing out especially in the environmental (4.67), economic (4.67), and governance (4.50) dimensions. This performance shows its direct alignment with SDGs 2 (*Zero Hunger*), 8 (*Decent Work and Economic Growth*), 12 (*Responsible Consumption and Production*), and 13 (*Climate Action*) [126].

In the social dimension, educational cooperation projects obtained the highest score (4.5), reflecting their effectiveness in promoting inclusion, collaborative learning, and capacity building, in line with SDGs 4 (*Quality Education*) and 17 (*Partnerships to Achieve the Goals*) [127]. However, this dimension was the least integrated in the set of projects, which indicates the need to strengthen community impact strategies.

From the environmental perspective, although research projects stood out (4.67), most of the initiatives showed medium or low scores, which reveals a gap in the application of agroecological practices and climate resilience, an aspect that should be a priority in future interventions [128].

The economic dimension was also better integrated into projects with research or production orientation. Research projects again stood out, highlighting the potential of linking knowledge production with sustainable economic strategies.

Finally, the governance dimension was the most consistently applied. SL: TFG-TFM projects obtained the maximum score (5.0), reflecting consolidated participatory structures, institutionalization of processes, and transparency, in line with SDGs 16 (*Peace, Justice, and Strong Institutions*) and 17 [129].

A relevant finding is that a high total score does not always imply balance among dimensions. Some projects showed significant internal imbalances, which reinforces the need to design interventions that ensure coherence among the social, environmental, economic, and institutional aspects.

The Service Learning WWP model (SL-WWP), developed over a decade, includes the CFS-IRA Principles and the SDGs, implementing them, e.g., in local learning activities integrating academic programs and focused on raising awareness about food waste and responsible consumption.

Key intermediate results highlight the importance of the model's overall framework for developing student competencies, applying the IPMA competence framework, and promoting the strengthening of partnerships. Furthermore, pilot waste reduction projects such as GIS-FESBAL Geoportal promote information monitoring and management. Among the directly measured results are reach, quality of design, and comparative standardized evaluation using a Likert scale. Inferred results include the contribution to improved food security, as evidenced by the benefits achieved in waste reduction and the application of the Principles.

In conclusion, the joint application of the CFS-IRA Principles and the SDGs has proven to be a useful tool for the analysis and improvement of projects with a sustainable approach. The most successful cases not only score well but also achieve synergies among dimensions, generating transformative, scalable, and inclusive impacts [68,88,129].

Considering the results, which reflect the need to strengthen the balance among the dimensions, a future evaluation using mixed methods is proposed. This approach would capture both efficacy and transformation by applying the following: A quantitative phase (Efficacy) prioritizing the inclusion of direct impact KPIs to measure waste reduction or the optimization of logistics would establish baselines and measure changes in the Food Banks' logistical carbon footprint. A qualitative phase (Transformation) assessing social impact and transformative learning through qualitative studies (interviews, analysis of reflections, etc.) would make it possible to evaluate the acquisition of socio-ethical competencies by students and community cohesion.

This integration would allow for an assessment of overall coherence, revealing whether the high performance in governance (4.0) effectively translates into synergies and the generation of scalable, inclusive, and transformative impacts in the social and environmental dimensions.

4. Conclusions

This research study demonstrates the originality and effectiveness of integrating Service-Learning projects with the Working with People (WWP) methodological model and the Responsible Investment in Agriculture and Food Systems (CFS-IRA) Principles. This combination constitutes an innovative educational approach that tangibly contributes to the advancement of the Sustainable Development Goals (SDGs), with special emphasis on responsible consumption and the reduction in food waste.

Among the most relevant findings is the validation of the SL-WWP model through a decade-long analysis of CBA-UPM-FESBAL, which has enabled the implementation of 212 projects across the lines of action of training and awareness, research, and knowledge transfer. The ability of this model to strengthen transversal competencies in students (technical, social, and ethical) while generating quantifiable positive impacts on vulnerable communities and third-sector actors is proven.

Furthermore, a structural alignment of the interventions with SDGs 2 (Zero Hunger), 4 (Quality Education), 12 (Responsible Consumption and Production), and 17 (Partnerships for the Goals) is identified, reinforcing the relevance of the approach within the framework of the 2030 Agenda. Concurrently, the multi-criterion evaluation based on the sustainability dimensions (social, environmental, economic, and governance) reveals outstanding performance in governance, although a heterogeneous integration of environmental and social criteria is detected in some cases, pointing to a priority area for improvement. This disparity responds to clear institutional mechanisms, such as organizational incentives, and points to the need to establish well-defined coordination structures and reporting systems to ensure the continuity of projects which have prioritized governance aspects. In contrast, the evaluation of environmental and social impacts requires more complex indicators and longer timeframes, gaps which have limited the systematic development of these effects.

Based on the findings obtained, it is recommended to institutionalize the SL-WWP model in the university system by integrating it into curricula and social responsibility programs, alongside the development of standardized evaluation protocols that ensure the balanced incorporation of the social, environmental, economic, and governance dimensions. Likewise, it is essential to promote the transferability of the model to diverse contexts through the systematic documentation of best practices and the establishment of inter-institutional cooperation. Finally, it is suggested to strengthen multisectoral collaboration mechanisms, incentivizing the participation of the public–private sector in the co-design, financing, and scaling of projects, to guarantee the sustainability and long-term impact of the initiatives.

In conclusion, this work consolidates the role of higher education as a transformative agent of change, demonstrating that the strategic cooperation of university, civil society, and public–private sectors through SL projects constitutes an effective pathway for addressing complex challenges, advancing towards a more just, supportive, and sustainable society.

Limitations and Research Perspectives

The methodological approach of this research study inherently imposes limitations on the external validity of the findings, as it focuses on the analysis of a specific case study: the ten-year trajectory of the Food Bank Chair (FBC). Nevertheless, the SL-WWP model implemented within the FBC is considered a replicable proposal for other higher education institutions. Likewise, initiatives derived from this framework, such as the CORAL Program, have demonstrated potential for scalability and transferability.

In terms of limitations identified in the project evaluation, the results revealed significant challenges in the cross-cutting integration of certain sustainability criteria, specifically the social (average score of 2.4) and environmental (2.3) dimensions, which were the least integrated across the evaluated projects, in contrast with the governance dimension (4.0). This disparity underscores the need to strengthen community impact strategies (social dimension) and to bridge the existing gap in the effective application of agroecological practices and climate resilience (environmental dimension) in future interventions.

Additionally, it was found that a high aggregate score does not constitute an indicator of balance among the different sustainability dimensions (social, environmental, eco-nomic, and governance). This finding reinforces the need to adopt a more coherent and comprehensive intervention design that ensures a balanced and synergistic incorporation of all evaluation criteria.

Overall, this research study emphasizes that the integration of the SL model with the CFS-IRA Principles and the SDGs is an effective educational strategy. A clear advantage resides in the governance dimension, which was consistently the highest-rated, demonstrating a strong institutional commitment to transparency and organizational management. Furthermore, as evidenced by the decade-long experience, the implementation of this model has succeeded in strengthening students' technical, social, and ethical competencies, and related projects have demonstrated progressive expansion.

Consequently, future lines of research should aim to strengthen the social and environmental dimensions, systematically addressing their inherent challenges. Such work could build upon existing technological projects, such as the GIS tool for calculating the carbon footprint, and upon community impact strategies that enhance inclusion and learning in alignment with Sustainable Development Goal 4. Finally, it is recommended to focus on assessing the replicability of the SL model in other higher education institutions and on designing interventions that ensure a coherent balance among the four dimensions to generate transformative impacts.

Author Contributions: Conceptualization: P.N.C. and I.J.F.E.; Methodology: P.N.C. and I.J.F.E.; Validation: P.N.C., I.J.F.E., G.A. and C.Z.; Formal analysis: P.N.C. and I.J.F.E.; Investigation: P.N.C., I.J.F.E., G.A. and C.Z.; Resources: P.N.C. and I.J.F.E.; Data curation: P.N.C., I.J.F.E., G.A. and C.Z.; Writing—original draft preparation: P.N.C. and I.J.F.E.; Writing—review and editing: P.N.C., I.J.F.E., G.A. and C.Z.; Visualization: P.N.C., I.J.F.E., G.A. and C.Z.; Supervision: P.N.C. and I.J.F.E.; Project administration: P.N.C.; Funding acquisition: P.N.C. All authors have read and agreed to the published version of the manuscript.

Funding: This research was supported by the collaboration agreement between the Food and Agriculture Organization of the United Nations (FAO) and the Polytechnic University of Madrid (UPM), within the framework of the project for the implementation of the CSA-RAI Principles in Latin America, the Caribbean, and Spain, and by the Food Bank Chair of the Polytechnic University of Madrid.

Institutional Review Board Statement: Ethical review and approval were waived for this study because it did not involve human participants or animals, and no personal data were collected. The Ethics Committee of the Polytechnic University of Madrid and the GESPLAN Research Group reviewed the documentation for this study and concluded that as no personal information from participants was used, the project did not require institutional ethical approval or review. Therefore, a formal waiver was granted.

Informed Consent Statement: Informed consent was obtained from all subjects involved in the study.

Data Availability Statement: All data regarding the results of this research study are available.

Acknowledgments: The authors gratefully acknowledge the support of the GESPLAN Research Group and the RU-RAI University Network, whose collaboration and contributions were essential to the development of this study. The authors also wish to thank FESBAL (Federación Española de Bancos de Alimentos) for its openness and for providing the opportunity to work collaboratively. During the preparation of this manuscript, the authors used Consensus and SciSpace to assist in research synthesis and verification. The authors reviewed and edited the output and take full responsibility for the final content of this publication. The authors gratefully acknowledge the support of the GESPLAN Research Group and the RU-RAI university network, whose collaboration and contributions were essential to the development of this study.

Conflicts of Interest: The authors declare no conflicts of interest.

References

1. Sibbel, A. Pathways towards Sustainability Through Higher Education. *Int. J. Sustain. High. Educ.* **2009**, *10*, 68–82. [CrossRef]
2. Žalėnienė, I.; Pereira, P. Higher Education for Sustainability: A Global Perspective. *Geogr. Sustain.* **2021**, *2*, 99–106. [CrossRef]
3. Global Footprint Network. Earth Overshoot Day 2024 Approaching. Available online: https://www.overshootday.org/ (accessed on 29 September 2025).
4. World Health Organization. Hunger Numbers Stubbornly High for Three Consecutive Years as Global Crises Deepen: UN Report. 2024. Available online: https://www.who.int/news/item/24-07-2024-hunger-numbers-stubbornly-high-for-three-consecutive-years-as-global-crises-deepen--un-report (accessed on 29 September 2025).
5. United Nations Environment Program. Food Waste Index Report 2024: Think Eat Save—Tracking Progress to Halve Global Food Waste. 2024. Available online: https://sdg2advocacyhub.org/wp-content/uploads/2024/03/food_waste_index_report_2024.pdf (accessed on 29 September 2025).
6. Ministry of Agriculture, Fisheries and Food (MAPA). The Ministry of Agriculture, Fisheries and Food Allocates an Additional 100 Million in Direct Aid from the CAP. 2025. Available online: https://www.tridge.com/ko/news/the-ministry-of-agriculture-allocates-an-add-dripwq (accessed on 29 September 2025).
7. Buzby, J.C.; Hyman, J. Total and Per Capita Value of Food Loss in the United States. *Food Policy* **2012**, *37*, 561–570. [CrossRef]

8. Beretta, C.; Stoessel, F.; Baier, U.; Hellweg, S. Quantifying Food Losses and the Potential for Reduction in Switzerland. *Waste Manag.* **2013**, *33*, 764–773. [CrossRef]
9. Mourad, M. Recycling, Recovering and Preventing "Food Waste": Competing Solutions for Food Systems Sustainability in the United States and France. *J. Clean. Prod.* **2016**, *126*, 461–477. [CrossRef]
10. Leal Filho, W.; Salvia, A.L.; Frankenberger, F.; Akib, N.A.M.; Sen, S.K.; Sivapalan, S.; Novo-Corti, I.; Venkatesan, M.; Emblen-Perry, K. Governance and Sustainable Development at Higher Education Institutions. *Environ. Dev. Sustain.* **2021**, *23*, 6002–6020. [CrossRef]
11. Young, B. Benefits of Volunteerism: From Extracurricular to Service Learning and Beyond. *Int. J. Art Des. Educ.* **2024**, *43*, 37–50. [CrossRef]
12. Principato, L.; Secondi, L.; Pratesi, C.A. Reducing Food Waste: An Investigation on the Behaviour of Italian Youths. *Br. Food J.* **2015**, *117*, 731–748. [CrossRef]
13. Hyun Kim, J. A Hyperlink and Semantic Network Analysis of the Triple Helix (University-Government-Industry): The Interorganizational Communication Structure of Nanotechnology. *J. Comput.-Mediat. Commun.* **2012**, *17*, 152–170. [CrossRef]
14. Larrán Jorge, M.; Herrera Madueño, J.; Calzado, Y.; Andrades, J. A Proposal for Measuring Sustainability in Universities: A Case Study of Spain. *Int. J. Sustain. High. Educ.* **2016**, *17*, 671–697. [CrossRef]
15. Morone, P.; Falcone, P.M.; Imbert, E.; Morone, A. Does Food Sharing Lead to Food Waste Reduction? An Experimental Analysis to Assess Challenges and Opportunities of a New Consumption Model. *J. Clean. Prod.* **2018**, *185*, 749–760. [CrossRef]
16. Ríos-Carmenado, I.D.L.; Mereles, M.L.A.; Deza, X.N.D.L. Connecting Sustainable Rural Development Projects and Principles of Responsible Investment in Agriculture and Food Systems from WWP Model: Lessons from Case Studies in Seven Countries. *arXiv* **2025**, arXiv:2025090058.
17. Rodríguez-Zurita, D.; Jaya-Montalvo, M.; Moreira-Arboleda, J.; Raya-Diez, E.; Carrión-Mero, P. Sustainable Development through Service Learning and Community Engagement in Higher Education: A Systematic Literature Review. *Int. J. Sustain. High. Educ.* **2024**, *26*, 158–201. [CrossRef]
18. Aramburuzabala, P.; Cerrillo, R. Service-Learning as an Approach to Educating for Sustainable Development. *Sustainability* **2023**, *15*, 11231. [CrossRef]
19. Molderez, I.; Fonseca, E. The Efficacy of Real-World Experiences and Service Learning for Fostering Competences for Sustainable Development in Higher Education. *J. Clean. Prod.* **2018**, *172*, 4397–4410. [CrossRef]
20. Firman, F.; Prianto, A.; Munawaroh, M.; Timan, A. Does Service-Based Learning Strengthen Students' Basic Skills? *Int. J. Learn. Teach. Educ. Res.* **2024**, *23*, 583–603. [CrossRef]
21. Rosen, M.A. Do Universities Contribute to Sustainable Development? *Eur. J. Sustain. Dev. Res.* **2019**, *4*, em0112. [CrossRef] [PubMed]
22. Jiménez Aliaga, R.; De los Ríos Carmenado, I.; Huamán Cristóbal, A.E. The CFS-IRA Principles as Instruments for the Management of Rural Development Projects: The Case of the Central Highlands of Peru. In Proceedings of the 26th International Congress on Project Management and Engineering, Terrassa, Spain, 5–8 July 2022.
23. Revilla-Cuesta, V.; Skaf, M.; Gentle-Black, J.; San-José, J.; Ortega-López, V. Train Future Agricultural Engineers at the University of Burgos, Spain, through a Service-Learning Project on Rural Depopulation and its Social Consequences. *Educ. Sci.* **2023**, *13*, 267. [CrossRef]
24. Martín Sánchez, A.; González Gómez, D.; Jeong, J.S. Service-Learning in the Education System: Evaluation of the Degree of Implementation as a Tool to Work on Education for Sustainability. *Apex. J. Sci. Educ.* **2024**, *8*, 38–52. [CrossRef]
25. Caspersz, D.; Olaru, D. Why Do Students Differ in the Value They Place on Pro-Social Activities? *J. Sociol.* **2013**, *51*, 1017–1038. [CrossRef]
26. Chenier, K.A.; Englebretson, E.; James-Barry, J.A.; Rigsby, A.N.; Rodolfich, A.E.; McQueen, E.P.; Sparks, E.L. Development of a Community-Driven Waste Reduction Education and Action Program. *Sustainability* **2024**, *16*, 4823. [CrossRef]
27. Martínez-Heredia, N.; Corral-Robles, S.; González-Gijón, G.; Sánchez-Martín, M. Exploring Inequality Through Service Learning in Higher Education: A Bibliometric Review Study. *Front. Psychol.* **2022**, *13*, 826341. [CrossRef]
28. Hunt, N.P. Project Learn: Supporting On-Campus Learning with On-Line Technologies. *Interact. Learn. Environ.* **1999**, *7*, 269–282. [CrossRef]
29. Berman, S. *Service Learning: A Guide to Planning, Implementing, and Assessing Student Projects*; Corwin Press: Thousand Oaks, CA, USA, 2006; ISBN 9781483364216.
30. Boru, N. The Effects of Service Learning and Volunteerism Activities on University Students in Turkey. *J. Educ. Train. Stud.* **2017**, *5*, 146. [CrossRef]
31. Kronick; Robert, F.; Robert, B.C.; Michele, G. *Experiencing Service-Learning*; ProQuest Ebook Central; University of Tennessee Press: Knoxville, TN, USA, 2011.
32. Salam, M.; Awang Iskandar, D.N.; Ibrahim, D.H.A.; Farooq, M.S. Service Learning in Higher Education: A Systematic Literature Review. *Asia Pac. Educ. Rev.* **2019**, *20*, 573–593. [CrossRef]

33. Farber, K. *Real and Relevant: A Guide for Service and Project-Based Learning*; Rowman & Littlefield Publishers: Lanham, MD, USA, 2017.
34. Smith, E.; Kayser, C.; Wallace, M.; Bosnake, A. Soar into STEMed: Examining the Impact of a Service-Learning Internship on a Pre-Service Teacher's Conceptions of Culturally Responsive Teaching. *Educ. Sci.* **2023**, *13*, 552. [CrossRef]
35. Johnson, A.M.; Notah, D.J. Service Learning: History, Literature Review, and a Pilot Study of Eighth Graders. *Elem. Sch. J.* **1999**, *99*, 453–467. [CrossRef]
36. Kenny, M.E.; Gallagher, L.A. Service-Learning: A History of Systems. In *Learning to Serve: Promoting Civil Society Through Service Learning*; Springer: Boston, MA, USA, 2002.
37. Rubio, L.; Escofet, A. *Service-Learning (SL): Keys to Its Development at University*; Octaedro: Barcelona, Spain, 2017.
38. Jacoby, B. Partnerships for Service Learning. *New Dir. Stud. Serv.* **1999**, *1999*, 19–35. [CrossRef]
39. Batlle, R. What Do We Talk About When We Talk About Service-Learning? *Rev. Reflect. Calmly Unravels Knots* **2011**, *972*, 49–54.
40. Bringle, R.G.; Hatcher, J.A. Implementing Service Learning in Higher Education. *J. High. Educ.* **1996**, *67*, 221–239. [CrossRef]
41. Eyler, J.; Giles, D.E., Jr. *Where's the Learning in Service-Learning*; Jossey-Bass: San Francisco, CA, USA, 1999; ISBN 978-0-7879-4687-1.
42. Astin, A.W.; Sax, L.J. How Undergraduates Are Affected by Service Participation. *J. Coll. Stud. Dev.* **1998**, *39*, 251–263.
43. Jacoby, B. (Ed.) *Service-Learning in Higher Education: Concepts and Practices*; Jossey-Bass: San Francisco, CA, USA, 1996; ISBN 978-0-7879-0275-4.
44. Naufal, W.N.A.D.M.; Aris, S.R.S.; Wahi, R. Key Success Factors for Implementation of Service-Learning Malaysia University for Society (SULAM) Projects at Higher Education Level: Community Perspectives. *Asian J. Univ. Educ.* **2024**, *20*, 152–172.
45. Vázquez, B.J.L.; Ledo, A.A.F.M.; Pazos, M.D.C.M.; Jiménez, S.S.; Chávez, V.R.; Castillo, M.Q. Community Social Service: Students Away from the Classroom Building Knowledge in Their Communities. *J. Serv. Sci. Res.* **2017**, *5*, 10–24. [CrossRef]
46. Bruce-Davis, M.N.; Gubbins, E.J.; Gilson, C.M.; Villanueva, M.; Foreman, J.L.; Rubenstein, L.D. STEM High School Research Experiences: Impact on Gifted Students' Science Process Skills and Interest in STEM Careers. *Gift. Child Q.* **2012**, *56*, 3–23. Available online: https://nrcgt.media.uconn.edu/wp-content/uploads/sites/953/2015/04/STEM_eBook.pdf (accessed on 29 September 2025).
47. Biedermann, A.; Muñoz López, N.; Serrano Tierz, A. Developing Students' Skills through Real Projects and Service Learning Methodology. In *Advances on Mechanics, Design Engineering and Manufacturing*; Lecture Notes in Mechanical Engineering; Springer: Cham, Switzerland, 2017; Volume 2, pp. 951–960. [CrossRef]
48. De los Ríos, I.; Cazorla, A.; Díaz-Puente, J.M.; Yagüe, J.L. Project-Based Learning in Engineering Higher Education: Two Decades of Teaching Competences in Real Environments. *Procedia Soc. Behav. Sci.* **2010**, *2*, 1368–1378. [CrossRef]
49. De Los Ríos Carmenado, I.; Afonso Gallegos, A.; Quintanero Lahoz, S.; Ortega Rincón, R.; Nole Correa, P.; Zuluaga, C.; Roncancio Burgos, M. FESBAL-UPM Food Bank Chair and the Service-Learning Projects from the 'Working with People' Perspective. In *Crisis After the Crisis: Economic Development in the New Normal*; Chivu, L., De Los Ríos Carmenado, I., Andrei, J.V., Eds.; Springer International Publishing: Cham, Switzerland, 2023; pp. 79–90. ISBN 978-3-031-30995-3.
50. Notarnicola, B.; Tassielli, G.; Renzulli, P.A.; Castellani, V.; Sala, S. Environmental Impacts of Food Consumption in Europe. *J. Clean. Prod.* **2017**, *140*, 753–765. [CrossRef]
51. Ivanova, D.; Stadler, K.; Steen-Olsen, K.; Wood, R.; Vita, G.; Tukker, A.; Hertwich, E.G. Environmental Impact Assessment of Household Consumption. *J. Ind. Ecol.* **2016**, *20*, 526–536. [CrossRef]
52. Steg, L.; Vlek, C. Encouraging Pro-Environmental Behaviour: An Integrative Review and Research Agenda. *J. Approx. Psychol.* **2009**, *29*, 309–317. [CrossRef]
53. Grilli, G.; Curtis, J. Encouraging Pro-Environmental Behaviours: A Review of Methods and Approaches. *Renew. Sustain. Energy Rev.* **2021**, *135*, 110039. [CrossRef]
54. Giesler, M.; Veresiu, E. Creating the Responsible Consumer: Moralistic Governance Regimes and Consumer Subjectivity. *J. Consum. Res.* **2014**, *41*, 840–857. [CrossRef]
55. Ulusoy, E. Experiential Responsible Consumption. *J. Bus. Res.* **2016**, *69*, 284–297. [CrossRef]
56. Aydin, A.E.; Yildirim, P. Understanding Food Waste Behavior: The Role of Morals, Habits and Knowledge. *J. Clean. Prod.* **2021**, *280*, 124250. [CrossRef]
57. Visschers, V.H.M.; Wickli, N.; Siegrist, M. Sorting out Food Waste Behaviour: A Survey on the Motivators and Barriers of Self-Reported Amounts of Food Waste in Households. *J. Approx. Psychol.* **2016**, *45*, 66–78. [CrossRef]
58. Chinie, A.-C. Challenges for Reducing Food Waste. *Proc. Int. Conf. Bus. Excell.* **2020**, *14*, 819–828. [CrossRef]
59. Zaichkowsky, J.L. Consumer Behavior: Yesterday, Today, and Tomorrow. *Bus. Horiz.* **1991**, *34*, 51–58. [CrossRef]
60. Dimitrieska, S.K.E. Understaning Children As Consumers. *TIJ's Res. J. Soc. Sci. Manag.-RJSSM* **2017**, *6*, 104–110.
61. Feinstein, L.; Bynner, J.; Duckworth, K. Young People's Leisure Contexts and Their Relation to Adult Outcomes. *J. Youth Stud.* **2006**, *9*, 305–327. [CrossRef]
62. Arıkan, I.; Üniversitesi, S.M.; Yüce, K.; Üniversitesi, K.M.; Tunçel, Ö.; Firat, A.; Kutucuoğlu, K.Y.; Saltik, I.A. Consumption, Consumer Culture And Consumer Society. *J. Community Posit.* **2013**, *13*, 182–203.

63. Sims, R.R. Changing an Organization's Culture Under New Leadership. *J. Bus. Ethics* **2000**, *25*, 65–78. [CrossRef]
64. Ahmed, S.; Byker Shanks, C.; Lewis, M.; Leitch, A.; Spencer, C.; Smith, E.M.; Hess, D. Meeting the Food Waste Challenge in Higher Education. *Int. J. Sustain. High. Educ.* **2018**, *19*, 1075–1094. [CrossRef]
65. Buerke, A.; Straatmann, T.; Lin-Hi, N.; Müller, K. Consumer Awareness and Sustainability-Focused Value Orientation as Motivating Factors of Responsible Consumer Behavior. *Rev. Manag. Sci.* **2017**, *11*, 959–991. [CrossRef]
66. United Nations Transforming Our World. *The 2030 Agenda for Sustainable Development. Resolution Adopted by the General Assembly on 25 September 2015 (A/RES/70/1)*; United Nations: New York, NY, USA, 2015.
67. FAO. *FAO and the Sustainable Development Goals: Achieving the 2030 Agenda Through Empowerment of Local Communities*; FAO: Rome, Italy, 2022.
68. Negrillo Deza, X.; Montero, A.C. The Ira Principles as a Basis For Research Between Universities and Companies. Available online: https://www.unae.edu.py/ojs/index.php/facem/article/view/357 (accessed on 29 September 2025).
69. Requelme, N.; Afonso, A. The Principles for Responsible Investment in Agriculture (CFS-RAI) and SDG 2 and SDG 12 in Agricultural Policies: Case Study of Ecuador. *Sustainability* **2023**, *15*, 15985. [CrossRef]
70. Acosta Mereles, M.L.; De los Ríos Carmenado, I.; Ávila Cerón, C.A.; Castañeda Sepulveda, R. Responsible Investment in the Processes of Substitution of Illicit Crops from the "Working with People" Model: Case Study Guaviare (Colombia). In Proceedings of the 27th International Congress on Project Management and Engineering, Donostia-San Sebastián, Spana, 10–13 July 2023.
71. Jiménez Aliaga, R.; De los Ríos Carmenado, I.; Cazorla Montero, A.; Huamán Cristóbal, A.E. Definition of Priority Objectives for the Implementation of the CFS-RAI Principles in the Central Highlands of Peru. In Proceedings of the 27th International Congress on Project Management and Engineering, Donostia-San Sebastián, Spana, 10–13 July 2023.
72. Federación Española de Bancos de Alimentos (FESBAL). 2025. Available online: https://www.fesbal.org/ (accessed on 29 September 2025).
73. The Global FoodBanking Network. How Food Banks Reduce Food Loss and Waste. Available online: https://www.foodbanking.org/reducing-food-loss-and-waste/ (accessed on 24 May 2024).
74. De los Ríos, I.; Cazorla, A.; Sastre, S.; Cadeddu, C. New University-Society Relationships for Rational Consumption and Solidarity: Actions from the Food Banks-UPM Chair. In *Envisioning a Future Without Food Waste and Food Poverty*; Brill: Leiden, ND, USA, 2018; pp. 181–190.
75. Nole, C.; Leonor, P.; Ccancce, A.; Los Ríos, I. Service Learning Projects for Responsible Consumption: 10 Years of Experience from the Food Bank Chair UPM. In Proceedings of the 27th International Congress on Project Management and Engineering, Donostia-San Sebastián, Spana, 10–13 July 2023.
76. Afonso, A.; Sastre, S. *Good Practices for Food Bank Management: Experience Capitalization*; Food Bank-UPM Chair; Polytechnic University of Madrid: Madrid, Spain, 2017.
77. Uceda, G.; De Afonso, A.; De los Ríos, I. *Challenges Facing the Post-2020 Horizon: The New Approach of Food Banks in Spain*; Food Bank Chair; Polytechnic University of Madrid: Madrid, Spain, 2018.
78. Bustamante, M.; Alfonso, A.; De los Ríos, I. Food Waste in School Cafeterias: Quantification and Identification of Possible Conditioning Factors. *Granja: Rev. Cienc. Vida* **2018**, *28*, 20–42.
79. Roncancio Burgos, M.; De los Ríos Carmenado, I.; Zuluaga, C. The SIG Applied in the Management of Projects for Awareness in Responsible Consumption. In Proceedings of the 27th International Congress on Project Management and Engineering, Donostia-San Sebastián, Spain, 12–14 July 2023; AEIPRO: Donostia-San Sebastián, Spain, 2023.
80. Farías Estrada, I.J.; De los Ríos Carmenado, I. Carbon Footprint of Food Banks: Conceptual Framework and Review of Experiences. 2023. Available online: https://portalcientifico.upm.es/es/ipublic/item/10136003 (accessed on 29 September 2025).
81. Farías Estrada, I.J.; De los Ríos, I.; Velilla, C.; Marín Ferrer, C. GIS as a Tool for Calculating the Carbon Footprint of Food Banks. 2024. Available online: https://portalcientifico.upm.es/es/ipublic/item/10307256 (accessed on 29 September 2025).
82. De Los Ríos, I.; Rodríguez, F.; Pérez, C. Promoting Professional Project Management Skills in Engineering Higher Education: Project-Based Learning (PBL) Strategy. *Int. J. Eng. Educ.* **2015**, *31*, 184–198.
83. Cerezo-Narváez, A.; de los Ríos Carmenado, I.; Pastor-Fernández, A.; Blanco, J.L.Y.; Otero-Mateo, M. Project Management Competences and Sustainable Development in Higher Education: Case Studies from Two Spanish Public Universities. preprints.org. 2018. Available online: https://www.preprints.org/frontend/manuscript/9dfdc50ea54eb817c191477c9380c242/download_pub (accessed on 29 September 2025).
84. Stratta Fernández, R.; De los Ríos Carmenado, I.; López González, M. Developing Competencies for Rural Development Project Management through Local Action Groups: The Punta Indio (Argentina) Experience. In *International Development*; Appiah-Opoku, S., Ed.; In tech Open: Rijeka, Croatia, 2017.
85. De los Ríos Carmenado, I.; Sastre-Merino, S.; Díaz Lantada, A.; García-Martín, J.; Nole, P.; Pérez-Martínez, J.E. Building World Class Universities through Innovative Teaching Governance. *Stud. Educ. Eval.* **2021**, *70*, 101031. [CrossRef]

86. Committee on World Food Security (CFS). *Principles for Responsible Investment in Agriculture and Food Systems*; Food and Agriculture Organization (FAO), World Food Program (WFP), International Fund for Agricultural Development (IFAD): Rome, Italy, 2014; Available online: http://www.fao.org/3/a-au866s.pdf (accessed on 29 September 2025).
87. Cazorla, A.; De los Ríos, I.; Afonso, A. *Principles for a Responsible Investment in Agriculture and Voluntary Guidelines on the Responsible Governance of Tenure of Land: Links with University*; Working Paper 2018/1; GESPLAN Research Group Polytechnic University of Madrid: Madrid, Spain, 2018.
88. Meyers, S.A. Service Learning as an Opportunity for Personal and Social Transformation. *Int. J. Teach. Learn. High. Educ.* **2009**, *21*, 373–381.
89. Mur Nuño, C.; De Los Ríos, I.; Mejía, J.; Nole, P. *Good Governance Practices in Food Banks: ESG Criteria from the APS-WWP Model*; University of Zaragoza: Zaragoza, Spain, 2024.
90. Caspersz, D.; Olaru, D. The Value of Service-Learning: The Student Perspective. *Stud. High. Educ.* **2017**, *42*, 685–700. [CrossRef]
91. De Los Ríos Carmenado, I.; Carlos, M.N.; Inclusive, C. Transparent Governance Structures from IRA Principles. Case FESBAL "Food Bank". In Proceedings of the 27th International Congress on Project Management and Engineering, Donostia-San Sebastian, Spain, 10–13 July 2023; pp. 10–13.
92. López, M.A.R. European Higher Education Area-Driven Educational Innovation. *Procedia Soc. Behav. Sci.* **2017**, *237*, 1505–1512. [CrossRef]
93. Herrero, L.M.J. *Circular-Spiral Economy: Transition to a Closed Economic Metabolism*; Ecobook: Madrid, Spain, 2020; Volume 2.
94. Cátedra Banco de Alimentos FESBAL-UPM. *ArcGIS Online GIS FESBAL-UPM: Contribution of Food Banks to Sustainable Development from the CFS-RAI Principles of the Committee on World Food Security*; GIS Application; Polytechnic University of Madrid: Madrid, Spain, 2023.
95. Yang, H.; Shi, X.; Bhutto, M.Y.; Ertz, M. Do Corporate Social Responsibility and Technological Innovation Get Along? A Systematic Review and Future Research Agenda. *J. Innov. Knowl.* **2024**, *9*, 100462. [CrossRef]
96. Salamon, L.M. Putting the Civil Society Sector on the Economic Map of the World. *Ann. Public Coop. Econ.* **2010**, *81*, 167–210. [CrossRef]
97. Sharp, G. Food Production and Consumption. *Capital. Nat. Social.* **2016**, *27*, 117–124. [CrossRef]
98. Food Banks Chair–UPM Food Banks Chair. Available online: https://blogs.upm.es/cba/ (accessed on 31 August 2025).
99. Sterling, S. Transformative Learning and Sustainability: Sketching the Conceptual Ground. *Learn. Teach. High. Educ.* **2010**, *5*, 17–33.
100. UNESCO. *Education for Sustainable Development Goals: Learning Objectives*; United Nations Educational, Scientific and Cultural Organization: Paris, France, 2017.
101. Coque, J. Food Banks as Organizational Networks: A Spanish Case Study. *Sustainability* **2017**, *9*, 643. [CrossRef]
102. FAO. *The State of Food Security and Nutrition in the World 2023: Urbanization, Agrifood Systems Transformation and Healthy Diets across the Rural–Urban Continuum*; FAO: Rome, Italy, 2023.
103. UNESCO. *Education in a Post-COVID World: Nine Ideas for Public Action*; UNESCO: Paris, France, 2020.
104. Barrón Torres, J.G.; Sánchez Limón, M.L. Organizational Resilience: A Theoretical Review of the Literature. *Manag. Stud.* **2022**, *38*, 235–249.
105. Eisner, E. Eight Important Conditions for Teaching and Learning in the Visual Arts. *Art Individ. Soc.* **2002**, *1*, 47–55.
106. Herrera, D.; Figuero, E.; Shapira, L.; Jin, L.; Sanz, M. The New Classification of Periodontal and Peri-implant Diseases. *Sci. J. Span. Soc. Periodontol.* **2018**, *1*, 94–110.
107. De los Ríos Carmenado, I.; Acosta Mereles, M.L.; Farias Estrada, I.J.; Nole Correa, P.; Zuluaga Zuluaga, C.L.; Mur Nuño, C.; García, J. *The Principles of Responsible Investment in Agriculture and Food Systems from the 'La Gran Recogida' Project of the Spanish Federation of Food Banks*; UPM/FESBAL; FAO: Rome, Italy, 2025.
108. De los Ríos Carmenado, I.; Nole Correa, P.; Zuluaga Zuluaga, C.L.; Roncancio Burgos, M. ApS projects from the Working with People–WWP–model in the framework of responsible food consumption: Experience from the UPM Food Bank Chair. In *Service-Learning Experiences at UPM: 2021 and 2022*; UPM: Kuala Lumpur, Malaysia, 2023.
109. Granados-Alós, L.; Catalán-Gregori, B. Application of the Service-Learning Methodology in the University Environment. *Eur. Public Soc. Innov. Rev.* **2025**, *10*, 1–18. [CrossRef]
110. Mungo, M.H. Closing the Gap: Can Service-Learning Enhance Retention, Graduation, and GPAs of Students of Color? *Mich. J. Community Serv. Learn.* **2017**, *23*, 42–52. [CrossRef]
111. Drewery, M.L.; Lollar, J. Undergraduates' Perceptions of the Value of Service-Learning. *Front. Educ.* **2024**, *9*, 1330456. [CrossRef]
112. Deans, T.; Meyer-Goncalves, Z. Service-Learning Projects in Composition and Beyond. *Coll. Teach.* **1998**, *46*, 12–15. [CrossRef]
113. Nuño, M.; De Los Ríos Carmenado, I. From the SDGs and ESG Criteria: The Case of Food Banks. 2024. Available online: https://portalcientifico.universidadeuropea.com/documentos/67bbb7a343384c18067962bc?lang=gl (accessed on 29 September 2025).

114. Ríos, I.d.L.R.; Sánchez, C.P.; Nuño, C.M. Learning from the Experience of the Madrid Food Bank: 30 Years Working on Volunteer Projects to Help Vulnerable People. *Braz. J. Bus.* **2024**, *6*, e70712. [CrossRef]
115. De los Ríos, I.; los Ríos, D. Madrid Food Bank: 20 Years Working on Volunteering Project in Favor of Needed People. 2016. Available online: https://oa.upm.es/46621/1/INVE_MEM_2016_251067.pdf (accessed on 29 September 2025).
116. Villena-Esponera, M.P.; Moreno-Ortega, A.; Baquero, R.A.; Ugarte-Gurrutxaga, M.I.; Moreno-Rojas, R.; de los Rios-Carmenado, I. COVID 19: Eating Behavior Changes Related to Individual and Household Factors during the COVID-19 Lockdown in Spain. *Arch. Latinoam. Nutr.* **2021**, *71*, 13–27. [CrossRef]
117. Sánchez Suárez, Y.; Pérez Castañeira, J.A.; Sangroni Laguardia, N.; Cruz Blanco, C.; Medina Nogueira, Y.E. Current Challenges of Logistics and the Supply Chain. *Ind. Eng.* **2021**, *42*, 169–184.
118. Shmelev, S.E.; Lefievre, N.; Saadi, N.; Shmeleva, I.A. Interdisciplinary Linkages among Sustainability Dimensions in the Context of European Cities and Regions Research. *Sustainability* **2023**, *15*, 14738. [CrossRef]
119. Klima Ronen, I. Action Research as a Methodology for Professional Development in Leading an Educational Process. *Stud. Educ. Eval.* **2020**, *64*, 100826. [CrossRef]
120. Blak Bernat, G.; Qualharini, E.L.; Castro, M.S. Enhancing Sustainability in Project Management: The Role of Stakeholder Engagement and Knowledge Management in Virtual Team Environments. *Sustainability* **2023**, *15*, 4896. [CrossRef]
121. Martín-Rubio, I.; Andina, D. University Knowledge Transfer Offices and Social Responsibility. *Adm. Sci.* **2016**, *6*, 20. [CrossRef]
122. Ballesteros-Ballesteros, V.A.; Zárate-Torres, R.A. Mapping the Conceptual Structure of University–Industry Knowledge Transfer: A Co-Word Analysis. *Publications* **2025**, *13*, 8. [CrossRef]
123. Foti, L.; Warwick, L.; Lyons, E.; Dhaliwal, S.; Alcorn, M. Knowledge Transfer and Innovation: Universities as Catalysts for Sustainable Decision Making in Industry. *Sustainability* **2023**, *15*, 11175. [CrossRef]
124. United Nations. 17 Goals to Transform Our World. Available online: https://www.un.org/en/climatechange/17-goals-to-transform-our-world (accessed on 29 September 2025).
125. FAO. *Biodiversity for Food and Agriculture—Frequently Asked Questions*; Updated Version; FAO: Rome, Italy, 2022.
126. FAO; IFAD; UNICEF; WFP; WHO. *The State of Food Security and Nutrition in the World 2023*; FAO: Rome, Italy, 2023.
127. United Nations Development Programme (UNDP). *Human Development Report 2022: Uncertain Times, Unsettled Lives—Shaping Our Future in a Transforming World*; United Nations Development Programme: New York, NY, USA, 2022.
128. FAO. *The State of The World's Biodiversity for Food and Agriculture FAO Commission On Genetic Resources for Food and Agriculture Assessments*; FAO: Rome, Italy, 2019.
129. United Nations. The Sustainable Development Goals (SDGs). Available online: https://sdgs.un.org/goals (accessed on 29 September 2025).

sustainability

MDPI

Article

Participatory Geographic Information Systems and the CFS-RAI: Experience from the FBC-UPM-FESBAL

Mayerly Roncancio-Burgos [1,2,*,†], Irely Joelia Farías Estrada [1,3,†], Cristina Velilla-Lucini [1,†] and Carmen Marín-Ferrer [1,†]

1 Planificación y Gestión Sostenible del Desarrollo Rural-Local (GESPLAN), Cátedra Banco de Alimentos, E.T.S.I. Agronómica, Alimentaria y de Biosistemas, Universidad Politécnica de Madrid, Avda, Puerta de Hierro 2, 28040 Madrid, Spain; ij.farias@alumnos.upm.es or ij.farias@ce.pucmm.edu.do (I.J.F.E.); cristina.velilla@upm.es (C.V.-L.); carmen.marin@upm.es (C.M.-F.)

2 Unidad de Planificación Rural Agropecuaria, Carrera 10 No. 28-49, Piso 11, Torre A, Bogotá 110321, Colombia

3 Facultad de Ciencias Sociales, Humanidades y Arte, Pontificia Universidad Católica Madre y Maestra, Autopista Duarte Km 1 1/2, Santiago 51000, Dominican Republic

* Correspondence: mayerly.rburgos@alumnos.upm.es

† These authors contributed equally to this work.

Abstract

This paper analyzes the implementation of the Geoportal SIG FESBAL–UPM, a Participatory Geographic Information System (PGIS) developed within the Master's and Doctorate programs in Rural Development Project Planning and Sustainable Management at UPM. The study introduces a model integrated with Project-Based Learning (PBL), the Working With People (WWP) framework, and the CFS-RAI principles to address challenges in responsible food systems. The geoportal designed to be applied at the Food Bank–UPM Chair–FESBAL, acts as an innovative instrument for participation among the different stakeholders enabling the spatialization and analysis of data across social, environmental, and governance dimensions. Functionally, it offers a robust foundation for evidence-based decision-making, systematizes geographic information, and visualizes data via the web, supporting research, training, and community engagement actions. Furthermore, this study details the specific projects and activities developed under the three involved action lines: research, training, and community engagement, identifying strengths and weaknesses in each. The findings affirm that this participatory approach ensures that the proposed solutions are aligned with local needs and priorities, increasing the sustainability and long-term success of the projects implemented through the geoportal.

Keywords: PGIS; project-based learning; sustainable development; responsible investment in agriculture and food systems; CFS-RAI principles; food banks; Working with People

1. Introduction

1.1. Context and Justification

The global agenda for sustainable development, materialized in the Sustainable Development Goals (SDGs) and reinforced by frameworks such as the Principles for Responsible Investment in Agriculture and Food Systems (CFS-RAI) [1], necessitates innovative solutions to complex socio-environmental challenges [2,3]. In this context, the Food Bank Chair (FBC) was established in 2013 on the initiative of the Spanish Federation of Food Banks (FESBAL) and the GESPLAN Research Group at UPM, with the firm purpose of promoting rational consumption through training, research, and knowledge transfer. Its mission closely aligns with mitigating the paradox of food waste and food insecurity, a critical situation where resource disposal coexists with the lack of access to adequate nutrition for vulnerable sectors [4–6].

To strengthen this work and effectively link knowledge with action, the FBC has implemented the WWP model [7]. Its three components (ethical–social, technical–business, and political–contextual) facilitate connection with society through pedagogical strategies such as PBL. These methodologies enable postgraduate students to acquire solid technical training while solving real-world problems, promoting values of sustainability, solidarity, and ethics. Given that the effectiveness of these initiatives depends on understanding the social and territorial context, the FBC has incorporated Geographic Information Systems (GISs), which are established instruments for addressing territorial problems under the premise that every phenomenon occurs in a specific place [8].

The capacity of GISs to represent and analyze information facilitates communities' ability to identify issues and design collaborative solutions, leading to their evolution into PGIS. Through the awareness-raising efforts of the Food Banks (FBs), this study analyzes the perception of utility, accessibility, and practical learning resulting from the construction of a geographic portal, as well as its contribution to the CFS-RAI principles and the SDGs, particularly SDG 2 (Zero Hunger) and SDG 12 (Responsible Consumption and Production). This methodological combination not only incentivizes students to develop solutions adapted to local problems [9] but also strengthens resilience by involving civil society actors and public–private entities in current challenges [10,11].

Integrating the WWP and PBL methodologies from a geographical perspective constitutes the central axis of this work. This integration has facilitated the design and implementation of the FESBAL–UPM Geoportal. A geoportal is essentially a web-based gateway that provides a single point of access to spatial information and services [12]; in this study, it is applied as a PGIS, designed to involve students in decision-making and foster community collaboration [10]. In this regard, the general objective of this article is to analyze how the integration of participatory methodologies in the geoportal's design strengthens learning and collaborative management among the stakeholders linked to the FBC.

To achieve this, the study pursues three specific objectives: first, to design a participatory geoportal that functions as a visualization and analysis tool for FBC projects, raising awareness about rational and responsible food consumption; secondly, to evaluate the tool's contribution across the strategic action lines of teaching, community engagement, and the promotion of applied research, while also analyzing evidence regarding the geoportal's contribution to the CFS-RAI principles and the SDGs; finally, to validate the geoportal as a replicable model of educational and territorial innovation based on PGIS, capable of systematizing the multidimensional impact of the 54 FBs, thereby transcending surplus management to encompass social, environmental, and governance dimensions.

This work contributes to the literature on GIS and sustainability by demonstrating that the integration of PGIS into collaborative frameworks, specifically through the WWP model and PBL, allows for a joint approach to social, environmental, and governance dimensions, overcoming fragmented applications of geospatial technology. The primary scientific contribution lies in providing a transferable methodological framework that operationalizes global ethical principles (CFS-RAI) into local action. Furthermore, it offers an evidence-based governance approach supported by geospatial transparency, which empowers stakeholders to move beyond data management toward sustainable decision-making.

https://doi.org/10.3390/su18031232

1.2. Theoretical Framework

1.2.1. Participatory Geographic Information Systems

GISs constitute fundamental technological tools in the information age, enabling the storage, analysis, and representation of spatially referenced data through the integration of hardware, software, and complex processes [10]. Their capacity to manage and cross-reference socio-economic and environmental variables allows the understanding of the spatial factors that influence territorial development, making them essential instruments for monitoring social change processes and advancing the achievement of the SDGs, whose success largely depends on the precise geographical localization of interventions [13, 14]. Nevertheless, traditional GISs were initially criticized for their technocratic nature, restricted to institutions with privileged access and excluding the local knowledge of communities, which is a key asset for generating contextualized and socially relevant data [14,15].

To overcome these limitations of exclusion and verticality, PGISs emerged. These systems integrate expert technical knowledge with the practical knowledge of society, functioning as dynamic spatial laboratories oriented toward social transformation [15–17]. This approach has strengthened the inclusion of civil society, allowing local perspectives to be incorporated into territorial planning and significantly improving the quality and relevance of the data generated [18]. PGISs have demonstrated their utility in diverse fields such as urban planning, natural resource management, and rural development, although their success depends on the meaningful participation of stakeholders, democratic access to technology, and institutional commitment [15]. In the educational field, they represent a unique opportunity for students and communities to develop geo-spatial skills and become actively involved in territorial development, thereby making the work of actors like the FBs visible and reinforcing their role in the governance of local food systems [7,14].

1.2.2. Project-Based Learning

PBL is defined as a comprehensive pedagogical methodology that fuses theory and practice, enabling students to develop technical, analytical, and professional competencies through the research and resolution of authentic territorial problems [19,20]. This approach substantially improves learning by compelling students to apply knowledge in real contexts, and, in addition to strengthening technical training in territorial planning and resource management, it promotes cooperative learning and teamwork, which are essential competencies for modern project management [20,21]. By confronting real-world problems, students transition from a passive role to an active one, generating contextualized viable proposals that are adjusted to available resources, thereby increasing their ethical commitment to the environment [22,23].

Within this framework, the combination of PBL with PGIS constitutes a highly effective, integrated methodology for addressing the complex challenges of sustainable development [7,15]. While PGIS provides the advanced tools necessary to collect, analyze, and visualize spatial data, PBL structures the learning process around applied projects that give social meaning and utility to that data. This synergy facilitates the design of solutions aligned with the CFS-RAI principles, establishing a fundamental framework for guiding projects with positive environmental, social, and economic impact, and promoting a more inclusive and equitable management of food systems [24].

1.2.3. Principles for Responsible Investment in Agriculture and Food Systems

The CFS-RAI principles have the fundamental objective of ensuring that investments in agriculture and food systems effectively contribute to sustainable development, promoting social equity, respect for the rights of local communities, and the generation of fair benefits

https://doi.org/10.3390/su18031232

for all involved stakeholders [1]. In parallel, the 2030 Agenda coherently articulates social, economic, and environmental targets through the 17 SDGs, considering them an "indivisible whole" that requires an integral approach [25]. In this context, the CFS-RAI principles are key to the sustainability of the FBC's projects, as they promote practices that respect human rights, foster environmental sustainability, and generate positive social impacts, addressing critical challenges such as hunger and food waste from a holistic perspective that considers social, environmental, economic, and governance dimensions [26,27].

The incorporation of technologies like PGIS significantly amplifies the scope and applicability of these principles, facilitating the necessary data collection and analysis to evaluate environmental impacts, conserve natural resources, and raise societal awareness [28]. The combination of local knowledge with robust technological tools strengthens communities' capacity to develop resilient projects, aligned with both the CFS-RAI principles and the SDGs. Ultimately, these principles emphasize innovation and research as drivers for strengthening food systems, promoting technologies that increase food security without compromising the resources of future generations. Table 1 details the intrinsic relationship between these principles and the SDGs.

Table 1. CFS-RAI principles.

Principles	Description	Related SDGs
1. Food Security and Nutrition	Improve food security and access to adequate nutrition.	SDG 2: Zero Hunger
2. Economic Sustainability	Promote viable investments that contribute to rural economic development.	SDG 8: Decent Work and Economic Growth
3. Environmental Sustainability	Protect natural resources and strengthen climate resilience.	SDG 13: Climate Action, SDG 15: Life on Land
4. Social Inclusion	Involve smallholder farmers, women, and indigenous communities.	SDG 5: Gender Equality, SDG 10: Reduced Inequalities
5. Human Rights	Respect and protect human, labor, and property rights.	SDG 16: Peace, Justice and Strong Institutions
6. Responsible Governance of Land and Natural Resources	Ensure equitable access and sustainable management of land and resources.	SDG 15: Life on Land
7. Transparency and Accountability	Clear participatory processes subject to monitoring and evaluation.	SDG 16: Peace, Justice and Strong Institutions
8. Strengthening of Food Systems	Improve supply chains, reduce losses, and increase resilience.	SDG 12: Responsible Consumption and Production
9. Innovation and Research	Develop technologies that improve productivity and reduce environmental impact.	SDG 9: Industry, Innovation and Infrastructure
10. International Cooperation	Foster global partnerships for sustainable investments in agriculture.	SDG 17: Partnerships for the Goals

Source: Authors' elaboration based on [1,2].

2. Materials and Methods

2.1. Methodological Approach

The research was developed under a qualitative, interpretive, and participatory methodological approach aimed at understanding how the integration of PGIS, PBL, and the WWP metamodel contributes to strengthening learning, collaboration, and project man-

https://doi.org/10.3390/su18031232

agement within the framework of the FBC UPM–FESBAL Chair. This approach is suitable for analyzing educational, social, and technological phenomena whose understanding depends on the interaction between stakeholders, local knowledge, spatial information, and organizational dynamics. The use of qualitative methods allowed for the identification of perceptions, experiences, and processes of technological appropriation that cannot be captured through conventional quantitative metrics.

Within this framework, the development of the FESBAL–UPM Geoportal was based on the WWP metamodel, which articulates community participation with training processes and collaborative planning. It integrates PGIS with the PBL approach and the CFS-RAI principles, positioning communities at the center of the decision-making process, ensuring that technological and educational solutions respond to their needs and aspirations [10]. This metamodel is synthesized in three fundamental components, closely aligned with the dimensions of sustainable development [26], as detailed below and illustrated in Figure 1:

- Ethical–Social: linked to the social dimension, this component emphasizes cooperation, solidarity, and social responsibility, guiding projects toward generating tangible benefits for vulnerable groups.
- Technical–Business: related to the environmental dimension, this connects technical knowledge and geo-spatial tools with social reality through collaboration with associations, donor companies, and FBs.
- Political–Contextual: coherent with the governance dimension, this integrates the institutional and normative environment (FESBAL, FAO, CFS-RAI, SDGs), ensuring the coherence, viability, and sustainability of the initiatives developed through participatory decision-making processes and clear regulatory frameworks.

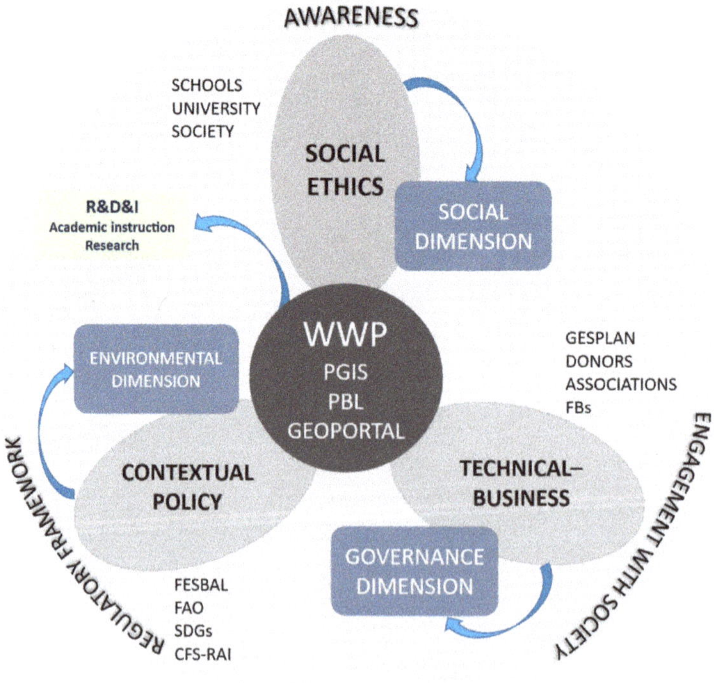

Figure 1. WWP model components applied to the dimensions of the geoportal.

https://doi.org/10.3390/su18031232

The methodological design was structured using qualitative triangulation, which combined three complementary sources of evidence:

- Documentary and Technical Analysis: analysis of the materials and products developed for the geoportal to define its structure and verify its coherence with the project objectives.
- Participatory Processes: processes conducted with faculty, students, and FESBAL volunteers through workshops and co-design sessions; these allowed for the identification of needs, the interpretation of use experiences, and the observation of how meaningful interactions with the technological tools occurred.
- Reflective Analysis: analysis derived from the implementation of PBL applied to the use of PGIS following the WWP model; this examined technological appropriation, spatial reasoning processes, and the integration of the geoportal into real learning and management activities.

The articulation of these three sources made it possible to obtain an integrated and robust vision of the geoportal's role in improving learning, participation, and collaborative management.

The evaluation of the results regarding improvements in governance, transparency, and sustainability was conducted through a qualitative approach, focusing on the analysis of the geoportal's design, implementation, and effective use. The assessment criteria included the degree of stakeholder engagement, the structure and accessibility of geospatial information, and its traceability. Furthermore, the alignment of these processes with international frameworks, such as the SDGs and the CFS-RAI principles, was analyzed. This allowed for an evaluation not only of the system's technical performance (outputs) but also of its impact on coordination dynamics, institutional trust, and shared information management (impacts).

2.2. Methodological Positioning Within the State of the Art

To situate the methodological approach developed in this work within the panorama of similar studies and reinforce its conceptual validity, a comparative analysis was conducted against research that combines participatory, technological, and educational approaches applied to sustainability and food security. The review of the state of the art evidenced that, although relevant studies exist on PGIS for community empowerment, such as those by Orban-Ferauge [15], or on environmental decision-making described by Jankowski [19], these lack explicit educational integration. Furthermore, more recent proposals, such as the Web-GIS modules by Luo and Park [29], while addressing service-learning, do not integrate global governance principles.

Table 2 synthesizes these differences, highlighting the originality of the FESBAL–UPM Geoportal by proposing a solution that not only manages spatial data but also operationalizes ethical (CFS-RAI) and educational (WWP) principles simultaneously.

The comparative analysis performed allows the FESBAL–UPM Geoportal proposal to be situated within the field of participatory and territorial studies, evidencing its added value through the simultaneous integration of educational frameworks (PBL and WWP), CFS-RAI principles, and PGIS tools. This articulation strengthens the study's methodological coherence and offers a conceptual basis for the interpretation of results in subsequent sections.

https://doi.org/10.3390/su18031232

Table 2. Comparison of related methodologies.

Study/Author	Methodology Used	Participatory Focus	Main Findings	Differences from the Present Study
Luo and Park (2020) [29]	Web-GIS modules for service-learning	Students and local communities	Improved digital skills and participation	Does not integrate governance principles or ethical evaluation
Jankowski (2009) [19]	PGIS in environmental decision-making	Citizens and local authorities	Strengthened legitimacy of decisions	Does not incorporate a normative framework for responsible investment
Orban-Ferauge (2011) [15]	PGIS and local empowerment	Community stakeholders	Reinforced territorial leadership	Does not articulate educational methodology or CFS-RAI principles
Current study (FBC-UPM-FESBAL)	WWP + CFS-RAI + PGIS Integration	Academic community, volunteers, and FBs	Geoportal as a tool for learning, management, and transparency	Combines educational dimension (PBL-WWP) with ethical and governance principles (CFS-RAI), linking technology and social responsibility

Source: Authors' elaboration based on [15,19,29].

2.3. Participatory Implementation and Stakeholder Roles

The social, environmental, and governance dimensions underpinning the WWP model are key to maximizing the impact of the FBC's actions. The social dimension drives awareness, volunteering, and collaboration to improve food security for vulnerable groups. The governance dimension promotes transparent and participatory management through collaboration between the university, the public sector, private companies, and civil society. The environmental dimension strengthens sustainability by reducing food waste and ensuring the responsible use of resources, in line with the SDGs.

The geoportal's operationalization was supported by the participation and coordination of the stakeholders linked to the FBC, under the guidance of the WWP model, which fostered horizontality and collective learning. Initially, an interdisciplinary focus group was established with researchers, students, and volunteers to diagnose needs and evaluate technological readiness, which allowed the PGIS objectives to be aligned with the real challenges of food security [9]. FESBAL played a crucial role by providing operational data and validating the content, facilitating interaction with volunteers, and promoting the systematization of best practices at the national level. Its participation was decisive in consolidating the project's social capital and collaborative structure.

Once the priority intervention areas were identified, students from the Master's and Doctoral programs in Rural Development Project Planning and Sustainable Management with GIS training were selected, applying the PBL approach to promote active learning and the resolution of real-world problems. The teams designed technological proposals aimed at managing food surpluses, raising social awareness, and strengthening community commitment [9]. They contributed to the collection, organization, and analysis of data, the design of thematic maps, dashboards, and story maps, and the documentation of best practices. Their participation strengthened both technical and digital competencies and social skills oriented toward responsible action, which constitutes a fundamental pillar of the WWP model's technical–business component.

https://doi.org/10.3390/su18031232

Table 3 details the specific roles and contributions, evidencing how this synergy strengthens human capital (SDG 4) and institutional transparency (CFS-RAI principle 7), ensuring that the technology responds to the expectations and values of the Food Bank network.

Table 3. Roles and contributions of stakeholders.

Stakeholder	Main Role	Contributions	Alignment with SDGs and CFS-RAI
Food Banks and FESBAL	Operational core and coordinator.	Data collection on surpluses and carbon footprint. Methodological coordination with GESPLAN. Georeferencing for project monitoring. Documentation of 37 Best Practices (24 social, 9 environmental, 4 governance).	CFS-RAI principle 7: transparency. Replicability of Best Practices. SDG 12: Responsible Consumption and Production.
Postgraduate Students	Agents of innovation and application.	Technical implementation of the PGIS applying the PBL methodology. Development of solutions for data management and territorial visualization.	SDG 4: Quality Education. CFS-RAI principle 6: strengthening human capital.
GESPLAN-UPM Researchers	Methodological leaders and conceptual architects.	Design of the conceptual architecture integrating PGIS, PBL, WWP, and CFS-RAI. Participatory workshops. Knowledge transfer and promotion of R&D&I. Environmental metrics (carbon footprint).	SDG 13: Climate Action. Ethical–Technical Framework: for social relevance.
Volunteers	Bearers of experiential and local knowledge.	Contribution of experiential knowledge in co-diagnosis workshops. Socio-economic and environmental characterization. Prioritization of problems using a bottom-up approach.	WWP Approach: for co-creation. Operational sustainability and social appropriation of technology.

The meetings and workshops organized by the FBC allowed for consensus on the geoportal's priorities, defining its implementation to achieve the following:

- Territorially analyze the beneficiaries of food aid.
- Map areas with high vulnerability to food insecurity (AROPE indicator).
- Monitor the CORAL program (Rational Food Consumption), active since 2014.
- Identify best practices implemented by the FBs.
- Evaluate and compare the carbon footprint derived from the FBs' activities with that associated with food waste.

Each of these projects was subsequently linked to its corresponding dimension within the WWP model (social, environmental, and governance), as shown in Table 4.

Table 4. Synthetic description of geoportal dimensions and associated products.

Dimension	Description	Output
Social	Analyzes human interactions and social well-being, including the satisfaction of basic needs and quality of life.	Monitoring of awareness actions on rational food consumption (CORAL). Control panel for the AROPE indicator and FBs beneficiaries. Interactive map of social best practices.

https://doi.org/10.3390/su18031232

Table 4. *Cont.*

Dimension	Description	Output
Environmental	Evaluates the environmental impact of food production and transport [3].	StoryMap (available online: https://storymaps.arcgis.com/stories/dc05cac275834828978eac294dd8e02e (accessed on 25 November 2025)) of the FBs' Carbon Footprint. Interactive map of environmental best practices.
Governance	Facilitates monitoring and the identification of areas of highest impact for the follow-up of initiatives [9,10].	Map of the location of the FBs. Control panel of historical beneficiary data. Interactive map of governance best practices.

2.4. Procedural Operationalization: WWP and Geoportal Lifecycle

To ensure methodological reproducibility and address the practical implementation of the WWP model, the process was structured in four operational phases that follow the flow illustrated in Figure 2. This structure ensures that the technical development of the geoportal is deeply rooted in social learning and institutional governance:

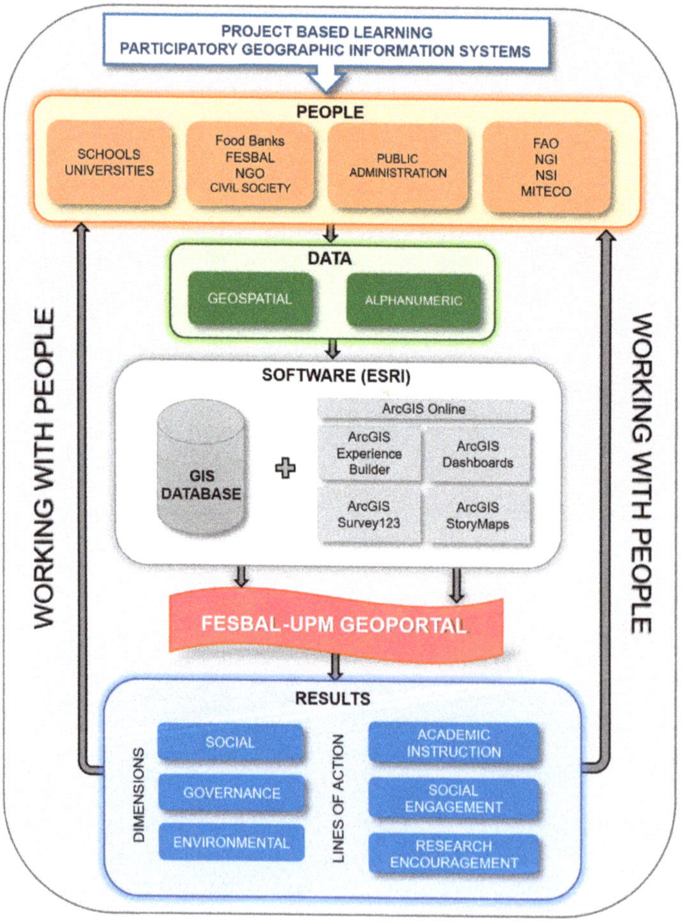

Figure 2. Implementation process of the FESBAL–UPM Geoportal.

https://doi.org/10.3390/su18031232

- Phase 1: Multisectoral Stakeholder Engagement (Ethical–Social Dimension): Following the creation of a multisectoral group (FBs, FESBAL, academia, and Non-Governmental Organizations (NGOs)), this phase operationalized the Ethical–Social dimension. It focused on co-diagnosis workshops to align the geoportal's objectives with the real needs and social values of the food aid network.
- Phase 2: Database Structuring and PBL (Technical–Business Dimension): In this stage, spatial and alphanumeric information was organized within a GIS environment. Students, applying the PBL approach, acted as technical links to ensure the database structure responded to the community's operational requirements.
- Phase 3: Technological Implementation (Technical–Business Dimension): Using Esri technology, raw data were transformed into accessible geospatial resources (maps, dashboards). This phase represents the technical–business dimension, where social learning occurs through the iterative validation of the tools by the stakeholders.
- Phase 4: Synthesis of Results and Governance (Political–Contextual Dimension): Finally, the results were organized into three strategic dimensions: social, environmental, and governance. This stage ensures that the geoportal's outputs are coherent with the institutional framework (CFS-RAI and SDGs), providing a tool for transparent decision-making.

To further clarify the procedural dynamics, Table 5 provides a synthetic matrix of the roles and responsibilities of each stakeholder throughout the geoportal's lifecycle. This matrix operationalizes the interaction between the PBL activities, where postgraduate students lead the technical development and broader WWP dimensions.

Table 5. Operational matrix of stakeholder roles and PBL interaction within the geoportal's lifecycle.

Project Phase (WWP Dimension)	Geoportal Lifecycle and PBL Activities	FESBAL	Vo_FBs/ Vo_Sch	NGOs/Civil Soc	Admin	FBC	Univ_Stu (PBL)
Phase 1: Planning (Ethical–Social)	Contextualization: Identification of social needs and stakeholder mapping.	X				X	
Phase 2: Design and Data (Technical–Business)	Data Creation: Collection and provision of primary/secondary data.	X	X		X	X	X
Phase 3: Implementation (Technical–Business)	Analysis and Validation: GIS processing, dashboard building, and iterative testing.	X	X			X	X
Phase 4: Synthesis (Political–Contextual)	Dissemination and Decision-making: Results sharing, policy alignment, and final use.	X		X		X	X

Abbreviations: FESBAL (Food Bank Federation), Vo_FBs (Food Bank Volunteers), Vo_Sch (School Volunteers), NGOs (Non-Governmental Organizations), Admin (Public Administration), FBC (Food Bank Chair), Univ_Stu (Postgraduate Students), X indicates the stakeholder(s) involved/participating in the corresponding phase and activity.

https://doi.org/10.3390/su18031232

2.5. *Technical Implementation of the Geoportal*

2.5.1. Geoportal Concept and Its Application at FBC FESBAL–UPM

While the general definition of a geoportal as a single point of access to spatial data [12] was introduced earlier, its conceptual and functional complexity warrants further detail in the context of this research. Maguire and Longley [30] describe it as a gateway on the World Wide Web that organizes content and services such as directories, search tools, community information, support resources, data, and applications. For Tait [31], it is simply a website where geographical content can be discovered. Based on these criteria, the FBC implemented the FESBAL–UPM Geoportal as a dissemination and awareness tool with the objective of raising consciousness and promoting rational food consumption through training, knowledge transfer, and the fostering of Research, Development, and Innovation (R&D&I).

Figure 2 synthesizes the geoportal implementation flow, which begins with the creation of a multisectoral group of stakeholders made up of the FBs, FESBAL, academic institutions, NGOs, public administrations, and civil society. The system's database, structured within a GIS environment, is fed by spatial and alphanumeric information. Using Esri technology, these data are transformed into geospatial resources accessible through the geoportal. Finally, the results are organized in three strategic dimensions: social, environmental, and governance, which reinforce the main action lines of the FBC: teaching, social engagement, and research promotion.

2.5.2. Construction and Validation of the Spatial Database

The data collection procedure was rigorously designed to ensure the traceability, reproducibility, and quality of the spatial and alphanumeric information, thus ensuring its coherence with the project objectives. The study area encompasses the entire national territory, integrating the 54 FBs that are part of FESBAL. This constitutes the complete universe of study, not a statistical sample, thereby reinforcing the territorial representativeness of the results. To ensure the reliability of this heterogeneous information, an iterative validation scheme was applied according to the WWP model: the instruments were preliminarily reviewed by the research team and subsequently subjected to cross validation in working sessions with volunteers. Inconsistencies detected were resolved proactively through direct communication with the personnel of each FB, guaranteeing the completeness of the database before its integration. It is important to note that no conflicts of interest existed, as all data were supplied voluntarily under the existing institutional collaboration framework, ensuring the protection of sensitive information.

Operational information primarily originated from the FBs, from whom structural, volunteer, and beneficiary entity data were collected via census surveys. Of particular note is the collection of empirical primary data for the calculation of the carbon footprint using a standardized instrument based on the GHG Protocol [32]. This procedure allowed for the rigorous quantification of emissions in three categories: Scope 1 (direct emissions from fuel and refrigerant consumption), Scope 2 (indirect emissions from electricity consumption), and Scope 3 (other indirect emissions, including donor mobility and waste management), which represents a significant innovation in the sector's environmental management. All this operational information was geocoded and linked to the provincial administrative boundaries layer of the NGI (National Geographic Institute) [33] for cartographic representation.

To provide the territorial and socio-economic context for the FBs' activity, official external indicators and data from complementary programs were integrated. From the NSI (National Statistics Institute), the annual population census [34] and the AROPE index (At Risk of Poverty and/or Social Exclusion) [35] were incorporated; these are essential

https://doi.org/10.3390/su18031232

indicators for spatially correlating the supply of food aid with social demand and territorial vulnerability. Additionally, within the framework of the CORAL awareness program, participating educational centers in the annual drawing contest were geolocated by postal address, allowing for the visualization of the penetration of awareness campaigns. Table 6 summarizes the data sources, typology, and technical description, evidencing the diversity of sources integrated into the system.

Table 6. Data sources, typology, and description.

Data Source	Data	Description
FESBAL	FBs' number of beneficiaries FBs' Good Practices Greenhouse gas (GHG) emissions	Table of the number of beneficiaries by year Classification of Good Practices by FBs Quantification of GHG emissions
CORAL Program	FBs participating in the drawing contest	Table of FBs' participation percentage
NSI	AROPE Index	Table of percentage of population at risk of poverty and/or social exclusion.
	Population census	Population census at the provincial and autonomous community level
Postal address	Geographic location of FBs and schools	Vector layer representing point geometry
NGI	Autonomous communities administrative boundaries Provinces administrative boundaries	Vector layer representing polygon geometry

2.5.3. Technical Architecture and Tools

Once the projects to be included in the database were selected, a series of software applications were used to configure the geoportal's user interface, defining the mode of access, visualization, and analysis of the geospatial information.

The FESBAL–UPM Geoportal was conceived as a dissemination and awareness tool, aligned with definitions of a geoportal as a unified access point to distributed geospatial resources [12,30,33]. The implementation flow, synthesized in Figure 2, starts with the integration of the multisectoral database described above and culminates in the visualization of results organized across the strategic dimensions (social, environmental, and governance).

The technological development was based on the Esri application suite, whose selection is justified by its capacity to guarantee the solidity, interoperability, and technical quality required for a national-scale project. The specific tools were integrated in a complementary manner to cover the entire data lifecycle:

- ArcGIS Online (available online: https://www.arcgis.com/ (accessed on 25 November 2025)): This served as the cloud-based infrastructure, allowing spatial information to be created, analyzed, and shared. Its architecture ensures that all members of the working group have simultaneous collaborative access for viewing, editing, and generating new data.
- ArcGIS Survey 123 (available online: https://survey123.arcgis.com/ (accessed on 16 November 2025)): This was implemented for the data collection phase, allowing for the generation of smart forms for the surveys. Its main contribution was the automatic recording of the geographical location of the information source, minimizing manual geocoding errors.
- ArcGIS Dashboard (available online: https://www.esri.com/en-us/arcgis/products/arcgis-dashboards/overview (accessed on 16 November 2025)): This is a key tool for business intelligence, used to create interactive control panels. These dashboards allow

https://doi.org/10.3390/su18031232

for the monitoring, analysis, and communication of critical information in real-time, facilitating informed decision-making based on spatial evidence.

- ArcGIS Experience Builder (available online: https://www.esri.com/en-us/arcgis/products/arcgis-experience-builder/overview (accessed on 16 No-vember 2025)): This facilitated the development of the final user interface, allowing for the creation of interactive and personalized web applications tailored to the needs of the FBs, without requiring complex code writing.
- ArcGIS StoryMaps (available online: https://storymaps.arcgis.com/ (accessed on 25 November 2025)): These were used as a narrative tool to combine heterogeneous content (maps, 3D scenes, multimedia, and text). Their use was fundamental for "telling the story" behind the data, making complex information (like the carbon footprint) comprehensible to a non-specialized audience.

The use of technology developed by an entity with the trajectory of Esri, especially its specialization towards the SDGs [31], ensures the technological sustainability of the project. Compliance with internationally recognized geospatial standards (such as those defined by the OGC) guarantees that the generated products present the levels of reliability, interoperability, and consistency necessary to achieve the stated objectives and allow for future scalability.

3. Results

3.1. Visualization, Dissemination, and Territorial Analysis

The geoportal's results are structured across the three strategic dimensions of sustainable development, demonstrating its capacity to transform operational data into territorial knowledge.

3.1.1. Social Dimension

In this dimension, the tool has proven effective for awareness-raising and vulnerability identification. Figure 3 illustrates the integration of the CORAL Program results, where the participation of educational centers in the fight against food waste and hunger. is georeferenced. This visualization allows not only for the recognition of school commitment but also for the analysis of the contest's temporal evolution, identifying participation patterns that serve to guide future promotional campaigns [36] and strengthening the institutional image of participating centers through their public testimony of social commitment [37].

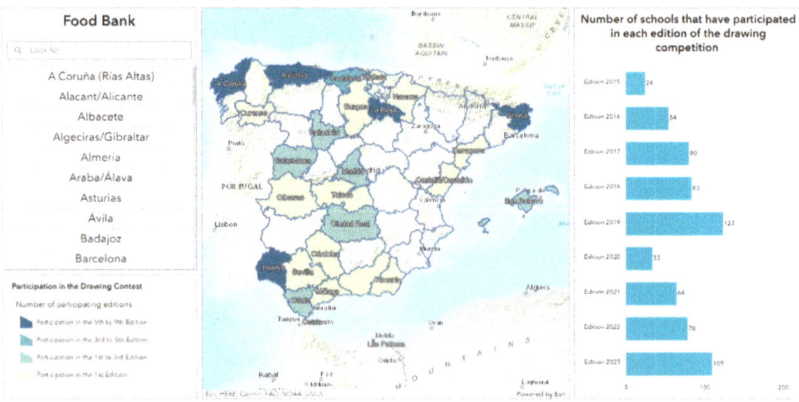

Figure 3. Implementation of the CORAL Program in the geoportal and territorial evolution of the contest. Note: The colors of Spain's provinces represent the map symbology for the level of school participation in CORAL Pro-gram food-waste awareness activities.

Furthermore, Figure 4 presents the spatial overlay between the distribution of FB aid and the AROPE Indicator. This territorial analysis allows for the correlation of solidarity actions with areas of higher risk of social exclusion, transforming abstract statistics into visual evidence of the humanitarian impact in the most needed areas [35].

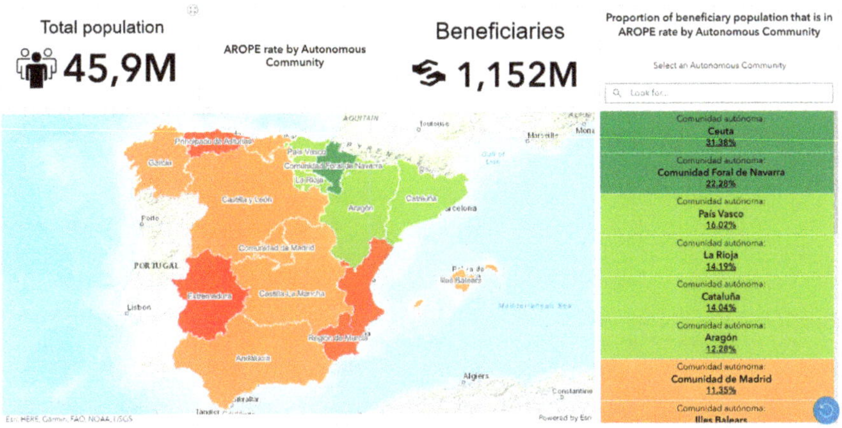

Figure 4. AROPE Indicator and percentage of population benefiting from FBs by autonomous community.

Complementarily, Figure 5 details the localization of the 24 identified Social Best Practices [37], functioning as a reference catalog for the replicability of successful initiatives.

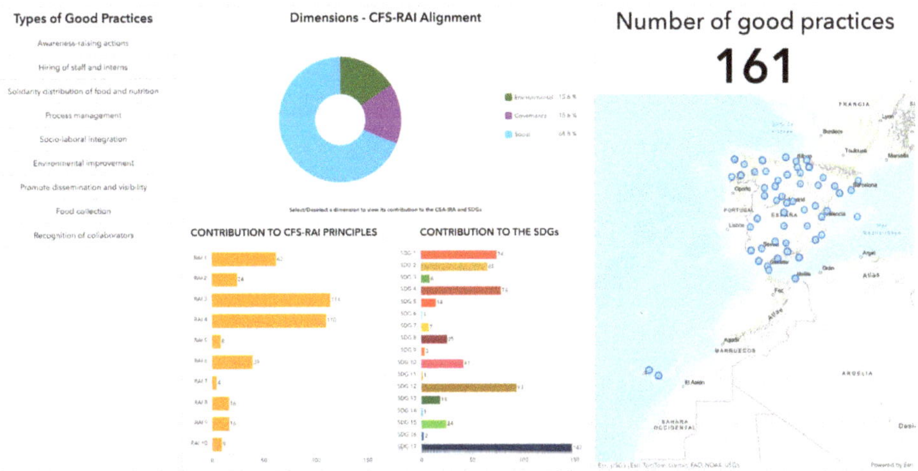

Figure 5. Results of the identification of Best Practices of the FBs.

3.1.2. Environmental Dimension

In the environmental dimension, the geoportal acts as a pioneer observatory for sustainability management through the "Carbon Footprint (CF) Reduction" project [38,39]. The environmental dimension measures the FBs' effect on the environment, as they play a relevant role in reducing the greenhouse gases (GHG) emitted into the atmosphere by taking advantage of goods and resources that would primarily be wasted. This Indicator exposes the amount of emissions released into the atmosphere as a result of any activity and, based on this knowledge, allows for the establishment of effective reduction measures [40].

https://doi.org/10.3390/su18031232

With this focus, the geoportal is perceived as an observatory of the FBs' CF, both in terms of knowledge generation and dissemination. Furthermore, the spatial analysis tools inherent to GIS and incorporated into the geoportal allow for the quantification of the scope of food transport by collaborating and charitable entities in the CF calculation using GIS network analysis tools [39].

The results obtained regarding the CF should not be interpreted as direct evidence of emission reductions; instead, they represent the establishment of a homogeneous and verifiable environmental baseline for the entire network FBs. The geoportal serves as a monitoring infrastructure that enables the systematic collection, validation, and representation of emissions associated with FB activities, creating the technical conditions necessary for future longitudinal comparative analysis.

It is also important to highlight that the geoportal includes nine Environmental Best Practices [41] linked to 15 FBs, offering a model for other FBs to replicate these initiatives and contribute to natural environment protection. This opportunity for replication in the fight against climate change allows FBs to receive support and participate in public calls, acknowledging their commitment to environmental responsibility in line with CFS-RAI principles 5 and 6. Within the educational environment, this initiative is framed as an innovative and transferable concept aimed at raising awareness among future generations [4–6]. Finally, the StoryMaps application has been used to display the project's evolution through maps and narratives (Figure 6).

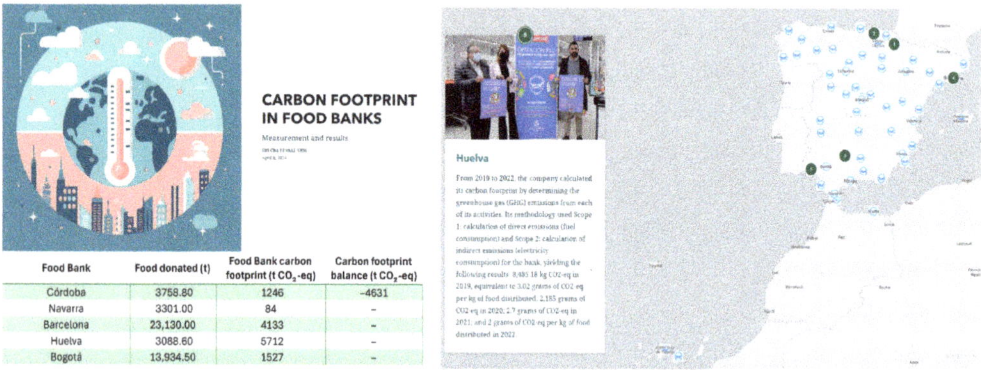

Food Bank	Food donated (t)	Food Bank carbon footprint (t CO_2-eq)	Carbon footprint balance (t CO_2-eq)
Córdoba	3758.80	1246	−4631
Navarra	3301.00	84	-
Barcelona	23,130.00	4133	-
Huelva	3088.60	5712	-
Bogotá	13,934.50	1527	-

(a) (b)

Figure 6. (**a**) Analysis of the FBs' carbon footprint; (**b**) Carbon Footprint StoryMap linked to the FBs' activity.

3.1.3. Governance Dimension

The geoportal transcends its technical function to be validated as a replicable model of territorial innovation that operationalizes the SDG. The governance dimension is linked to the management of the FBs as entities that operate with transparency and independence, promoting efficiency and strengthening trust in their work. This contributes to generating a more positive image of the FBs in society and fostering a better perception of their labor.

This dimension is aligned with CFS-RAI principle 7 (respect heritage rights and provide for transparency), as the geoportal acts as a mechanism for public disclosure and legal–administrative clarity regarding food aid distribution. Furthermore, it integrates principles 9 (innovation) and 10 (cooperation) by enhancing the organization's capacity to adapt to changes and innovate in its processes, such as incorporating new technologies and promoting food culture [41]. It also considers the FBs' ability to effectively involve the different stakeholders in the food chain, such as producers, distributors, retailers, and consumers. These processes of collaborative governance contribute to better coordination of activities and the definition of joint policies and strategies [9,42,43].

The geoportal, therefore, becomes a tool that facilitates the linkage and participation of these interest groups in the FBs' governance and represents an opportunity for all involved stakeholders to work collaboratively toward a more inclusive and sustainable food security. The linking of FESBAL's annual reports with the geolocation of the FBs, donors, and beneficiary entities allows for the monitoring of distributed food quantities over time, effectively operationalizing the transparency and accountability requirements of principle 7, while promoting the dissemination of the banks' actions.

The geoportal incorporates the four Governance Best Practices [41], which are linked to awareness campaigns and dissemination actions on social networks. These contribute to bringing the reality of the FBs and their volunteers closer to society, acting with complete transparency and solidarity.

Figure 7 shows how the geoportal facilitates the visibility of the FBs' transparent and solidarity-based management through the implementation of the Governance Best Practices and the number of beneficiaries of the aid.

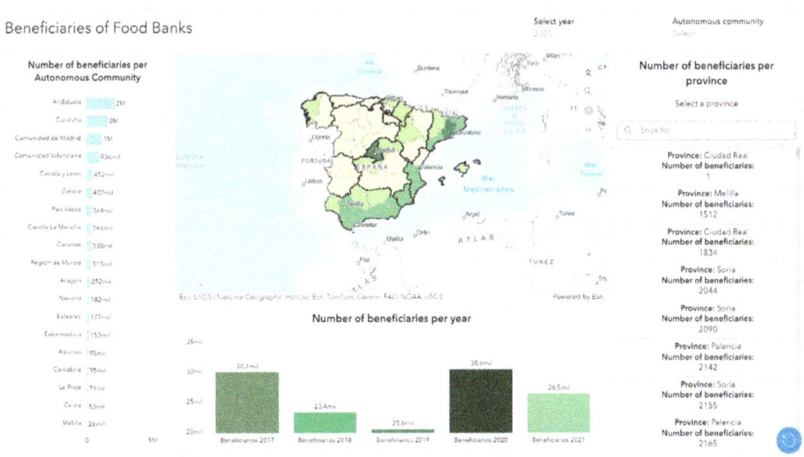

Figure 7. Implementation of Governance Best Practices in the geoportal.

3.2. *Strengthening Teaching, Societal Engagement, and Research*

The FBC's mission is to reinforce actions aimed at combating hunger, poverty, and food waste through awareness and sensitization projects developed by GESPLAN in collaboration with FESBAL. Its initiatives are articulated around three fundamental axes: teaching, societal engagement, and the promotion of research. In this framework, the FESBAL–UPM Geoportal is consolidated as a strategic tool that supports territorial planning and management, integrating academic knowledge with social action to generate real and sustainable impact.

3.2.1. Teaching and Educational Transfer

The geoportal compiles the final Master's degree projects developed from the PBL educational innovation strategy [44], the volunteer actions of university students, and the workshops conducted in schools, thereby reinforcing training and educational cooperation from school to university. This allows for the connection of different educational levels and learning modalities, generating a holistic vision of the FBC's work.

Furthermore, the geoportal incorporates into its database the reports and conclusions derived from conferences, seminars, and roundtables involving professionals and volun-

https://doi.org/10.3390/su18031232

teers linked to the FBs. This documentary integration makes it possible to identify and analyze the relationship between the FBC's work and competencies such as leadership, sustainability, social responsibility, and community cohesion [6,45].

Collectively, these functionalities transform the geoportal into an educational innovation transfer tool, while simultaneously acting as a shared knowledge repository that facilitates the systematization of practices and the strengthening of educational cooperation among stakeholders.

3.2.2. Societal Engagement

Societal engagement materializes through the dissemination and transfer of knowledge from the university to the citizenry. In this sense, the geoportal has been designed as a collaborative tool to provide visibility to the FBC's activities, facilitating content organization, information retrieval, community participation, and data access [30]. When designing the geoportal interface using interactive panels and dashboards, care was taken to ensure that the system is accessible to all users, including those with limited technical knowledge, thereby promoting equitable and transparent participation [29] and greater inclusion in food security-related decision-making processes. By extending access to data and tools, transparency and collaboration among all involved stakeholders are reinforced.

The FESBAL–UPM Geoportal increases the visibility of food security initiatives, attracting the interest of society, public and private entities, as well as potential collaborators who wish to participate in the FBs' activities. The operability offered by the geoportal facilitates the understanding of projects both over time and across space, showing their impact in different geographical locations. The platform encourages citizen participation by motivating their collaboration and serving as a gateway for new partners, not only from the university sphere but also NGOs, private companies, public administrations, and committed citizens.

Throughout the years of the geoportal's development, various dissemination activities have been carried out to highlight its utility and operation, such as: "The FESBAL-UPM Food Bank Chair uses smart maps to fight against waste," "Geospatial technology against food waste," and "The FBC FESBAL-UPM uses smart maps in the Great Food Collection to fight against waste" [44–46].

Table 7 illustrates this evolution by selecting four key temporal milestones: the beginning (2014), model consolidation (2018), the pandemic period (2021), and the present (2024). The data evidence an exponential growth in annual participation, rising from barely 19 entities in the first year to 1454 in the last year, especially highlighting the mobilization of civil society. Analyzing the complete period, the geoportal has managed to integrate an accumulated historical volume of 4988 collaborating entities. This massive increase empirically validates the tool's capacity to mobilize social capital, transform transparency into trust, and strengthen SDG 17 (Partnerships for the Goals), consolidating a robust multisectoral network around food security.

Table 7. Historical evolution of FBC collaborating entities (key milestones).

Type of Entity	2014 (Inception)	2018 (Consolidation)	2021 (Pandemic)	2025 (Current)
Public	6	184	153	272
Civil Society	11	10	8	1.013
Private	2	188	75	169
Annual Total	19	382	236	1.454

Source: own elaboration based on internal FBC data.

https://doi.org/10.3390/su18031232

3.2.3. Promotion of Research

The geoportal's contribution to promoting R&D&I is reflected in the research projects and doctoral theses currently underway. These objectives are directly related to the analytical capacity of GIS, such as the identification of spatial patterns and trends of specific socio-economic indicators, the measurement of the FBs' carbon footprint (incorporating the digitalization of food and volunteer transport networks), or the territorial study of aid to individuals at risk of social exclusion. Effectively, the geoportal's ability to display information that might have otherwise gone unnoticed without map visualization, alongside the incorporation of socio-temporal variables, has allowed for a more complete and detailed vision of reality, especially in the context of food security [45].

The FESBAL–UPM Geoportal not only facilitates the visibility, analysis, and dissemination of the FBC's projects but also plays a crucial role in strengthening ties between the university and society, supporting education, research, innovation, and citizen collaboration [4]. Table 8 illustrates the implemented projects in the geoportal, differentiated by the FBC's action line to which they are linked.

Table 8. Distribution of geoportal projects by action line and strategic dimension.

Action Line	No. Items	Social Dimension	Environmental Dimension	Economics Dimension	Governance Dimension
Dissemination and Transparency	73	38.6%	40.7%	45%	47.6%
Training and Awareness	108	50.5%	41.4%	40.7%	41.1%
Research	31	10.9%	17.9%	14.3%	11.3%
Overall Total	212	100%	100%	100%	100%

3.3. *Model Validation and Contribution to the SDGs*

The geoportal, as a result of the PGIS approach, not only facilitates access to the results of projects in research, training, and societal engagement but also becomes an innovative tool that drives the fulfillment of several SDGs. Although it originated aligned with SDG 2 (Zero Hunger) and SDG 12 (Responsible Consumption) by supporting food security projects and responsible consumption awareness, its scope extends to other SDGs. It strengthens SDG 4 (Quality Education) by providing practical project-based learning, developing students' digital, analytical, and critical skills, especially in the use of geospatial technologies, and fostering applied research and knowledge transfer. Similarly, it contributes to SDG 11 (Sustainable Cities and Communities) by facilitating more informed territorial decision-making through the integration of data on phenomena occurring within the territory, promoting participatory and transparent planning. Finally, it drives SDG 17 (Partnerships for the Goals) by strengthening collaboration among universities, governments, and civil society.

4. Discussion

The results confirm that this study proposes and validates a methodological framework based on the WWP metamodel. By integrating PGIS for the monitoring and evaluation of projects implemented through the PBL strategy, this model focuses on responsible consumption and food security. It is guided by the CFS-RAI principles as a normative and value-based reference, showing clear potential for transferability to other sustainable development contexts. This approach combines methodological innovation, participatory technology, and ethical criteria in project management. Ultimately, the study advances

https://doi.org/10.3390/su18031232

holistic governance, aligning with the SDGs and overcoming fragmented applications of geospatial technology.

From the perspective of the CFS-RAI principles, the FESBAL–UPM Geoportal improves food security and nutrition by generating territorialized evidence. This evidence guides food distribution and fosters transparent participatory governance by integrating multiple stakeholders in data analysis. Moreover, it strengthens human capital by serving as a pedagogical instrument for technical capacity building. It also contributes to environmental and socioeconomic sustainability (principles 4 and 7) by enabling effective balanced planning with reduced environmental impact. Beyond a technical tool, the geoportal is an academic infrastructure for transformative action, strengthening the link between knowledge, participation, and sustainability.

In contrast to traditional static approaches, the geoportal transforms data management into a dynamic collaborative process. The centralization and standardization of operational data for the complete universe of 54 FBs within an interoperable environment is a significant contribution. Unlike previous studies focusing on partial samples, this model enables operational coherence and an emergency response capacity unprecedented in the sector. By providing a unified territorial vision, it ensures that decision-making is based on real-time evidence rather than fragmented reports, facilitating a rapid and coordinated institutional response.

Regarding environmental sustainability, the systematization of CF calculation represents a significant methodological advance. While current literature demands immediate quantitative indicators to evaluate climate impact, this study demonstrates that—particularly within the third-sector context, characterized by high operational heterogeneity and scarce standardized historical data—the construction of an environmental baseline constitutes a necessary scientific contribution. In this sense, the geoportal acts as an enabling infrastructure for environmental monitoring by establishing a standardized methodology based on the GHG protocol and a system for homogeneous data collection and validation across the entire FB network.

This framework lays the technical foundation required for the future development of rigorous longitudinal quantitative analyses. The value of the geoportal resides in its capacity to establish a reliable baseline, without which it would be impossible to accurately measure future emission reductions [38–40]. By prioritizing the creation of this monitoring infrastructure, the model addresses the specific challenges of non-profit organizations, ensuring that future environmental interventions are supported by a solid and traceable data structure. This methodological reflection justifies the study's focus on baseline construction as a prerequisite for any meaningful climate impact assessment within the sector.

However, the current lack of consolidated longitudinal metrics to evaluate CF reduction is a relevant limitation. While the system is operational, its recent implementation precludes temporal trend analysis at this stage. Nevertheless, this work establishes the methodological foundations for future evaluation by defining a replicable baseline for all 54 FBs. This will eventually allow for the assessment of the medium- and long-term impact of environmental interventions through periodic data updates. By acknowledging this limitation, the study provides a structured plan for future research, aligning methodological innovation with actual environmental performance monitoring over time.

From a governance perspective, the study addresses consistency and quality challenges in data generated by non-specialized volunteers. The risk of heterogeneity was effectively mitigated through the WWP collaborative model. The results demonstrate that GIS technology alone is insufficient without cross-validation protocols and social appropriation of information. The FESBAL–UPM Geoportal sets a PGIS standard by linking validation procedures directly to trust-building and institutional transparency (CFS-RAI

https://doi.org/10.3390/su18031232

principle 7; SDGs 16/17). Unlike other citizen science initiatives, this model ensures that reliability stems from social consensus and technical rigor, emphasizing human capital as the core of the system.

Regarding transferability, the study validates the geoportal as a replicable model provided a university–society alliance exists to sustain the pedagogical dimension. Unlike projects focused on isolated technical aspects, this proposal integrates human capital as a structural component, ensuring innovation is social and educational as well as technological. However, challenges such as the digital divide and varying competencies among volunteers must be addressed. Strengthening continuous training and exploring hybrid architectures are necessary for long-term resilience. While developed with Esri tools, the model is adaptable to open-source solutions like QGIS Cloud, enhancing its viability in resource-limited contexts.

Beyond immediate operational outputs, potential impacts transcend the system's initial effects. In governance, a shared environment of validated information strengthens coordination and accountability across the network. Regarding transparency, a single traceable source reduces information asymmetries, fostering a culture of data-driven management. Integrating social, environmental, and governance dimensions allows the geoportal to monitor progress toward the 2030 Agenda, particularly SDG 12. This approach reinforces strategic coherence in the third sector and its international replicability regardless of territorial scales or technological availability, facilitating complex systems analysis through transdisciplinary knowledge.

5. Conclusions

The implementation of the FESBAL–UPM Geoportal demonstrates the value of integrating geospatial tools with participatory approaches to strengthen social initiatives. By articulating the PGIS strategy with the WWP metamodel, PBL, and CFS-RAI principles, this research transformed dispersed and heterogeneous data from 54 Food Banks (FBs) into a unified territorial strategy. This process certified the model's pedagogical capacity, where the PBL methodology acted as a driver of practical innovation, developing essential geospatial and social management skills in students and volunteers, thereby strengthening the organization's human capital.

From an academic perspective, this study expands the PGIS literature by demonstrating its potential as an integrative infrastructure for learning and social action. The integration of methodological frameworks with a global normative ethical framework (CFS-RAI) provides a holistic approach that addresses the limitations of previous research, which often focused on isolated technological or participatory dimensions. This multidimensional model serves as a validated transferable framework for sustainable project management, effectively moving beyond a traditional purely technical application of GIS.

The geoportal has become a structural pillar for FESBAL's accountability and transparency strategy. By operationalizing CFS-RAI principle 7, the system facilitates public disclosure and legal-administrative clarity regarding food aid distribution. This multidisciplinary approach not only promotes participatory governance but also contributes to raising public awareness of poverty and food waste. Through accessible cartographic resources, the project has successfully institutionalized transparency, allowing for the replication of territorial initiatives and expanding the social impact of the national FB network.

Furthermore, the structured architecture (PGIS–PBL–WWP) provides a robust academic infrastructure capable of aggregating information at a national level without losing the link to local realities. The systematization of indicators, such as the carbon footprint (CF), establishes a replicable framework for sustainable monitoring and longitudinal evaluation. As a policy-relevant tool, this research bridges the gap between digital innovation and

https://doi.org/10.3390/su18031232

social accountability, offering a roadmap for policymakers to optimize resource allocation and align institutional governance with the global sustainability agenda and the SDGs.

Finally, the study acknowledges that long-term viability transcends technology, requiring permanent stakeholder collaboration to ensure data validation and effective use. Future research lines are oriented toward developing predictive demand models and logistics optimization through Artificial Intelligence (AI). Additionally, exploring hybrid data architectures will be crucial to reinforcing system resilience against the digital divide. Ultimately, this model positions geospatial technology as an essential driver of social inclusion, ensuring that digital transformation remains an ethical engine for sustainable development and a structural pillar for evidence-based public policy.

Author Contributions: M.R.-B., I.J.F.E., C.V.-L. and C.M.-F.; methodology, M.R.-B., I.J.F.E., C.V.-L. and C.M.-F.; validation, M.R.-B., I.J.F.E., C.V.-L. and C.M.-F.; formal analysis, M.R.-B., I.J.F.E., C.V.-L. and C.M.-F.; investigation, M.R.-B. and I.J.F.E.; resources, M.R.-B. and I.J.F.E.; data curation, M.R.-B., I.J.F.E., C.V.-L. and C.M.-F.; writing—original draft preparation, M.R.-B. and I.J.F.E.; writing—review and editing, C.V.-L. and C.M.-F.; visualization, M.R.-B., I.J.F.E., C.V.-L. and C.M.-F.; supervision, C.V.-L. and C.M.-F.; project administration, I.J.F.E. and C.V.-L.; funding acquisition, I.J.F.E. All authors have read and agreed to the published version of the manuscript.

Funding: This research received no external funding.

Institutional Review Board Statement: Ethical review and approval were waived for this study because it did not involve human participants or animals, and no personal data were collected. The Ethics Committee of the Polytechnic University of Madrid and the GESPLAN Research Group reviewed the documentation for this study and concluded that as no personal information from participants was used, the project did not require institutional ethical approval or review. Therefore, a formal waiver was granted.

Informed Consent Statement: Informed consent was obtained from all subjects involved in the study.

Data Availability Statement: Publicly available datasets were analyzed in this study. These data can be found here: https://experience.arcgis.com/experience/1593455f5d3d46c898860e1d43674e68/page/Inicio (accessed on 29 November 2025). The Geoportal SIG FESBAL–UPM provides open access to the spatial data, dashboards, and indicators discussed in this research.

Acknowledgments: The authors would like to thank the Universidad Politécnica de Madrid (UPM) for providing access to the ArcGIS Online platform used in this study. We also express our gratitude to the volunteers and staff of the 54 Spanish Food Banks for their active participation, as well as to the managers of FESBAL for their collaboration and outstanding social work in support of the most disadvantaged populations.

Conflicts of Interest: The authors declare no conflicts of interest.

References

1. Committee on World Food Security (CFS). *Principles for Responsible Investment in Agriculture and Food Systems*; Food and Agriculture Organization (FAO): Rome, Italy, 2014. Available online: http://www.fao.org/3/a-au866s.pdf (accessed on 30 December 2025).
2. United Nations. *Transforming Our World: The 2030 Agenda for Sustainable Development*; United Nations: New York, NY, USA, 2015. Available online: https://sustainabledevelopment.un.org/content/documents/21252030%20Agenda%20for%20Sustainable%20Development%20web.pdf (accessed on 30 December 2025).
3. FAO. *Food Wastage Footprint. Impacts on Natural Resources*; Natural Resources Management and Environment Department: Rome, Italy, 2013. Available online: https://www.fao.org/3/i3347e/i3347e.pdf (accessed on 30 December 2025).
4. De los Ríos-Carmenado, I.; Afonso-Gallegos, A.; Quintanero-Lahoz, S.; Ortega-Rincón, R.; Nole-Correa, P.; Zuluaga, L.C.; Roncancio-Burgos, M. FESBAL-UPM Food Bank Chair and the Service-Learning Projects from the 'Working with People' Perspective. In *Proceedings of the International Conference on Economic Scientific Research—Theoretical, Empirical and Practical Approaches, Bucharest, Romania, 9–10 December 2021*; Springer International Publishing: Cham, Switzerland, 2021; pp. 79–90. [CrossRef]

https://doi.org/10.3390/su18031232

5. Nole Correa, P.L. Proyectos de Aprendizaje Servicio Desde el Modelo Working with People: Aplicación al Consumo Responsable Desde la Cátedra Bancos de Alimentos. Doctoral Thesis, Universidad Politécnica de Madrid (UPM), Madrid, España, 2025. [CrossRef]
6. Nole Correa, P.; Farias Estrada, I.J.; Aliaga, G.; Zuluaga, C. Service Learning Projects and CFS-IRA Principles: Application to the Food Bank Chair from the Working with People Model. *Sustainability* **2025**, *17*, 10212. [CrossRef]
7. Cazorla, A.; De los Ríos, I.; Salvo, M. Working with People (WWP) in Rural Development Projects: A Proposal from Social Learning. *Cuad. Desarro. Rural.* **2013**, *10*, 131–157.
8. Longley, P.A.; Goodchild, M.F.; Maguire, D.J.; Rhind, D.W. *Geographic Information Science and Systems*, 4th ed.; John Wiley & Sons: Hoboken, NJ, USA, 2015.
9. Hammelman, C.; Levkoe, C.; Agyeman, J.; Kharod, S.; Moragues Faus, A.; Munoz, E.; Oliva, J.; Wilson, A. Integrated Food Systems Governance: Scaling Equitable and Transformative Food Initiatives through Scholar-Activist Engagement. *J. Agric. Food Syst. Community Dev.* **2020**, *9*, 71–86. [CrossRef]
10. McCall, M.K.; Dunn, C.E. Geo-Information Tools for Participatory Spatial Planning: Fulfilling the Criteria for 'Good' Governance? *Geoforum* **2012**, *43*, 81–94. [CrossRef]
11. Nyerges, T. Understanding the Scope of GIS: Its Relationship to Environmental Modeling. In *Environmental Modeling with Geographic Information Systems*; Goodchild, M., Steyaert, L., Parks, B., Eds.; Oxford University Press: London, UK, 1993; pp. 75–93.
12. Esri. What Is a Geoportal? Esri Support. Available online: https://enterprise.arcgis.com/en/inspire/10.8/get-started/introduction-to-geoportals.htm (accessed on 20 November 2024).
13. Sastre-Merino, S.; Dorado-Martín, G.; De los Ríos-Carmenado, I. Los Sistemas de Información Geográfica Participativos como Herramientas para el Desarrollo Rural Sostenible: Análisis Conceptual y Revisión de Experiencias. In *Actas del XIV Congreso Internacional de Ingeniería de Proyectos*; AEIPRO: Madrid, España, 2010. Available online: https://oa.upm.es/39271/1/INVE_MEM_2010_216170.pdf (accessed on 30 December 2025).
14. United Nations Committee of Experts on Global Geospatial Information Management (UN-GGIM). *The SDGs Geospatial Roadmap*; United Nations: New York, NY, USA, 2022. Available online: https://ggim.un.org/meetings/GGIM-committee/11th-Session/documents/The_Geospatial_SDGs_Roadmap_WGGI_IAEG_SDGs_20210804.pdf (accessed on 5 January 2026).
15. Orban-Ferauge, F. *Participatory Geographic Information Systems and Land Planning: Life Experiences for People Empowerment and Community Transformation*; Presses Universitaires de Namur: Brussels, Belgium, 2011. Available online: https://www.iapad.org/wp-content/uploads/2015/07/orban.pdf (accessed on 5 January 2026).
16. Rambaldi, G.; Kyem, P.; McCall, M.; Weiner, D. Participatory Spatial Information Management and Communication in Developing Countries. *Electron. J. Inf. Syst. Dev. Ctries.* **2006**, *25*, 1–9. [CrossRef]
17. Corbett, J.; Rambaldi, G.; Kyem, P.; Olson, R.; Wiener, D.; Muchemi, J.; McCall, M.; Chambers, R. Overview: Mapping for Change-the emergence of a new pratice. In *Mappirng for Change: Practice, Technologies and Communication*; IIED Publications: London, UK, 2006. Available online: https://www.iied.org/sites/default/files/pdfs/migrate/14507IIED.pdf (accessed on 5 January 2026).
18. Dunn, C. Participatory GIS: A People's GIS? *Prog. Hum. Geogr.* **2007**, *31*, 616–637. [CrossRef]
19. Jankowski, P. Towards Participatory Geographic Information Systems for Community-Based Environmental Decision Making. *J. Environ. Manag.* **2009**, *90*, 1966–1971. [CrossRef]
20. Blumenfeld, P.; Soloway, E.; Marx, R.; Krajcik, J.; Guzdial, M.; Palincsar, A. Motivating Project-Based Learning: Sustaining the Doing, Supporting the Learning. *Educ. Psychol.* **1991**, *26*, 369–398. [CrossRef]
21. De los Ríos, I.; Rodriguez, F.; Pérez, C. Promoting Professional Project Management Skills in Engineering Higher Education: Project-Based Learning Strategy. *Int. J. Eng. Educ.* **2015**, *31*, 184–198.
22. Martí, J.A.; Heydrich, M.; Rojas, M.; Hernández, A. Aprendizaje basado en proyectos: Una experiencia de innovación docente. *Rev. Univ. EAFIT* **2010**, *46*, 11–21.
23. Cobo Gonzales, G.; Valdivia-Cañotte, S.M. *Aprendizaje Basado en Proyectos; Materiales de Apoyo a la Docencia. Colección Metodologías de Enseñanza y Aprendizaje*; Pontificia Universidad Católica del Perú: Lima, Perú, 2017. Available online: https://repositorio.pucp.edu.pe/server/api/core/bitstreams/8e87d1d4-0655-4336-9add-6afa7e2679ee/content (accessed on 5 January 2026).
24. Villanueva Morales, C.; Ortega Sánchez, G.; Díaz Sepúlveda, L. Aprendizaje Basado en Proyectos: Metodología para Fortalecer tres habilidades transversales. *Rev. Estud. Exp. Educ.* **2022**, *21*, 433–445. [CrossRef]
25. Kim, M. Project-based community participatory action research using geographic information technologies. *J. Geogr. High. Edu.* **2018**, *42*, 61–79. [CrossRef]
26. Nilsson, M.; Griggs, D.; Visbeck, M. Policy: Map the Interactions between Sustainable Development Goals. *Nature* **2016**, *534*, 320–322. [CrossRef] [PubMed]

27. Cazorla, A.; De los Ríos, I. Hacia una Metauniversidad para un desarrollo sostenible: Inversión Responsable en agricultura y Sistemas Alimentarios. In *Actas del XXVIII Congreso Internacional de Dirección e Ingeniería de Proyectos*; AEIPRO: Jaén, España, 2024. [CrossRef]
28. Creswell, J.W. *Research Design: Qualitative, Quantitative, and Mixed Methods Approaches*, 5th ed.; Sage Publications: Thousand Oaks, CA, USA, 2018.
29. Luo, X.; Park, W. Development of a Web-Based GIS Learning Module for Community Asset Mapping to Enhance Service Learning in Social Work Education. *Int. J. Librariansh.* **2020**, *5*, 36–52. [CrossRef]
30. Maguire, D.J.; Longley, P.A. The emergence of geoportals and their role in spatial data infrastructures. *Comput. Environ. Urban Syst.* **2005**, *29*, 3–14. [CrossRef]
31. Tait, M.G. Implementing geoportals: Applications of distributed GIS. *Comput. Environ. Urban Syst.* **2005**, *29*, 33–47. [CrossRef]
32. World Resources Institute (WRI); World Business Council for Sustainable Development (WBCSD). *The Greenhouse Gas Protocol: A Corporate Accounting and Reporting Standard, Revised Edition*; WRI/WBCSD: Washington, DC, USA, 2014. Available online: http://pdf.wri.org/ghg_protocol_2004.pdf (accessed on 5 January 2026).
33. Instituto Geográfico Nacional (IGN). Cartografía Base de España del Atlas Nacional de España. Centro de Descargas. Available online: https://centrodedescargas.cnig.es (accessed on 15 November 2024).
34. Instituto Nacional de Estadística (INE). Censo Anual de Población. Available online: https://www.ine.es/jaxiT3/Tabla.htm?t=67988&L=0 (accessed on 15 November 2024).
35. Instituto Nacional de Estadística (INE). Riesgo de Pobreza y/o Exclusión Social (Estrategia Europa 2020): Indicador AROPE. Available online: https://www.ine.es/jaxiT3/Tabla.htm?t=10011 (accessed on 15 November 2024).
36. Esri. GIS for SDGs. Available online: https://www.esri.com/en-us/industries/sustainable-development/sdg (accessed on 20 November 2024).
37. Delpish, R.; Jiang, S. Visualization of Food Bank/Agency Distribution. *Proc. Hum. Factors Ergon. Soc. Annu. Meet.* **2019**, *63*, 1879–1882. [CrossRef]
38. Farías Estrada, I.; De Los Ríos Carmenado, I. The Carbon Footprint of the Food Banks: Conceptual Framework and Review of Experiences. In Proceedings of the 27 th International Congress on Project Management and Engineering, San Sebastián, España, 10–13 July 2023; pp. 1219–1235. [CrossRef]
39. Farías Estrada, I.; De los Ríos, I.; Velilla, C.; Marín Ferrer, C. Los SIG como herramienta para el cálculo de la Huella de Carbono de los Bancos de Alimentos. *AEIPRO* **2024**. [CrossRef]
40. MITECO. Guía Para el Cálculo de la Huella de Carbono y para la Elaboración de un Plan de Mejora de una Organización. Ministerio de Transición Ecológica, España. 2025. Available online: https://www.miteco.gob.es/es/cambio-climatico/temas/mitigacion-politicas-y-medidas/guia_huella_carbono_tcm30-479093.pdf (accessed on 5 January 2026).
41. Mur Nuño, C.; De los Ríos, I.; Mejía, J.; Nole, P. Buenas Prácticas de Gobernanza en los Bancos de Alimentos: Criterios. In Proceedings of the Actas del VII Congreso Internacional Sobre Aprendizaje, Innovación y Cooperación, CINAIC 2023, Madrid, España, 18–20 October 2023. Available online: https://zaguan.unizar.es/record/131992/files/084.pdf (accessed on 5 January 2026).
42. Aiking, H.; de Boer, J. Food sustainability: Diverging interpretations. *Br. Food J.* **2004**, *106*, 359–365. [CrossRef]
43. Cárcaba, A.; González, E.; Ventura, J.; Arrondo, R. How Does Good Governance Relate to Quality of Life? *Sustainability* **2017**, *9*, 631. [CrossRef]
44. Cazorla-Montero, A.; De los Ríos-Carmenado, I.; Pasten, J.I. Sustainable development planning: Master's based on a project-based learning approach. *Sustainability* **2019**, *11*, 6384. [CrossRef]
45. Rocancio Burgos, M.; De Los Ríos Carmenado, I.; Zuluaga, C. GIS Applied in the Management of Responsible Consumption Awareness Projects. In Proceedings of the 27 th International Congress on Project Management and Engineering, San Sebastián, España, 10–13 July 2023.
46. Esri España. La Tecnología Geoespacial Contra el Desperdicio Alimentario. FESBAL & Cátedra Banco de Alimentos. Available online: https://www.esri.es/es-es/descubre-los-gis/casos-de-exito/otros/cs-bancoalimentos-desperdicio (accessed on 5 January 2026).

https://doi.org/10.3390/su18031232

sustainability

Article

Strengthening of the Rural Community and Corn Food Chain Through the Application of the WWP Model and the Integration of CFS-RAI Principles in Puebla, México

José Regalado-López, José Antonio Maimone-Celorio * and Nicolás Pérez-Ramírez *

Colegio de Postgraduados, Campus Puebla, Puebla 72760, CP, Mexico; josere@colpos.mx
* Correspondence: maimone.antonio@colpos.mx (J.A.M.-C.); perezn@colpos.mx (N.P.-R.)

Abstract: Strengthening producer groups, the rural community, and agri-food chains are important actions to help solve the problem of food poverty, improve the living conditions of producers and promote sustainable development in rural México. It is necessary to seek new ways to improve decision-making by producer groups and establish some principles to strengthen the different links in agri-food chains. The objective of this study was to analyze the integration of the Principles for Responsible Investment in Agriculture (PRIA) in the corn agri-food chain in order to assess its strengthening. A study was carried out in three cases based on the application of the "Working With People" (WWP) model as well as interviews with key actors. It was found that cooperating groups with a higher degree of application of the WWP model and PRIAs have a higher degree of stability and sustainable development and strengthen the integration and cooperation of local action groups. These groups have the technical component better organize the agri-food processes and better incorporate the PRIAs and improve their economic, social, and environmental development compared to other groups that do it in a traditional way.

Keywords: small producers; rainfed agriculture; local action groups; maize cultivation; responsible agricultural investment; CFS-RAI principles; food systems

1. Introduction

Based on data from the Agricultural Census (2022) of the National Institute of Statistics and Geography of Mexico (Spanish acronym INEGI) [1,2], shown in Table 1, it is possible to demonstrate that agricultural activity both nationally and in the state of Puebla are important activities, especially for producers living in agricultural and very agricultural regions in these two areas. The most predominant crops in these regions are corn and beans, planted under rainfed conditions, whose production is destined for self-consumption and a surplus that is sold in the local market. The production levels obtained in these crops are low, although there is evidence, as has been documented among corn producers in Tlaltenango, Puebla [3], that shows high levels of grain and forage yields in corn, when the available technology is applied with greater precision. However, it is important to highlight the poor relationship between the area planted and the low use of agricultural machinery, insurance, and agricultural credit, strategic components for regional agricultural development.

In addition to the low use of machinery and factors related to financing, Mexican agriculture faces a set of specific problems that further affect the income levels of the population engaged in agricultural activities. Based on Figure 1, a group of problems detected by the production units can be observed; however, there are some that, according to the opinion of the production units, have a greater impact on food production. Among these are (1) climatic and biological factors, (2) the high costs of inputs and services, (3) difficulties in marketing products after the pandemic, and (4) insecurity problems. The perception of the production units with respect to this problem indicated that more than 88.8% considered that the high costs of inputs and services and climatic and biological problems were the factors that most affected Mexican agriculture.

Table 1. Characteristics of the Production Units at the national level and in Puebla State.

Elements	National, México	State, Puebla
Total production units	4,440,265	440,752
Surface area for agricultural use (ha)	26,104,422.68	1,015,173.77
Sown area (ha)	20,547,097.17	926,890.58
Receiving "senior adults" support	895,837	85,728
Tractor use	1,793,338	187,816
Credit use	265,508	14,437
Insurance use	78,140	621

Source: INEGI, 2022. [1,2].

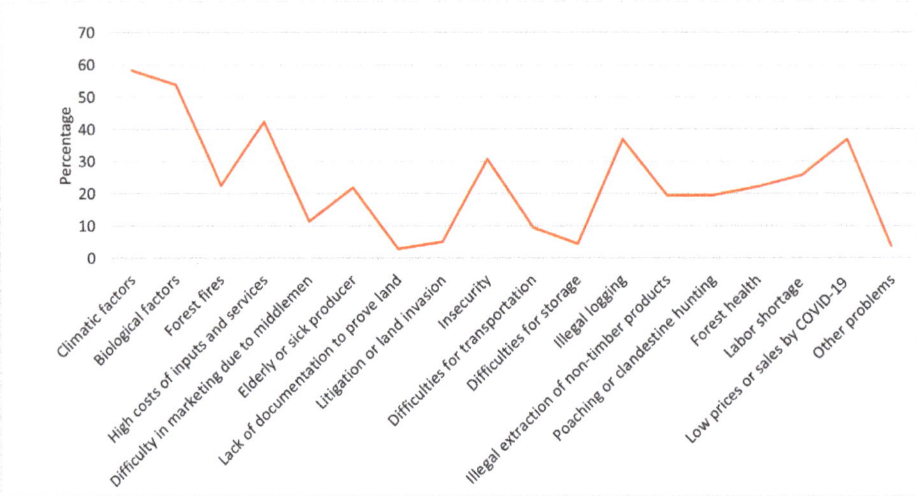

Figure 1. Main problems affecting agricultural and livestock production units. Source: INEGI, 2022 [1,2].

The limitation that most affects the production units cannot be addressed from an individual perspective; it requires a group effort. In this sense, the experience generated in the processes of strengthening producers and rural communities indicates that the "Working White People" (WWP) model and the proposal developed in the territory under study, considering associativity and organization, the latter considered as a company, are the most appropriate tools to reduce the intensity of this limitation on production and towards the strengthening of the rural community. De los Ríos [4] and Cazorla [5] point out that the WWP model is presented as a conceptual proposal to address the integrated approach to rural prosperity, based on the consideration of three dimensions: political-contextual, technical–entrepreneurial, and ethical–social. These three dimensions of the model can be synthesized into two central concepts: partnership and enterprise. The origin and integration of these concepts go back to the work developed by Smith [6] in 1962. In his discussion on the importance of separating associations and institutions, he points out that, in 1949, MacIver [7] mentioned that when "man creates associations he must also create rules and procedures for the handling of common affairs and for the regulation of the activity of the members among themselves, such forms are: indistinctly institutions". In the work developed by Hodgson [8], it is mentioned that organizations are a special type of institutions that stand out for establishing chains of orders that define the responsibilities within the organization, an important feature of a company.

On the other hand, Niño [9] points out that in order to understand the role of both concepts and their application in the process of social development, it is necessary to define them. In this sense, and through an intellectual process, he established that an association is an "instrument to establish purposes, to constitute and develop as a collective subject and to join forces or individual powers to form a greater power, sufficient to relate positively and successfully with other subjects on the way to the realization of the purposes established by the partners themselves". Instead, "an organization is just a company, that is, a space that can have different dimensions but that does not vary in terms of its generic objective: the maximization of economic profit". The integration of both concepts to build social development processes at the territorial level makes it possible to consolidate the strengthening of rural communities in the territory through the application of Principles of Responsible Investment in Agriculture and Food Systems (CFS-RIA) [10]. With this integration, it is possible to improve the production, transformation and consumption stages of the maize agri-food chain in rainfed conditions [11]. In this process of improvement, the principles of (1) food security and nutrition, (2) economic development, and (9) governance and accountability [10] are considered. Albisu [12] mentions that agri-food chains perform their functions from the moment a product starts from its raw materials until it is consumed, along with what are known as the links.

From the point of view of socioeconomic reality, the agri-food chain is understood as a system that groups interrelated economic and social actors who participate in activities that add value to a good from its production until it reaches the consumers [11]. The Catalan Food Safety Agency [13] considers that, in the food chain, the production stage is a very important initial phase; however, once the product is obtained, the transformation and distribution phases begin. In the latter, the food industry plays a relevant role, emphasizing a retail structure, such as bars, restaurants, flea markets, and markets on wheels, which come into direct contact with the consumer phase (transport, handling, preservation). These processes constitute the agri-food chain models In México, the "Secretaría de Agricultura" [14] has identified three models developed within the framework of neoliberal policy: (1) the Mexican Food System (SAM), (2) the neoliberal agri-food regime, and (3) the agri-food chain based on production, transformation, and commercialization in family agriculture, which provides income and food for consumption and the market, contributing to food security and sovereignty; the study in question is located in the latter model.

Regardless of the food chain model, different actors participate in each of its phases. In the specific case of the corn food chain, which is practiced in rainfed conditions by small producers, a group of actors has been identified, which, according to Moctezuma-López [15], is made up of agricultural input suppliers, production systems, storage, processing, distributors, and final consumers.

The research questions posed in this study are as follows: (1) What are the existing maize and other crop agri-food systems in the Izta-Popo territory? (2) How has the WWP used in conjunction with the PRIAs strengthened the integration and efficiency of the farmer producer associations that make up the cases studied? The objectives of this research were to identify three agri-food models of maize and other crops in the Izta-Popo territory in order to apply PRIAs in these cases using the WWP model as a support method.

1.1. Intervention Territory

The main problems affecting active agricultural use units in the state of Puebla are climatic factors, high costs of inputs and services, and low prices of agricultural products [1]. Thus, the intervention territory is located in the east of the state of Puebla (Figure 2). It is composed of 11 municipalities that comprise part of the natural regions of the Puebla Valleys and the Sierra Nevada. This space has an area of more than 800 km^2. This territory has a flat topography in the area of the Puebla Valley but a rugged topography in the foothills of the volcanoes.

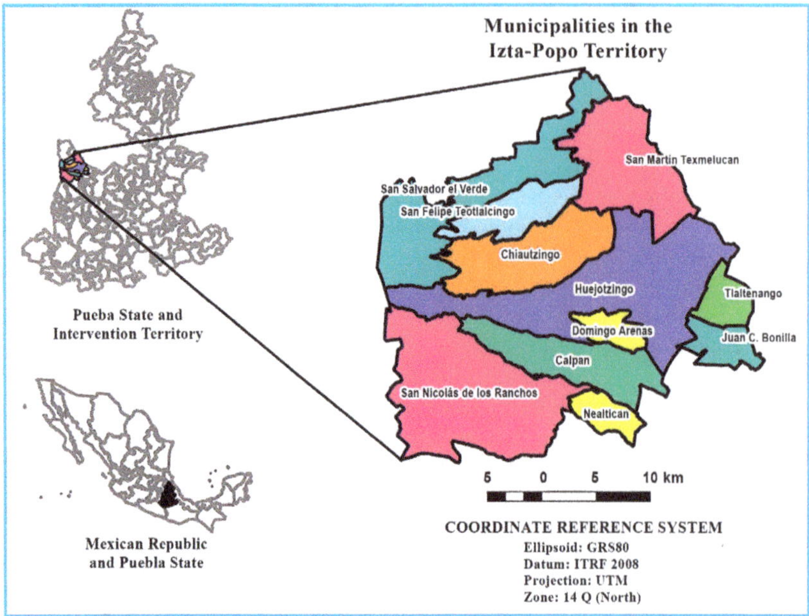

Figure 2. Location of the intervention territory, composed of 11 municipalities. Source: Own elaboration with spatial information from the Digital Map of México [16]. The Spanish words in the Figures 2–4 correspond to the names of the municipalities of the Izta-Popo territory of the state of Puebla in Mexico. A municipality is the basic political-administrative unit in Mexico where a state is subdivided.

The altitude of this region ranges from 2100 m in Juan C Bonilla to 5465 m in San Nicolás. Due to this altitudinal gradient, the most important climates are temperate and semi-cold sub-humid with summer rains, and in the higher parts of the volcanoes, there are cold and very cold climates. The average temperature is around 15 °C, but in some municipalities, such as San Salvador, Chiautzingo, and Calpan, there are average temperatures between 2 and 8 °C. The most representative soils are arenosols, andosols, phaeozems, and cambisols, which have favorable characteristics for agricultural development [17–23].

These physical conditions of the territory allow for the production of different annual crops and fruit trees in temperate climates; most rainfed agriculture takes place in the flat part of the territory and on the lower slopes of the volcanoes (Figure 3). The annual crops grown in the flat area are mainly corn and beans. On the slopes of the volcanoes, fruit trees and annual crops are grown interspersed with fruit trees.

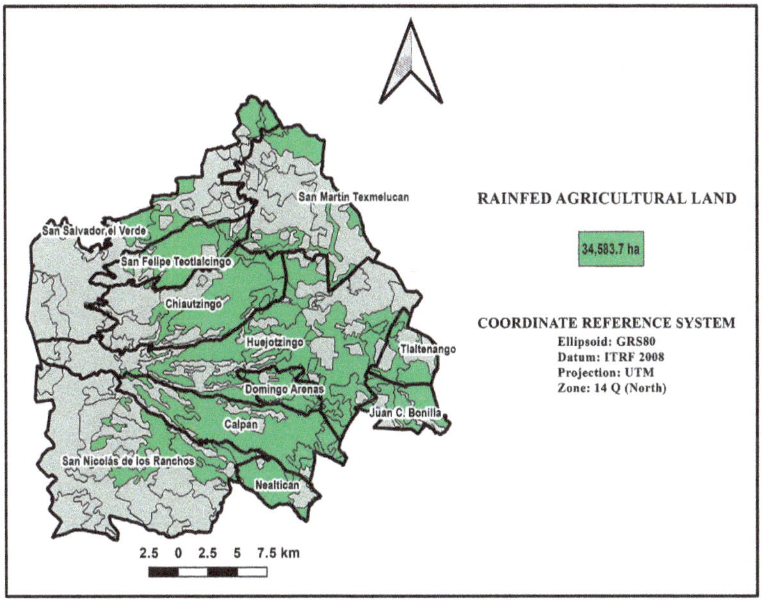

Figure 3. Distribution of rainfed agriculture in the territory. Source: Own elaboration with spatial information from the Digital Map of México [16].

Corn is the most important crop in terms of area planted and volume and value of production. Beans are also important in terms of area planted. However, both crops have low yields, which means that there is significant potential to increase their production. These grains are the basis of the local population's diet, and producers are applying strategies to add value to them. The inclusion of the Principles for Responsible Investment in Agriculture (PRIA) in these strategies will strengthen agri-food chains.

Regarding fruit trees, peach and pear are the most important in terms of area planted and volume and value of production. Walnut and apple are less important in terms of area but have an important production value (Table 2). These fruit trees are of great culinary relevance, since they are the raw material for making "chiles en nogada" (chilis in walnut sauce), which is a stew that is highly appreciated by the population and visitors at "chiles en nogada" fairs. Currently, the establishment of plantations of these trees is being promoted to cover the great demand for their fruits.

Table 2. Main crops and fruit trees that support the agri-food chains in the territory.

Agricultural Species	Planted Surface (ha)	Production	
		Volume (tons)	Value (Thousands of Mexican Pesos)
Corn	21,178.7	38,579.1	269,672.8
Beans	1791.2	1767.9	26,323.3
Peach	543.1	3775.1	31,607.4
Apple tree	191.8	1203.5	9525.9
Pear tree	796.4	6646.2	22,895.3
Hickory	127.0	534.7	15,673.1

Source: [24].

Regarding socio-demographic conditions, the total population of the territory is more than 400,000 inhabitants. The municipalities with the largest populations are San Martin Texmelucan and Huejotzingo, with more than 100,000 inhabitants. The least populated municipalities are Domingo Arenas and Tlaltenango, with around 7000 people [25]. The average population density is 500 inhabitants/km^2, although the municipalities of San Martin and Juan C. Bonilla exceed 1000 inhabitants/km^2.

On the other hand, the population 15 years of age and older without basic education is 41%, and illiteracy is 50%. The marginalization index varies from 54.4 to 58.2%, i.e., most of the municipalities have a low and very low degree of marginalization [26]. Likewise, 68.4% of the population lives in poverty (Figure 4), but the most outstanding aspect is that 18.5% of this percentage lives in extreme poverty and 81.5% in moderate poverty [27].

These are the socio-demographic, marginalization, and poverty conditions in which the population of this territory lives and which, for its development, depends on the different agri-food chains based on small-scale rainfed family farming. It is necessary to seek strategies or elements that contribute to strengthening the agri-food chains, such as the RIA principles, so that the chains can better contribute to improving the poverty and marginalization conditions of the territory and, in turn, promote more sustainable human development in all aspects.

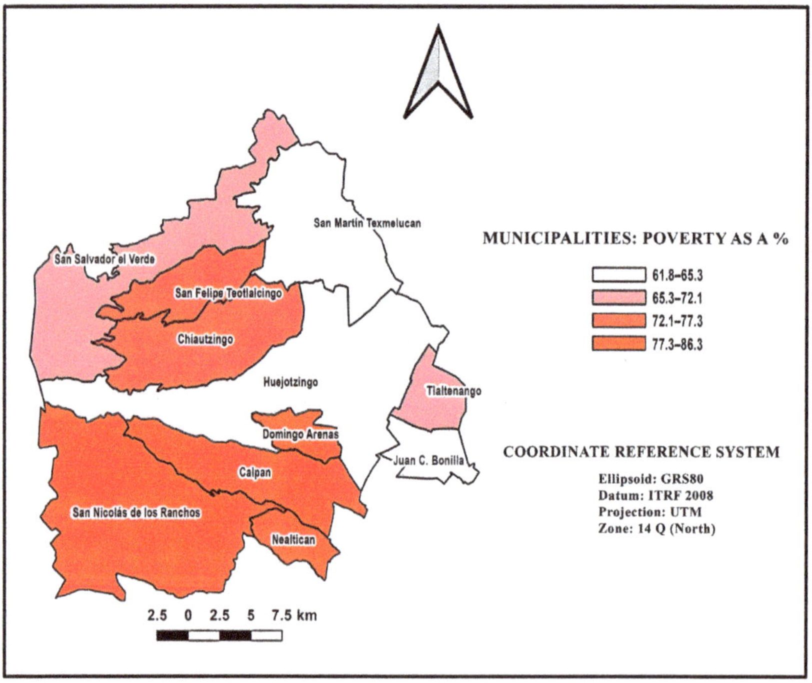

Figure 4. Distribution of poverty in the intervention territory. Source: Own elaboration with spatial information from the Digital Map of México [16] and poverty information from CONEVAL [27].

In the intervention territory, called "izta-popo", the environmental and socio-cultural conditions and food patterns have influenced the planting of corn and beans in monoculture or associated systems, in addition to intercropping systems with fruit trees, where the agri-food model of family farming prevails. This model takes into account the resources available for the exploitation of the production units, which has allowed the construction of specific models of agri-food chains whose agglutinating axis is the cultivation of corn. These models are as follows. Model (1): The agri-food chain of corn production is associated with beans at the family level that transforms corn into different by-products that are marketed in tianguis in México State [28]. Model (2): The agri-food chain of corn in monoculture is produced at the family level. In this model, corn is transformed and marketed in the form of tortillas and tostadas. The product is made using a tortilla machine obtained by a group of producers [29]. Model (3): The chain of corn and chili is produced at the family level. Corn and chili are marketed through a cooperative with a technical commerce component [30]. Knowledge about the integration of PRIA-FAO in these models is the objective of this article. In Table 3, we visualize the local agri-food conditions and characteristics similar and corresponding to each model mentioned and the selected municipalities of the intervention territory.

Table 3. Local characteristics similar to each agri-food system.

		Model 1	**Model 2**	**Model 3**
Territory	Municipality	Huejotzingo	Tlaltenango	Calpan
	Locality	Miguel Tianguizolco	No location	San Andrés
Production	Product	Corn, beans, ayocote and pumpkin.	Corn	Fruits, poblano peppers and corn
	Farmers	50	500	250
	Total Surface	100 ha	1200 ha	208 ha
Transformation	Product	Tlacoyo and tortilla	Tortilla and tostada	Chile en Nogada
	Families	600	100	80
Commercialization	Points of sale	50 markets in CDMX y Edo. de México	280 places, e.g., miscellaneous, schools, and others.	150 restaurants and 5 events.

Source: Prepared by the authors with approximate data from key informants.

1.2. Criteria for Case Selection

Among the other relevant aspects considered for the first case are the following points. (1) The rural family production unit studied belongs to the community of San Miguel Tianguizolco, located between the communities of San Juan Pancoac and Santa María Tianguistengo in the municipality of Huejotzingo. Within this area, there are more than 1000 family production units dedicated to the transformation of corn and beans into other by-products. In other words, it is a representative unit of the agri-food model addressed in the case in question. (2) Since 2023, researchers from the Colegio de Postgraduados, linked to the University Network for the application of PRIAs, selected this rural-family production unit to participate in the establishment of a demonstration module to develop an economic development project (PRIA 1) based on peach or Creole peaches and food security (PRIA 2) through the planting of corn associated with guide beans, pumpkin, and inoculation of Huitlacoche in 1000 corn plants. (3) The members of the family unit are willing to participate in activities related to linkage and research processes of professors and students of the Colegio de Postgraduados.

The second case justified its inclusion because it is a unique experience in the territory in terms of integrating the association and organization, the latter understood as a company. Its construction dates from the year 2010, starting its integration with the parents. Currently, the association is integrated by the children of the first initiators of a technological application project to raise the levels of corn yields, considering the principles of the WWP model at that time. This group has observed the application of PRIA 2, food security and nutrition, and the current principles of 8, 9, and 10, promoting safe food systems, governance, and accountability.

The third case is very interesting because it has a structure that is not observed in many associations under the associative and legal figure of a "cooperative". In this case, they manage the market through a manager with a specialty in marketing. Their experience of integrating processed agricultural products into local, regional, and international markets motivated them to link this association even more to generate experiences in the application of PRIAs and because of the willingness of the board of directors to be involved in these activities as a result of the collaboration with researchers from the Colegio de Postgraduados.

2. Materials and Methods

From the empirical component, this research was based on the case study that was used to learn, in greater depth, the characteristics and differences presented by each of the agri-food models identified in the territory. Due to the nature of the study and the approach to the process of constructing the different agri-food models in the territory in question, a multiple-case study was applied [31,32]. The first case consists of an agri-food model based on production, transformation, and commercialization by the members of one family unit. The second case analyzes production, transformation of grain to tortilla, and its commercialization through a partnership called "Sociedad de Producción Rural de R. L.". Finally, the third case is a cooperative in which the members are in charge of the production and transformation process; however, in the commercialization phase, they have a technical component, an expert in the market who is in charge of carrying out all the management related to this phase. To gather information, questionnaires were applied to all members of family units and partnerships working in the different models of the maize agri-food chain. In addition, in-depth interviews were conducted with one or two key informants per case in order to delve more deeply into some specific issues of the chain, complemented by the experiences generated with the application of the WWP model. Six participatory workshops were also held with corn producers, technicians, public officials, and other key stakeholders to identify needs, priorities, and development opportunities [33]. The workshops promoted a reflection on the Principles for Responsible Investment in Agriculture and Food Systems (PRIA) and their applicability in the local context as well as the monitoring of plots established on the land of the partners that make up the groups that constitute the tools used to carry out demonstrations and to directly observe the application of the principles in each of the cases [4]. These PRIA were developed by the FAO Committee on World Food Security (CFS).

The process of development and weighting of the qualitative and quantitative indicators and variables of this research are based on consultations with university experts from the GESPLAN Group of the "Universidad Politécnica de Madrid" and the "Colegio de Postgraduados" of Mexico. as Also, a joint design is applied, involving the community of the three cases of the Izta-Popo territory in Puebla, using the WWP model. The following software was used for data processing: IBM SPSS Statistics version 24.0, Excel Build 16.0.19003.42309, Word Build 20250606.7 from Office 365 A1 and QGIS 3.34 Prizren. Another part of the analytical framework is based on the references shown at the end of each table in the Results section. A detailed explanation is added in the Supplementary Materials "Research Methods Instruments and Construction of Indicators".

2.1. Methodology

2.1.1. The WWP Model, PRIAs and Food Systems

The research from the conceptual part was based on the WWP model that focuses on social learning, community participation, and the integration of ethical–social, technical–entrepreneurial, and political–contextual dimensions, seeking a balance between technical, behavioral, and contextual competencies. In the process of social learning, the concept of local action groups (LAGs) has shown to be an effective tool to promote rural development from a territorial perspective [34]. LAGs facilitate interaction between local actors and promote participatory development management [35]. The use of the components of the WWP model has contributed to linking the principles and their application with the different groups involved in the different components of the maize agri-food chain.

It is important to point out that the WWP model was highlighted as a central concept of this study, and the process of development and application of its components has resulted in the integration of rural production units in the form of associations. These are understood as the concentration of power through the union of individual forces to achieve common objectives. The association constitutes the social–ethical component of the WWP, but experience has shown that this component is not sufficient to address the challenges faced by rural family units in the production, transformation, and commercialization processes, since a technical–entrepreneurial component that integrates the concept of business organization is required. Associative (social) and organizational (economic) processes are fundamental pieces in agricultural and rural development processes. These variables have been used to observe the behavior of the cases and are the concepts developed in this article. The other theoretical part that complements this research is based on the framework of the ten PRIAs, which were created by the UN Committee on World Food Security.

As a component of the materials and methods, RIA principles and food systems were included, which constitute a range of activities related to production, transformation, and market [10]; while food systems comprise another wide range of individual, collective, and institutional actors that carry out the components of the agri-food chain, in a socio-political, economic, technological, and natural framework, the chain activities are carried out. The five RIA principles most closely related to the agri-food chain models managed by family units in the intervention territory are as follows: (1) contribute to food security and nutrition; (2) contribute to inclusive and sustainable economic development and poverty eradication; (3) respect cultural heritage and traditional knowledge; (4) support diversity and innovation; and (5) incorporate inclusive and transparent governance structures, processes, and grievance mechanisms.

2.1.2. Agri-Food Chain Models

According to INEGI [36], México has an area of 2.2 million hectares, of which 1.9 million correspond to rural areas and 0.3 million to population centers, roads, and bodies of water, among other geographic features. In the rural areas during the period 2021–2022, an agricultural use of one million hectares was recorded, representing 53% of the total rural area. Of this area, 329,761 hectares were cultivated with white corn. However, in the central, southern, and southeastern regions of México, small production units have less than 5 hectaes for cultivation and are mostly managed under rainfed conditions where in addition to white corn, more than 60 native corn breeds are cultivated [37]. When the "Project Puebla" started in the current area of influence of the Postgraduate College in Puebla State, it had 116,000 hectares of agricultural land, of which corn represented 69%, beans 15.9%, and the rest in other crops. The most important production systems prevalent at the beginning of the project are maize alone, association with guide beans (*Phaseolus vulgaris*), bush beans (*Phaseolus vulgaris*), intercropped maize in fruit orchards, and ayocote

(*Phaseolus coccineas*) [38]. In the study area, maize monoculture continues to prevail as an important crop in terms of area planted as well as the production systems of association with guide bean and intercropped maize in fruit orchards. Agri-food chain models were generated around these production systems [30].

3. Results

3.1. Case 1: Agri-Food Chain Maize-Bean Guide Bean Association

3.1.1. Relationship of the Family Unit to the CFS-RAI Principle

Before addressing the details of the chain, it is important to point out that the first international meeting on CFS-RAI principles at the "Colegio de Posgraduados", Campus Puebla, México, was held from 7 to 11 July 2024. On 7 June, a field trip was conducted in the community of San Miguel Tianguizolco, which belongs to Huejotzingo, one of the municipalities that make up the Izta-Popo territory. This territory is the area of intervention for the application of the economic development and food security project within the framework of the principles of Responsible Investment in Agriculture and Food Systems, all based on the production, transformation, and commercialization of native fruit trees, as well as the planting of native seeds of corn, beans, squash, and huitlacoche [39]. The project is using the Intercropped "Milpa" model with fruit trees under dry conditions as a unifying axis [40,41]. The owner of the plot and project participant, as well as academics from the Technological Institutes of Atlixco and San Martin Texmelucan, were present during the field tour of the meeting (Figure 5). These institutes are part of the Network of Higher Education Institutions, formed within the framework of the implementation of the RIAs in Latin America. During the tour, the owner of the plot indicated that his family unit has six hectares of dry land in his production unit, and with the support of his members, they transform corn and beans into products that are marketed in México City, such as tlacoyos, gorditas, and tortillas. They also sell fruits [28].

Figure 5. Field and sowing tour of the Izta-Popo territory of intervention and application of the PRIAs.

3.1.2. Characteristics of the Family Production Unit (FPU)

The climatic and soil conditions as well as social and economic factors allowed the family unit to generate, conserve, and improve the agri-food chain. This unit produces, transforms, and markets products and different by-products derived from the different crops grown in the FPU. The FPU is located in San Miguel Tianguizolco. The unit is made up of three men and three women. The average age of the members is 49 years, and the level of schooling is up to 6 years. They have 6 hectares of farmland, in rainfed conditions, and the family has its own machinery, including a tractor with implements and farming tools that allow them to carry out soil preparation and cultural activities in a timely manner. Even though decision-making and labor come from the members of the FPU, at certain times of the year, local laborers are hired to carry out cultural work (weed and pest control) and harvesting. Approximately 86-day laborers are hired each agricultural cycle at a total cost of MXN 25,800.0.

3.1.3. Production, a Component of the Corn–Bean and Ayocote Agri-Food Chain

The FPU is organized and uses its own resources to produce the necessary raw materials. It produces corn and beans for self-consumption and as inputs to transform them into various by-products. The productive activity begins with the preparation of the land during the months of February and March, and then, depending on the available humidity in the soil, planting can take place in April and May; however, due to the adverse effects of climate change, planting has been delayed until June, as happened during the 2024 agricultural cycle. According to the representative of this FPU, production involves an average investment of MXN 12,000.0 per hectare and yields 3.0 tons of corn and 500 kg of guide beans per hectare. With the sale of these products, a gross income of MXN 42,000.0 is obtained; with this cost and income, a benefit–cost ratio of 3.5 is obtained; for every peso invested, a profit of 2.5 pesos is obtained. This means that in the production phase, this agri-food system is profitable.

3.1.4. Processing, Transformation

The FPU transforms its production into various by-products made from corn and beans. Tlacoyos are the main by-product produced in the transformation process in the corn–bean agri-food chain. Pérez-Hernández et al. [42] documented the process of transformation of maize and beans into tlacoyo in the community of "San Miguel Tianguizolco", and, in this case, the same process was observed, which includes the following stages: (1) reception of raw material; (2) cleaning of corn and beans; (3) nixtamalization of corn and cooking beans; (4) grinding; (5) conditioning on a table; (6) manual or semi-manual molding; and (7) cooking on an iron comal. Considering these stages of the transformation process and the available labor, FPU manufactures tlacoyos, tortillas with nopal and wheat additives, and memelas, among other products. Ttlacoyos are made from nixtamalized corn and beans and are packaged by the dozen.

3.1.5. Commercialization, Marketing

The transformed products are marketed in the CDMX in the so-called "markets on wheels" on each Friday. The products sold in these public spaces are tlacoyos, wheat and nopal tortillas, memelas, and other products, which are transported in their own vehicle to the place of sale, travel on Fridays, stay on Saturdays, and return to their place of origin on Sundays. This has been the route commonly used by this family unit and other producers for more than 40 years. The people in charge of the FPU are responsible for marketing these products. Marketing is not delegated to the children or other people; it is the exclusive responsibility of the heads of the family. The construction of this market scheme was carried out by people living in Mexico City who visited the town of San Miguel Tianguizolco and interacted and agreed with the family members of the FPU to market local corn and bean products. A transportation network has been developed around the commercialization process that is responsible for transporting the tlacoyo producers from their place of origin to the points of sale and vice versa. The prices are set based on supply and demand. The FPU under study receives MXN 40.0 for each dozen tlacoyos it sells and MXN 25.0 for each dozen tortillas. Likewise, the production cost for making a dozen tlacoyos is MXN 15.0. With these costs and incomes, a benefit–cost ratio of MXN 2.60 for each dozen tlacoyos is estimated, which indicates that it is a profitable activity [43]. Tlacoyo can contribute to improve the income of FPUs engaged in its production.

3.2. *Case 2: Corn Agri-Food Chain and Its Transformation into Tortillas Through a Rural Production Society*

This case is located in the municipality of Tlaltenango, Puebla [44]. The agri-food model most commonly practiced in this locality is monoculture corn combined with dairy cattle farming. Regarding milk production, in verbal communication with key informants in the community, an estimated production volume of 20,000 L of milk per day is marketed to intermediaries in the neighboring towns of the municipality.

3.2.1. Personal Characteristics of the Members of the Association

The "Sociedad de Producción Rural" (SPR) Campo Lima began its integration process in 2010; however, it was in 2015 when it was legally constituted as a legal entity within the framework of the agrarian law of Title Four, which deals with the operation and formation of rural societies [45]. The SPR is made up of seven partners whose average age is 46 years and have an average of 8 years of schooling. Among the members of the society, they have a total area of 45 hectares of arable land in irrigated and rainfed conditions. All the partners have their own machinery (tractor and implements) and implement to carry out agricultural activities.

3.2.2. Production, Transformation and Commercialization as Elements of the Maize Agri-Food Chain of the Campo Lima Society and Its Relationship to the CFS-RIA Principles

In the background to the application of the CFS-RIA principles, it is important to note that the president of the Campo Lima Rural Production Society participated in the second version of the business course organized by the GESPLAN research group of "Universidad Politécnica de Madrid" (UPM).

The SPR Campo Lima is developing a model within the framework of the production, transforming, and commercialization process, which consists of having greater participation in these components of the chain, especially in processing and marketing. The main elements of the model are (1) production of certified seed of the HS-2 hybrid from the "Colegio de Postgraduados"; (2) marketing of the seed for commercial planting; (3) grain storage; (4) processing of the grain into tortillas; and (5) tortilla marketing strategy (Figure 6). The figure shows two important components for the structuring of the associative processes: the integration of partners for decision-making and the configuration of the technical–entrepreneurial component for the execution of decisions. From this component, the actions related to production, transformation, and commercialization will be carried out.

3.2.3. Seed Production and Commercial Plantings

The SPR Campo Lima and the "Colegio de Postgraduados" signed a licensing agreement to produce HS-2 corn seed, a patent generated by researchers at this Public Research Center to produce certified seed. SPR plans to produce a volume of 5 tons of HS-2 corn seed in 2025, with the supervision of the National Seed Inspection and Certification System (Spanish SNICS), the federal government institution in charge of seed certification in México. With this volume of certified seed production, the SPR will design a strategy for its distribution among local producers and, in some cases, outside the municipality. A record will be kept of the producers benefiting from the seed with the purpose of stockpiling the volume of harvested grain for its transformation into tortillas.

3.2.4. Transformation and Commercialization

The SPR acquired a tortilla machine at the end of 2024 to begin the process of transforming corn grain into tortillas and to be able to face two problems that affect them, namely (1) an unfavorable market, since the owners of the small-scale tortilla factories set the prices of the grain, so the prices are generally low, and (2) the deferred payment for the sale of the grain that is generally made very late and without paying an interest rate due to the ease of credit. This problem was the reason for the group of small producers that make up the SPR to venture into the processing of their corn harvest. This experience is unprecedented in the locality and in the region. In other words, it was the first time that a group of small producers ventured into a process of transforming their raw material through an entrepreneurial approach [29].

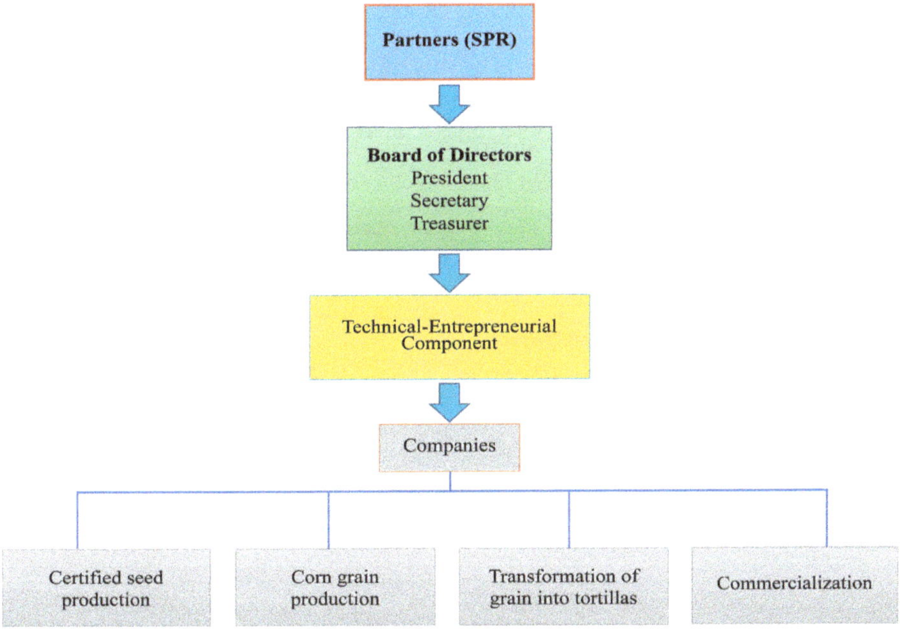

Figure 6. Governance structure for improving the maize agri-food system.

According to the Apolo Company [46], the tortilla machine purchased by the group has the following characteristics. (It produces 30 tortillas per minute or 1800 tortillas per hour, processing between 45 and 55 kg of dough per hour with tortillas of 25 g each. The total cost of the equipment was MXN 392,899.34. The current demand for corn grain from the tortilla machine is 250 kg per day; with this milling volume, the SPR requires 84 tons of corn annually to supply this demand. With the corn yield levels obtained by the group and considering the available surface area, they are able to produce 60 tons, with a shortfall of 20 tons to complete the grain demand; the shortfall can be solved by stockpiling grain from other local producers who plant H-S2 corn.

The tortilla transformation company (Figure 7) is operated by two full-time employees, one with experience in handling the equipment and in the processing method, and the other is an assistant in activities related to this activity and sales on the premises (delivery person). The delivery person is in charge of distributing tortillas to the group of clients who buy them from the company.

3.3. Case 3: Agri-Food Chain of Maize Produced at the Family Level but Transformed and Marketed Through a Cooperative

3.3.1. Relationship of the Cooperative's Activities with the CFS-RIA

Before describing this third case, it is interesting to highlight the participation of the cooperative's representatives in the second version of the training course aimed at entrepreneurs, which was given online by the GESPLAN Research Group of the Universidad Politécnica de Madrid (UPM) in 2024 [47]. They are currently linked with the Colegio de Postgraduados to build living laboratories whilst applying the WWP model and emphasizing the planning model as social learning.

Figure 7. Opening of the SPR Campo Lima small-scale tortilla factory.

In this agri-food chain, the main actor is a cooperative society called Guardianes de Calpan, which is similar to the other two models in that the members produce and transform corn at the family level. However, it has a feature that makes it different from the other two cases: Its integration into the market is through technical personnel specialized in the subject. The group develops the three stages of the corn agri-food system: production, transformation, and commercialization, as well as other complementary productive activities such as rural gastronomic experiences and the promotion of handicrafts made with materials from the region. The society was integrated in 2003 and was legally constituted in 2017, developing the three components of the agri-food chain. The group is composed of 17 partners, of whom 6 are men and 13 are women (Figure 8). The members of the cooperative have an average age of 55 years, although the age of some members is 20 years. The level of schooling is 9 years; this data coincides with the years of studies that the general population has.

Figure 8. Meeting with Guardianes de Calpan cooperative to promote PRIAs.

3.3.2. Production

The cooperative is located in the municipality of Calpan. Agriculture in this municipality is based on a traditional production system that combines staple crops (corn and beans) with various species of fruit trees, including peach (*Prunus pérsica*), capulín (*Prunus salicifolia*), plum (*Prunus domestica*), pear (*Pyrus communis*), walnut (*Juglans regia*), and apricot (*Prunus armeniaca*). Corn and fruit production is carried out using a combination of traditional soil preparation, planting, fertilization, and harvesting practices. Maize is planted with colored creole corn, with the blue seed being the most important because it is used to make pinole, a very traditional drink among the inhabitants of the territory [30,48,49].

3.3.3. Transformation

The productive base of the members of the cooperative allows them to obtain a diversity of products between the months of May and December, which are the raw materials for making mole poblano, chiles en nogada (Figure 9), tamales, tlapaxtamales, atoles, tlatlapas, cacamas, chileatoles, stews with quelites, huazontles, or nopales, tlacoyos, tortillas, pipianes, huaxmoles, breads, liquors, and sweets (ates, jams, syrups, pinoles, muéganos, and jamoncillos). However, of all the diversity of derivatives obtained from agricultural and fruit production, the cooperative places greater emphasis on the processing of blue corn into flour and the preparation of chiles en nogada; this dish was named Intangible Cultural Heritage of Humanity by UNESCO in 2010, and is made with local fruit products, including panochera apple, pear, peach, walnut, pomegranate, and sour pomegranate. The work developed in this transformation process allows the cooperative to link up with a network of companies and restaurants located in different parts of México. In this way, the cooperative is certified and recognized as a member of the network by providing information and promoting the consumption of its products, which are made with local ingredients and sold at fair prices.

Figure 9. *Chile en nogada* made by the Guardianes de Calpan cooperative at the restaurant and terrace Panchita y Genaro.

3.3.4. Commercialization

The incorporation of the cooperative to the network of companies and restaurants was made possible by the presence of a technical member in the work team who is in charge of marketing the different products that the members produce, mainly pinole and chile en nogada, which are derived from corn and fruit trees. This marketing component is a process that consists of the following aspects: (1) depending on the demand towards the inside of the cooperative, the available volume of each product is quantified; (2) with this data, the entrepreneurs are interviewed and more information is provided about the origin of the products; (3) for the case of corn flour, it is highlighted that it comes from native blue corn; (4) it is ensured that the production is obtained with limited use of mineralized fertilizers and (5) they are produced by small producers who practice family farming [50].

The production of fruits, especially native species, are the most important elements in the production of chile en nogada; in this activity, again, the technical member of the cooperative team plays an important role in the marketing process. Holding a fair between the months of August and September (lasting approximately nine weeks), the technical member promotes the arrival of tourists from Mexico City and Puebla; these visitors generate an important economic benefit for the cooperative and the family units that participate in this regional event. The technical member is in charge of organizing field trips with tourists to the cooperative members' plots to show them firsthand how they manage their fruit trees. The cooperative produced and sold 450 chiles en nogada in 2024, each one was priced at MXN 300.0. This amount generated a production value of MXN 135,000.0. Also, the company marketed fresh chiles poblanos and delivered 3556 kg of them; on average, eight chiles weigh 1 kg. A comparison of the three agri-food cases of the Izta-Popo Valley in Puebla is shown on Table 4.

Table 4. Comparison of the three agri-food chain cases.

	A. Maíz-Frijol Guía	B. SPR Campo Lima	C. Guardianes Calpan
Location	Huejotzingo	Tlaltenango	Calpan
Irrigation	Rainfed and well	Well and rainfed	Rainfed and well
Surface area	6 ha	45 ha	8 ha
Members	5	7	17
Average age	49 years	46 years	55 years
Schooling	6 years	8 years	9 years
Production	MIAF *	seed, grain and corn	Fruits, chile poblano
Transformation	Tlacoyos and gorditas	tortilla de nixtamal	Chiles en nogada, others
Marketing	Local markets	tortillería shop, stores	Restaurants, fairs
Production cycles	2 to 6 months	2 to 6 months	3 to 4 months
Contract labor	86	90	40
cost–benefit analysis	3.5	3.4	4
Associative figure	Family Unit	Rural Production Society	Cooperative

* Milpa interspersed with corn, beans, squash, and fruit trees. Source: Prepared by the authors based on data collected in interviews, field visits, and surveys of the groups in the three cases.

3.4. Relationship Between the Application of the CFS-RIA Principles and Agri-Food Models

Based on the empirical evidence documented with the people who manage these agri-food systems, the level of progress in the application of the CFS-RIA principles in each model was categorized as High, Medium, and Developing (Table 5). The analysis indicated that the three agri-food models present an average level of high progress in the application of PRIA 1, 2 and 9.

Table 5. Level of advancement of CFS-RIA principles in agri-food models.

CFS-RIA Principle	Progress Indicators *	AM α	SL β	GC θ	Identified Strengths	Identified Challenges
1. Contribute to food security and nutrition	▪ Increase in corn yields per hectare.	84	90	80	There has been an increase in the production of corn and other staple foods, as well as an increase in crop diversification, local consumption, and self-consumption of nutritious foods.	Dependence on external inputs (fertilizers and seeds) for corn. Limited access to fair markets.
	▪ Crop diversification through polyculture strategies.	80	74	88		
	▪ Increased availability of food for family consumption.	88	88	94		
	▪ Promote local consumption.	86	96	94		
	▪ Include simple and nutritionally balanced foods in the market.	94	94	94		
	GLP **	86	88	90		

Table 5. *Cont.*

CFS-RIA Principle	Progress Indicators *	AM α	SL β	GC θ	Identified Strengths	Identified Challenges
2. Contribute to sustainable and inclusive economic development and the eradication of poverty	▪ Diversification of products derived from or transformed from crops (innovation).	90	90	96	Family income is strengthened through crop diversification and the sale of surpluses and by-products. The action groups optimize the use of resources and improve profitability in corn and polycrop production. Increased adaptation of technologies and practices to local needs.	Prioritization of economic profitability over sustainability. Lack of access to credit or savings banks. Difficulty competing with large producers and companies or switching to a high value-added market through product transformation.
	▪ Promotion of action groups.	84	86	96		
	▪ Encourage local business relationships and improve business and production practices.	80	88	94		
	▪ Adoption and awareness of basic sustainable practices.	82	80	90		
	GLP **	84	86	94		
9. Incorporate inclusive and transparent governance structures, processes, and grievance mechanisms	▪ Creation of Territorial Action Groups (TAGs) for participatory development management.	80	84	90	The association of small producers is the main objective of the interventions, encouraging their participation in decision-making and their association in cooperatives or societies. The relationship with academic institutions is strengthened through research projects and field training.	Limitations of governmental structures. Lack of formalization due to the absence of specific laws.
	▪ Strengthening producer organizations and strategic collaboration networks.	78	92	94		
	▪ Sharing useful and relevant information.	82	88	96		
	▪ Promoting mediation mechanisms to provide solutions.	84	86	94		
	GLP **	82	88	94		

* The PRIA Progress Indicator was calculated as a percentage according to a weighting of quantitative data and the average valuation obtained from interviews, PRIA courses, and perception surveys. AM α = Case 1, SL β = Case 2, GC θ = Case 3. ** General Level of Progress (GLP) is the rounded average of PRIAs progress indicators, where less than 50% is Basic or Developing, 50% to 70% is Medium, 71% to 90% is High, and 91% to 100% is Very High. Source: Own elaboration based on information collected and [3,4,10,30,35,47,51–56].

In the three agri-food cases, a medium or moderate level of implementation progress was generally observed in PRIAs 3, 7, and 8; while in PRIAs 4, 5, 6, and 10, a basic or developing level was observed.

3.5. Relationship of the Application of the Working with People Model in Each Agri-Food Model

Working With People is a strategy to carry out linkage projects considering the processes of associativity and organization. This process is supported by the following components of the WWP model: (1) Technical–entrepreneurial; (2) ethical–social; (3) political–contextual. The process-approaches are (4) social learning [57]; (5) project management [58]; and (6) asset building [59]. Based on the results derived from the field work, it was detected that the technical–entrepreneurial, ethical–social, and social learning components present a high level of application (Table 6). However, the rest of the components are in the moderate and developing categories.

Table 6. Level of application of the WWP model in agri-food cases.

WWP Model	Application Indicator *	AM α	SL β	GC θ	Identified Strengths	Identified Challenges
1. Technical–Entrepreneurial Component	▪ The participation of the technical market component in the stages of the production chain.	74	82	98	The application of this component is high, especially in the Campo Lima and Guardianes de Saberes Sabores de Calpan groups, because they have a high level of associativity within their groups.	Consolidate the entrepreneurial component of the "Campo Lima" group. Replicate the experience of the Guardianes de Saberes y Sabores de Calpan cooperative in the other two agri-food cases. Generate commercial synergies among the three groups.
	▪ Structuring and business organization for transformation.	80	88	92		
	▪ Acceptance of strategic support from universities and their academic integration with rural groups.	90	90	90		
	▪ Coordination of meetings as an incentive for joint actions.	80	92	94		
	▪ Quality and standardization of the groups' products.	88	96	94		
	GLP **	82	90	94		
2. Ethical–Social Component	▪ Creation of local action groups (LAGs) and rural development councils.	80	84	90	The participation of rural groups in decision-making is promoted and gender equity is fostered, although inequalities persist.	Inequality in access to resources and opportunities. Persistence of traditional gender roles. Difficulty in achieving a balance in the participation of all stakeholders. Lack of trust in institutions and agents for dialog. Abuse of intermediaries that affect fair trade.
	▪ Implementation of participatory planning processes.	82	90	92		
	▪ Inclusion of women and young people in workshops and productive activities.	80	80	96		
	▪ Promotion of dialog and negotiation between stakeholders.	82	84	86		
	▪ Approach and mediation with rural groups to promote and strengthen organizational work behavior and attitudes.	88	90	94		
	GLP **	82	86	92		
4. Social Learning process-approach	▪ Use of participatory methodologies such as workshops and focus groups for group and stakeholder development.	78	82	88	Collective learning and the adaptation of strategies based on experience are encouraged, as well as an interdisciplinary reflection and analysis of the actions undertaken.	Difficulty in systematizing information and learning. Need to strengthen monitoring and evaluation capacities. Resistance to evaluation and feedback.
	▪ Exchange of knowledge and experiences to promote social dynamics between producers and universities.	92	90	90		
	▪ Participatory monitoring and evaluation of projects.	88	88	90		
	▪ Adaptation of development models to local conditions through innovation.	84	92	90		
	GLP **	86	88	90		

* The WWP Application Indicator was calculated as a percentage based on a weighting of quantitative data and the average value obtained from interviews, WWP workshops, and perception surveys. AM α = Case 1, SL β = Case 2, GC θ = Case 3. ** General Level of Progress (GLP) is the rounded average of the PRIAs progress indicators, where less than 50% is Basic, 50% to 70% is Medium, 71% to 90% is High, and 91% to 100% is Very High. Source: Prepared by the authors based on the information collected and [4,5,10,30,34,35,47,51,56,59–62].

4. Discussion

One of the most important findings of this study is the identification of different models of agri-food systems based on the level of associativity of the actors that manage these models. It is important to mention that one of the agri-food models (Case 3), in addition to being immersed in a process of association, includes, in its structure, the technical component of market competencies to carry out the commercialization process. In this model, more and better results are observed in terms of economic prosperity and in the application of RIA principles. These results are consistent with other studies in which the integration of the social and economic components constitute the elements favoring the economic income and social cohesion of the members, as in the case of the Camposeven cooperative [4,63]. They also correspond to other studies that highlight the importance of participation, innovation, and territorial management for sustainable rural development [4,59]. The application of the WWP model, with its emphasis on social learning, active participation, and respect for local knowledge, proved to be an appropriate framework for the integration of the CFS-RIA principles in the maize chain [64].

The findings made it possible to learn the use made by these three cases of the tools provided by the Working White People (WWW) model and the degree to which the Principles of Responsible Investment in Agriculture and Food Systems are applied, that is, to put the motto of working with people first to create joint actions that integrate the knowledge of experience and expert knowledge that allows solving problems by integrating both kinds of knowledge [62]. The integration of the agri-food models with the components of the WWW made it possible to observe the degree of application of the PRIAs in the production, transformation, and marketing processes; in this sense, in principles 1 and 9, their application was high for the three cases studied. These results agree with the findings of Acosta et al. (2023) [60] and Jiménez et al. (2023) [53], who, when weighing the perception of the population regarding the importance and prioritization of objectives to select the PRIAs, found that more than 91% of the population interviewed fixed their attention on principle 1 ("Contribute to food security and nutrition"), and in its prioritization, it was considered among the first places. In the case of principle 9, although its application was high, it was more relevant in two cases where the agri-food model is developed through associative processes. For principle 2 ("Contribute to sustainable economic development and poverty eradication"), congruence was found between the high degree of application in the Mexican cases and the findings of Requelme and Afonso (2023) [65], who highlight that this PRIA is incorporated in Ecuador's agricultural policies with the highest percentage of appearance in the policy guidelines and is aligned with the SDGs.

According to Fontana et al. (2025) [66], the application of the RAI principles and the WWP Model in Paucar del Sara Peru allowed them to carry out an integrative and multidimensional analysis considering the territory as the object of study. In the case of the Izta-Popo territory in Puebla, three cases of agri-food models were considered that used maize as the binding axis. These cases develop processes based on the components of the WWP model: political–contextual, technical–business, and ethical–social, to improve the processes that integrate the entire corn agri-food chain, applying some of the principles of responsible investment. Cachipuendo et al. (2025) [67] found that the development of agroecological technologies and the recovery of ancestral knowledge have been fundamental to improve the sustainability of the local agri-food system among other factors, which allow food availability and improve profitability, in line with the SDGs and the CFS-RIA principles. The three case studies in México provide elements that indicate the importance of generating processes to improve ancestral agri-food models; such is the case of the classic milpa, which integrates a diversity of crops among which corn, beans, and squash stand out. When technical and social innovations are used in the production process of this agri-food system, especially in the transformation and commercialization processes, there is greater availability of food and improved income, as was found in the results.

On the other hand, this study integrated the concept of living laboratories through the establishment of demonstration plots, which function as spaces for experimentation and learning; in these plots, producers validated innovative technologies and practices [3,30]. In this sense, knowledge transfer was not limited to a linear model but was based on a process of dialog and interaction between different actors [61].

Although significant progress has been made in strengthening rural communities, it is important to recognize that major challenges remain. Lack of access to financing, market uncertainty, and the effects of climate change are some of the obstacles that must be addressed to ensure the sustainability of development processes [59]. It is necessary to continue strengthening the capacities of producers, consolidating local organizational structures, and promoting public policies that support family farming and rural development with a territorial approach [68].

The cultivation of corn in the territory of action continues to be a priority for the production units that, through innovation processes, have developed agri-food systems that seek not only to produce their own food but also to generate new ways of participating in the stages of the food chain of transformation and commercialization, a scheme that was not observed before. The identification of the three agri-food systems is an important finding of this research because it highlights the efforts made by farming families, either at the family level or through associative processes, to make better use of available resources and improve their relationship with other actors within and outside the territory.

5. Conclusions

The incorporation of PRIAs in the maize agri-food chain using the WWP model is an effective strategy for strengthening rural communities in Puebla, México. Its application in the three Puebla agri-food cases demonstrates that it is possible to achieve a fairer, more equitable, and sustainable rural development, as long as active participation, respect for local knowledge, innovation, and territorial management are prioritized. The comparative case study demonstrated that the food chain is better organized when PRIA 1 and 2 are applied together with the technical–entrepreneurial and ethical–social components of the WWP model, which led to improvements in crop production and diversification, as well as in local consumption and self-consumption of nutritious food. A better strengthening of the association in rural groups was achieved with the application of PRIA 9, and the integration of stakeholders in the agri-food models was reached through the application of the social learning approach of the WWP model, which favors the generation of sustainable development projects that are more effective and relevant in meeting local needs, based on the convergence of knowledge, resources, action, and experience. It is necessary to continue with teaching, research, and linkage with study groups through the continuous improvement of projects based on the PRIAs and WWP, leading to the addition and optimization of more responsible investment principles in their practices and the creation of more initiatives that promote balanced sustainable development in its three dimensions: economic, social, and environmental. Thus, ensuring that projects remain relevant, effective, long-lasting, and that they promote and strengthen safe, healthy, and sustainable agri-food systems is important.

This study provides valuable information for the formulation of public policies and the implementation of rural development projects with a territorial approach in Mexico and other countries with similar challenges. It is essential to continue researching, adapting, and improving the strategies and models used in this study in order to achieve an even greater impact on strengthening rural communities and building a more prosperous and sustainable future. For example, as part it efforts to strengthen public policies in Mexico, the Polytechnic University of Madrid has offered a training course on PRIAs in international cooperation since 2021. The course is aimed at professional managers interested in working with people in rural territories and applying PRIAs, and it is a public policy recommendation designed to improve the work of stakeholders such as the government and public universities in rural territories. Additionally, public policy recommends identifying and consolidating the primary agri-food systems in the rural Izta-Popo territory of Puebla. These systems should be approached as living laboratories to facilitate the transmission of knowledge to decision-makers at local and regional levels as well as to technicians and association representatives interested in responsible investment in agriculture and food systems.

Supplementary Materials: The following supporting information can be downloaded at https://www.mdpi.com/article/10.3390/su17125442/s1. Research Methods Instruments and Construction of Indicators.

Author Contributions: Conceptualization: J.R.-L., J.A.M.-C., and N.P.-R.; methodology: J.R.-L., J.A.M.-C., and N.P.-R.; validation: J.R.-L. and J.A.M.-C.; formal analysis: J.R.-L., J.A.M.-C., and N.P.-R.; investigation: J.R.-L., J.A.M.-C., and N.P.-R.; resources: J.R.-L.; data curation: J.A.M.-C. and N.P.-R.; writing—original draft preparation: J.R.-L., J.A.M.-C., and N.P.-R.; writing—review and editing: J.A.M.-C. and N.P.-R.; visualization: J.A.M.-C. and N.P.-R.; supervision: J.R.-L.; project administration: J.R.-L., J.A.M.-C.; internal funding acquisition: J.R.-L. All authors have read and agreed to the published version of the manuscript.

Funding: This research received no external funding.

Institutional Review Board Statement: The internal institutional ethical regulations and norms indicate that in these cases no review by an Ethics Committee is required for this type of research, since it is non-interventionist, and therefore does not represent any psychological or physical harm to the participants.

Informed Consent Statement: Informed consent was obtained from all study participants, who were informed of the study's purpose and gave their consent to take part.

Data Availability Statement: All data relevant to the results are contained within the article; further data can be obtained upon request.

Conflicts of Interest: The authors report no conflicts of interest.

References

1. Instituto Nacional de Geografía, Estadística e Informática, INEGI. *Resultados Oportunos del Censo Agropecuario 2022 Puebla, Boletín de Comunicado de Prensa*; INEGI: Aguascalientes, México, 2023; pp. 1–2.
2. Instituto Nacional de Geografía, Estadística e Informática, INEGI. *Resultados Oportunos del Censo Agropecuario 2022 Puebla, Presentación*; INEGI: Aguascalientes, México, 2023; pp. 13–30.
3. Regalado, J.; Pérez-Ramírez, N.; Ramírez Juárez, J.; Méndez Espinoza, J.A. Los Grupos de Acción y La Aplicación de Tecnología de Alta Productividad Para Maíz de Secano En Localidades Del Plan Puebla, México. *La Granja* **2021**, *34*, 91–104. [CrossRef]
4. Ríos-Carmenado, I.D.L.; Becerril-Hernández, H.; Rivera, M. La agricultura ecológica y su influencia en la prosperidad rural: Visión desde una sociedad agraria (Murcia, España). *Agrociencia* **2016**, *50*, 375–389.
5. Cazorla, A.; de los Ríos, I.; Salvo, M. Working With People (WWP) in Rural Development Projects: A Proposal from Social Learning. *Cuad. De Desarro. Rural* **2013**, *10*, 131–157.
6. Smith, H.E. El concepto de "institución": Usos y tendencias. *Rev. De Estud. Políticos* **1962**, *1*, 93–104.
7. MacIver, R.M.; Page, C.H. *Society: An Introductory Analysis*, 1957th ed.; Macmillan & Co Ltd.: London, UK, 1949; Volume 1.
8. Hodgson, G.M. ¿Qué son las instituciones? *Rev.CS* **2011**, *8*, 17–53. [CrossRef]
9. Niño Velásquez, E.; Regalado López, J.; Hernández González, T. La Asociación Campesina Independiente y Sus Relaciones Con El Estado e Instituciones. *Reg. Rev. Interdiscip. En Estud. Reg.* **1998**, 62–72.
10. Comité de Seguridad Alimentaria Mundial (CSA). *Principios para la Inversión Responsable en la Agricultura y los Sistemas Alimentarios*; Comité de Seguridad Alimentaria Mundial: Rome, Italy, 2014.

11. García Winder, M.; Riveros, H.; Pavez, I.; Rodríguez, D.; Lam, F.; Arias Segura, J.; Herrera, D.; Agricultura (IICA). *Cadenas Agroalimentarias: Un Instrumento Para Fortalecer La Institucionalidad Del Sector Agrícola y Rural*; COMUNIICA, 5; Instituto Interamericano de Cooperación para la Agricultura (IICA): San José, Costa Rica, 2009; ISBN 978-92-9248-146-9.
12. Albisu, L.M. Las cadenas agroalimentarias como elementos fundamentales para la competitividad de los productos en los mercados. *Rev. Mex. De Agronegocios* **2011**, *28*, 451–452.
13. Agencia Catalana de Seguridad Alimentaria La Cadena Alimentaria. Principios y los Requisitos Generales de la Legislación Alimentaria. Reglamento (CE) No 178/2002. Available online: http://acsa.gencat.cat/es/seguretat_alimentaria/ (accessed on 24 March 2025).
14. Ramírez-Juárez, J. Régimen Alimentario y Agricultura Familiar. Elementos Para La Soberanía Alimentaria. *Remexca* **2023**, *14*, e3529. [CrossRef]
15. Moctezuma-López, G.; Espinosa-García, J.A.; Cuevas-Reyes, V.; Jolalpa-Barrera, J.L.; Romero-Santillán, F.; Vélez-Izquierdo, A.; Bustos Contreras, D.E. Innovación tecnológica de la cadena agroalimentaria de maíz para mejorar su competitividad: Estudio de caso en el estado de Hidalgo. *Rev. Mex. De Cienc. Agrícolas* **2010**, *1*, 101–110.
16. Instituto Nacional de Geografía, Estadística e Informática, INEGI. Mapa digital de México Para Escritorio Versión 6.3. Proyectos. Proyecto Básico de Información 2020. Available online: https://www.inegi.org.mx/temas/mapadigital/#Descargas (accessed on 24 March 2025).
17. Instituto Nacional de Geografía, Estadística e Informática, INEGI. *Compendio de Información Geográfica Municipal 2010, Huejotzingo, Puebla*; INEGI: Aguascalientes, México, 2010.
18. Instituto Nacional de Geografía, Estadística e Informática, INEGI. *Compendio de Información Geográfica Municipal 2010, Chiautzingo, Puebla*; INEGI: Aguascalientes, México, 2010.
19. Instituto Nacional de Geografía, Estadística e Informática, INEGI. *Compendio de Información Geográfica Municipal 2010, Tlaltenango, Puebla*; INEGI: Aguascalientes, México, 2010.
20. Instituto Nacional de Geografía, Estadística e Informática, INEGI. *Compendio de Información Geográfica Municipal 2010, San Salvador el Verde, Puebla*; INEGI: Aguascalientes, México, 2010.
21. Instituto Nacional de Geografía, Estadística e Informática, INEGI. *Compendio de Información Geográfica Municipal 2010, Juan C. Bonilla, Puebla*; INEGI: Aguascalientes, México, 2010.
22. Instituto Nacional de Geografía, Estadística e Informática, INEGI. *Compendio de Información Geográfica Municipal 2010, Calpan, Puebla*; INEGI: Aguascalientes, México, 2010.
23. ADICAP. *Atlas de Riesgos Del Municipio de San Martín Texmelucan*; ADICAP & Gobierno Municipal de San Martín Texmelucan: San Martín Texmelucan, México, 2018.
24. Secretaría de Agricultura y Desarrollo Rural del Gobierno de México Sistema de Información Agroalimentaria de Consulta, Siacon. Available online: http://www.gob.mx/siap/documentos/siacon-ng-161430 (accessed on 27 March 2025).
25. Instituto Nacional de Geografía, Estadística e Informática, INEGI. Censo de Población y Vivienda 2020. Available online: https://www.inegi.org.mx/programas/ccpv/2020/#datos_abiertos (accessed on 27 March 2025).
26. Consejo Nacional de Población del Gobierno de México, CONAPO. Índice de Marginación (Carencias Poblacionales) Por Localidad, Municipio y Entidad. Available online: https://datos.gob.mx/busca/dataset/indice-de-marginacion-carencias-poblacionales-por-localidad-municipio-y-entidad (accessed on 27 March 2025).
27. Consejo Nacional de Evaluación de la Política de Desarrollo Social, CONEVAL. Medición de La Pobreza Por Grupos Poblacionales a Escala Municipal 2010, 2015 y 2020. Available online: https://www.coneval.org.mx/Medicion/Paginas/Pobreza_grupos_poblacionales_municipal_2010_2020.aspx (accessed on 27 March 2025).
28. Méndez Espinoza, J.A.; Valencia Bastida, I.; Ramírez Juárez, J.; Pérez Ramírez, N.; Regalado López, J.; Hernández Flores, J.A. Caracterización y Tipificación de Las Unidades Domésticas Que Participan En La Cadena Agroalimentaria Maíz-Tlacoyo. *Agric. Soc. y Desarro.* **2024**, *21*, 389–401. [CrossRef]
29. Maimone-Celorio, J.A.; Regalado-López, J.; Gallego-Moreno, F.J.; Pérez-Ramírez, N.; Méndez-Espinoza, J.A. Comparative Analysis of Four Corn (Zea Mays L.) Varieties, Transformed from Grain Corn into Tortilla. *Agro Prod.* **2024**, *17*, 3–13. [CrossRef]
30. Regalado-López, J.; Castellanos-Alanis, A.; Pérez-Ramírez, N.; Méndez-Espinoza, J.A.; Hernández-Romero, E. Modelo Asociativo y de Organización Para Transferir La Tecnología Milpa Intercalada En Árboles Frutales (MIAF). *Estud. Soc.* **2020**, *30*, 2–27. [CrossRef]
31. Coller, X. *Estudio de Casos, Madrid, Cuadernos Metodológicos*; Centro de Investigaciones Sociológicas: Madrid, España, 2005; Volume 30, ISBN 978-84-7476-387-4.
32. Ponce Andrade, A.L. El Estudio de Caso Múltiple. Una estrategia de Investigación en el ámbito de la Administración. *Rev. Publicando* **2018**, *5*, 21–34.
33. Bulman, A.; Cordes, K.Y.; Mehranvar, L.; Merrill, E.; Fiedler, Y. *Guía Sobre Incentivos Para la Inversión Responsable en la Agricultura y Los Sistemas Alimentarios. Roma, FAO y Centro Columbia sobre Inversión Sostenible*, 1st ed.; FAO: Roma, Italia, 2021; ISBN 978-92-5-134575-7.

34. Regalado López, J.; Mendoza, R.; Ríos Carmenado, I.; Díaz Puente, J.M. Adaptación del modelo leader en el territorio huejotzingo, puebla: Una nueva propuesta de desarrollo rural local. In Proceedings of the Actas del XV Congreso Internacional de Ingeniería de Proyectos, Huesca, España, 6–8 July 2011; E.T.S.I. Agrónomos (UPM): Madrid, España, 2011; pp. 1533–1545.
35. Cazorla Montero, A.; De los Ríos Carmenado, I.; Yagüe, B. Trabajando con la gente en los proyectos de desarrollo rural: Una conceptualización desde el Aprendizaje Social. In *Modelos Para el Desarrollo Rural Con Enfoque Territorial en México*; Altres Costa-Amic Editores, S.A. de C.V.: Puebla, México, 2011; ISBN 978-968-839-585-1.
36. Instituto Nacional de Geografía, Estadística e Informática, INEGI. *Panorama Sociodemográfico de Puebla. Censo de Población y Vivienda 2020*, 1st ed.; Instituto Nacional de Estadística y Geografía de México: Aguascalientes, México, 2021; Volume 1, ISBN 304.601072.
37. Turrent Fernández, A.; Cortés Flores, J.I.; Espinosa Calderón, A.; Hernández Romero, E.; Camas Gómez, R.; Torres Zambrano, J.P.; Zambada Martínez, A. MasAgro o MIAF ¿Cuál Es La Opción Para Modernizar Sustentablemente La Agricultura Tradicional de México. *Remexca* **2017**, *8*, 1169–1185. [CrossRef]
38. Centro Internacional de maíz y Trigo; CIMMYT; Ministry of Agriculture, Government of Mexico; Government of the State of Puebla. *The Puebla Project: Seven Years of Experience, 1967–1973*, 1st ed.; CIMMYT: Puebla, México, 1974; Volume 1.
39. Madrigal-Rodríguez, J.; Villanueva-Verduzco, C.; Sahagún-Castellanos, J.; Acosta Ramos, M.; Martínez Martínez, L.; Espinosa Solares, T. Ensayos de producción de huitlacoche (Ustilago maydis Cda.) hidropónico en invernadero. *Rev. Chapingo Ser. Hortic.* **2010**, *16*, 177–182. [CrossRef]
40. Cortés Flores, J.I.; Turrent Fernández, A. *La Milpa Intercalada en Árboles Frutales (MIAF) Tecnología Multiobjetivo Para el Desarrollo de la Agricultura en Laderas*; Colegio de Postgraduados: Texcoco de Mora, México, 2021; p. 2.
41. Ordoñez-Ovalle, J.; Gómez-Martínez, E.; Soto-Pinto, L.; González-Santiago, M.V. El sistema milpa intercalado con árboles frutales (MIAF): Evaluación agroecológica a diez años de su implementación en Chamula, Chiapas, México. *Rev. Campo-Territ.* **2022**, *17*, 109–136. [CrossRef]
42. Pérez-Hernández, L.M.; Almeraya-Quintero, S.X.; Guajardo-Hernández, L. La producción de tlacoyos como alternativa de desarrollo en San Miguel Tianguizolco, Puebla, México. *Agro Product.* **2017**, *10*, 71–77.
43. Vargas-Cárdenas, T.; Thomé-Ortiz, H.; Ávalos De La Cruz, D.A.; Escalona-Maurice, M.; Gómez-Merino, F.C. El Tlacoyo Como Recurso Alimentario, y Su Relación Con La Oferta Turística Local: Casos Texcoco y Chiconcuac, Estado de México En México. *Agro Prod.* **2020**, *13*, 67–73. [CrossRef]
44. Secretaría de Economía del Gobierno de México Tlaltenango: Economía, Empleo, Equidad, Calidad de Vida, Educación, Salud y Seguridad Pública. Available online: https://www.economia.gob.mx/datamexico/es/profile/geo/tlaltenango (accessed on 28 March 2025).
45. Cabral Martell, A. Las figuras asociativas como alternativa en los Agronegocios. *Rev. Mex. De Agronegocios* **2004**, *VIII*, 378–389.
46. Apolo 30 Tipo 8–Tortilladoras Apolo. Tortilladoras Apolo. 2021. Available online: https://tortilladorasapolo.com/apolo-30-basica/ (accessed on 20 June 2024).
47. CFS-FAO-UN. (2024). PROGRAMA INTERNACIONAL SOBRE LOS PRINCIPIOS DE INVERSION RESPONSABLE EN LA AGRICULTURA Y LOS SISTEMAS ALIMENTARIOS, PRINCIPIOS CSA-IRA. Available online: https://www.colpos.mx/cp_pdf/eventos/2024/guia_curso_principios_ira.pdf (accessed on 20 June 2024).
48. Secretaría de Economía del Gobierno de México Calpan: Economía, Empleo, Equidad, Calidad de Vida, Educación, Salud y Seguridad Pública. Available online: https://www.economia.gob.mx/datamexico/es/profile/geo/calpan (accessed on 28 March 2025).
49. Littaye, A. The Role of the Ark of Taste in Promoting Pinole, a Mexican Heritage Food. *J. Rural Stud.* **2015**, *42*, 144–153. [CrossRef]
50. Jaleta, M.; Gebremedhin, B.; Hoekstra, D. Smallholder Commercialization: Processes, Determinants and Impact. In *Discussion Paper No. 18. Improving Productivity and Market Success (IPMS) of Ethiopian Farmers Project*; International Livestock Research Institute: Nairobi, Kenya, 2009; 55p.
51. Cazorla Adolfo, A.; De Los Ríos Carmenado, I. *Jornadas de Diálogo Para la inclusión de los Principios Para la Inversión Responsable en la Agricultura (IAR) y las Directrices Voluntarias de Gobernanza de la Tierra (DVGT) en el Ecosistema Universitario de Latinoamérica*; Conclusiones; Grupo GESPLAN, Universidad Politécnica de Madrid: Madrid, España, 2018; p. 31.
52. GESPLAN, Universidad Politécnica de Madrid. *Programa Internacional sobre los Principios de Inversión Responsable en la Agricultura y los Sistemas Alimentarios FAO*; Valoración de Objetivos de los Principios CSA-IRA en la Tercera Edición del Programa; Universidad Politécnica de Madrid: Madrid, Spain, 2024.
53. Jiménez Aliaga, R.; De Los Rios Carmenado, I.; Cazorla Montero, A.; Huamán, C.; Amparo, E. Definición de objetivos prioritarios para la implementación de los Principios CSA-IRA en la sierra central de Perú. In Proceedings of the 27th International Congress on Project Management and Engineering, Donostia-San Sebastián, Spain, 10–13 July 2023.
54. Jiménez Aliaga, R.; De Los Ríos-Carmenado, I.; San Martín Howard, F.; Calle Espinoza, S.; Huamán Cristóbal, A. Integration of the Principles of Responsible Investment in Agriculture and Food Systems CFS-RAI from the Local Action Groups: Towards a Model of Sustainable Rural Development in Jauja, Peru. *Sustainability* **2022**, *14*, 9663. [CrossRef]

55. Jiménez Aliaga, R.; De Los Ríos-Carmenado, I.; Huamán Cristóbal, A.E.; Aliaga Balbín, H.; Marroquín Heros, A.M. Competencies and Capabilities for the Management of Sustainable Rural Development Projects in the Value Chain: Perception from Small and Medium-Sized Business Agents in Jauja, Peru. *Sustainability* **2023**, *15*, 15580. [CrossRef]
56. FAO. Inversión Responsable en la Agricultura y los Sistemas Alimentarios—Dirigido al Sector Privado. Academia de Aprendizaje Electrónico de la FAO. 2022. Available online: https://elearning.fao.org/course/view.php?id=706 (accessed on 20 June 2024).
57. Friedmann, J. Planning as Social Learning. *Work. Pap.* **1981**, 1–7.
58. Winch, G. Rethinking Project Management: Project Organizations as Information Processing Systems? In Proceedings of the PMI Research Conference, London, UK, 11–14 July 2004; Project Management Institute: Newtown Square, PA, USA.
59. Midgley, J. *Social Development: Theory and Practice*; SAGE Publications Ltd.: London, UK, 2014; ISBN 978-1-4129-4778-7.
60. Acosta Mereles, M.L.; De los Ríos Carmenado, I.; Ávila Cerón, C.A.; Castañeda Sepulveda, R. Inversión responsable en procesos de sustitución de cultivos ilícitos desde el modelo "Trabajando con Personas": Estudio de caso Guaviare (Colombia). In Proceedings of the Comunicaciones presentadas al XXVII Congreso Internacional de Dirección e Ingeniería de Proyectos, Donostia-San Sebastián, Spain, 10–13 June 2023; AEIPRO, Asociación Española de Dirección e Ingeniería de Proyectos: Donostia-San Sebastián, Spain, 2023; Volume 197.
61. Cazorla Montero, A.; De Los Ríos Carmenado, I. *Rural Development as "Working With People": A Proposal for Policy Management in Public Domain*, 1st ed.; E.T.S.I. Agrónomos (UPM): Madrid, España, 2012; ISBN 978-84-615-7154-3.
62. Cazorla-Montero, A.; De Los Ríos-Carmenado, I. From "Putting the Last First" to "Working with People" in Rural Development Planning: A Bibliometric Analysis of 50 Years of Research. *Sustainability* **2023**, *15*, 10117. [CrossRef]
63. De Los Ríos Carmenado, I.; Rivera, M.; García, C. Redefining Rural Prosperity through Social Learning in the Cooperative Sector: 25 Years of Experience from Organic Agriculture in Spain. *Land Use Policy* **2016**, *54*, 85–94. [CrossRef]
64. Mur Nuño, C.; De los Ríos Carmenado, I. Hacia una "Gobernanza basada proyectos" desde los ODS y los criterios ASG: El caso de los Bancos de Alimentos. In Proceedings of the 28th International Congress on Project Management and Engineering, Jaén, España, 3–4 July 2024; pp. 1775–1793.
65. Requelme, N.; Afonso, A. The Principles for Responsible Investment in Agriculture (CFS-RAI) and SDG 2 and SDG 12 in Agricultural Policies: Case Study of Ecuador. *Sustainability* **2023**, *15*, 15985. [CrossRef]
66. Fontana, A.; Velasquez-Fernandez, A.; Rodriguez-Vasquez, M.I.; Cuervo-Guerrero, G. Territorial Analysis Through the Integration of CFS-RAI Principles and the Working with People Model: An Application in the Andean Highlands of Peru. *Sustainability* **2025**, *17*, 1380. [CrossRef]
67. Cachipuendo, C.; Requelme, N.; Sandoval, C.; Afonso, A. Sustainable Rural Development Based on CFS-RAI Principles in the Production of Healthy Food: The Case of the Kayambi People (Ecuador). *Sustainability* **2025**, *17*, 2958. [CrossRef]
68. Dixon, J.; Gibbon, D.P.; Gulliver, A.; Hall, M. *Sistemas de Producción Agropecuaria y Pobreza: Como Mejorar Los Medios de Subsistencia de Los Pequeños Agricultores En Un Mundo Cambiante*; FAO: Rome, Italy; World Bank: Washington, DC, USA, 2001; ISBN 978-92-5-104627-2.

sustainability

MDPI

Article

Territorial Analysis Through the Integration of CFS-RAI Principles and the Working with People Model: An Application in the Andean Highlands of Peru

Alejandro Fontana, Antonio Velasquez-Fernandez, Maria Isabel Rodriguez-Vasquez and Grecia Cuervo-Guerrero *

PAD Escuela de Dirección, Universidad de Piura, Piura 20001, Peru; alejandro.fontana@pad.edu (A.F.); antonio.velasquez.f@pad.edu.pe (A.V.-F.); proyec.investigacion@pad.edu.pe (M.I.R.-V.)
* Correspondence: carla.cuervo@alum.udep.edu.pe; Tel.: +51-980-848-939

Abstract: The characterization of territory in Peru's Andean regions faces significant challenges due to a lack of comprehensive methodologies capable of addressing the complexity of these contexts. This research aims to bridge that gap by developing a methodology that integrates the CFS-RAI Principles with the Working with People (WWP) model to provide a detailed, contextualized framework for territorial analysis. The framework leverages the CFS-RAI Principles' focus on sustainable agriculture and the WWP model's proven effectiveness in fostering social transformation in Aymara communities in southern Peru. The research centers on Paucar del Sara Sara, a province in Ayacucho, Peru, characterized by a Human Development Index (HDI) of 0.42 and significant development potential rooted in opportunities for organic agriculture and collaboration with the mining sector. Employing geographic analysis and qualitative methods, this study draws comparisons with existing literature and presents insights from the case study to develop a matrix of key variables for territorial analysis in Andean regions. Additionally, the research introduces a methodology for defining mining companies' areas of influence while addressing prevalent socioeconomic challenges in these territories.

Keywords: territorial analysis; territorial development; areas of direct influence; social conflicts

1. Introduction

The concept of territorial development has been explored since the late 1990s as a response to the limited success of strategies aimed at addressing rural poverty [1]. This approach has been extensively studied in the social and economic sciences, particularly within rural contexts where the interactions among space, social actors, and resources play a crucial role. Territorial development is defined as a process involving both productive transformations and institutional changes, aimed at building new capacities and opportunities for local actors [2]. Torre [3] emphasizes that this approach enhances social well-being and prosperity by balancing competition, cooperation, initiatives, and opposition among diverse stakeholders.

Despite advancements in sustainable territorial policies, significant challenges remain in implementing these strategies, especially in high Andean and rural regions [4]. These areas face unique obstacles, including geographic isolation, reliance on fragile local economies, and limited population integration into development processes [5,6]. In Peru's Andean region, these challenges are reflected in lower Human Development Index scores compared to other Peruvian areas—0.769 on average versus 0.881 in Moquegua, as reported by the UNDP in 2022—and a poverty rate of 34.3%, significantly higher than Moquegua's 13.1%.

Furthermore, while 90% of regional governments in the Andean region in Peru have regional development plans, only 17% of provinces have a local development plan. If this analysis is extended to the district level, a mere 2.8% possess local development plans.

Addressing these challenges requires not only the formulation of territorial policies but also the alignment with the unique economic, environmental, social, and institutional dimensions of each local territory [7,8]. Additionally, further research is needed to effectively adapt international frameworks to high Andean contexts, facilitating sustainable and equitable development in these regions [9].

The CFS-RAI Principles, designed to guide investments in agriculture and food systems, offer a conceptual framework for planning development in Peru's Andean region. Here, economic activities are closely linked to agriculture, operating within fragile ecosystems where smallholder farmers—typically managing plots averaging 1.0 hectare—or community members control most arable land. The successful establishment of Local Action Groups (LAG), affiliated with the Mantaro Regional Development Center (RDC) at the Universidad Nacional Mayor de San Marcos (UNMSM) and implemented under the CFS-RAI Principles in the town of Jauja, an Andean region of Peru, demonstrates the practical applicability of these principles in this study [10].

The WWP model, with its multidimensional approach to development and emphasis on integrating human values, expert knowledge, and experiential knowledge, has initiated social transformation in various Andean communities. The implementation of the WWP model in the Andean region of Puno has enabled the Coordinadora de Mujeres Aymaras, an association of 400 Aymara women across four provinces surrounding Lake Titicaca, to transform a rudimentary production of alpaca wool knitwear into an artisanal enterprise with export potential to ten countries. This transformation has been driven by the women themselves, through a process of "social learning", and supported by a technical team that effectively integrated their indigenous experiential knowledge with the expert insights of its advisors [11]. Studies have shown that the application of this model significantly increases social, relational, economic, and human capital [11].

The territorial development program in Paucar del Sara Sara, Ayacucho, Peru, provided a unique case study for refining criteria to analyze high Andean territories where a territorial development strategy is being implemented. This research aimed to develop a methodology for conducting a detailed and contextualized analysis of the territory by integrating the CFS-RAI Principles with the WWP model [12,13]. Prior studies underscore the utility of such integrated frameworks in developing holistic development strategies [14].

Development strategies are unlikely to succeed without a comprehensive analysis that integrates economic, institutional, social, and environmental dimensions. Incorporating territorial policies that prioritize local participation is essential for sustainable resource management and the effective implementation of these strategies [15].

The research also examined the contributions of various academic scholars regarding the variables required to effectively understand the territory in which a territorial development strategy is to be implemented. While these authors generally agree on the need for a multidimensional analysis, the specific dimensions and variables they propose differ significantly. Consequently, this study seeks to develop a methodology for analyzing a territory with the aim of implementing local territorial development programs in Peru's Andean region. This objective forms the core of the research, which was prompted by the implementation of a territorial development program in Paucar del Sara Sara, a locality situated in Peru's Andean Highlands.

Specifically, this study introduces a matrix that integrates dimensions from both functional and territorial approaches. The proposed multidimensional territorial analysis has also facilitated the delineation of planning units within the study area, enabling a planning process that better aligns with local realities.

Implementing this methodology to systematize territorial analysis in Peru's high Andean regions holds the potential to significantly enhance the effectiveness of development programs, foster greater community participation, and optimize resource allocation. Moreover, recognizing the social, economic, historical, and cultural relationships among the vertical life zones in these regions underscores the importance of incorporating these interconnections when defining territories for development programs or establishing areas of interaction with mining companies.

This approach could significantly influence the delineation of direct influence areas for mining operations in the Andean region. A well-defined area of interaction that considers these relationships can play a pivotal role in mitigating social conflicts, which often jeopardize the sustainability of mining activities in these zones.

2. Conceptual Framework for Territorial Development and the Andean Highland Areas

The conceptual framework for this research is grounded in definitions of territory and territorial development, as well as a synthesis of dimensions and variables proposed for territorial analysis. It also outlines the integration of the CFS-RAI Principles and the WWP model [12,13], presenting a cohesive framework for advancing territorial development and serving as a robust basis for territorial analysis. Finally, it provides an overview of the Andean Highland areas of Peru.

To define the concepts of territory, territorial development, and the compilation of dimensions and variables proposed for territorial analysis, case studies or research projects that proposed or evaluated territorial development strategies and simultaneously described the study area were reviewed. A total of 107 articles published between 2004 and 2024 in journals such as *Sustainability, Social Sciences & Humanities Open, Journal of Rural Studies, Mundo Agrario, Revista Brasileira de Ciências Sociais, Scientific Papers Management, Economic Engineering in Agriculture and Rural Development, City, Territory and Architecture, Economies, RIMISP-Latin American Centre for Rural Development, Springer Geography, Journal of Environmental Management and Tourism,* and *International Journal of Economics and Business Administration* were analyzed. Primary keywords used for the literature search included "rural development" and "territorial development", while secondary terms were "Peru", "diagnosis", "cases", "management", "strategy", "variables", and "dimension".

2.1. Territory

Etymologically, the term territory originates from the Latin territorium, with the root terra signifying the space where human beings live and interact. Territorium highlights the connection between a physical space and human activity, referring to a spatial area defined by the control exerted over it by a particular group [16].

According to Renault, "territory is a social construction, a set of historical-cultural, economic, geo-environmental, and institutional relations that give rise to and, at the same time, express an identity, a sense of belonging, and a purpose shared by social actors" [17] (p. 16). Torre [18] adds a forward-looking perspective, defining territory as a dynamic reality in constant construction, shaped by a set of projects and aspirations. Thus, the territory reflects the evolving relationship between a group of people and a physical space, or more precisely, the interactions among individuals connected through that space.

Fuini [19] identifies four dimensions of human interactions within the territory: political, economic, cultural, and natural. The political dimension pertains to the control and delimitation of space through power dynamics, often tied to the authority of the state. The economic dimension regards the territory as a source, distribution center, or object of economic resources. The cultural dimension encompasses the symbolic connection and appropriation of space by a group, fostering a shared identity. Lastly, the natural dimension addresses the interactions between society and the environment, emphasizing efforts to balance human activity with ecological sustainability.

From this analysis, it can be deduced that the territory, as a planning unit, is not a pre-existing reality but rather a construct shaped by the outcomes of multidimensional analysis.

2.2. Territorial Development

Torre [18] highlights the growing importance of territorial development, spurred by increased interest among researchers in designing effective territorial policies. He argues that traditional development analyses have often prioritized production and innovation, overlooking governance. The territorial approach seeks to address this gap by emphasizing the critical role of local population participation in the success of development projects.

Estrada and Olivares [20] posit that the essence of the territorial approach lies in integrating sectoral and disciplinary perspectives, achieved when space itself becomes the unifying entity. Similarly, a report by ECLAC [21] underscores the necessity of stakeholder unity in the development process. Henriques [22] supports this view, advocating for the territorial approach as a sustainable strategy for poverty prevention.

The concept of multidimensional territorial development has its roots in Europe. Drawing from earlier successful experiences, the Common Market Council established the European Regional Development Fund (ERDF) in 1975 as a financial tool to address territorial imbalances. Subsequent modifications in 1985 and 1988 reinforced the territorial orientation of regional development as the primary path for social inclusion, abandoning functional criteria in regional development [13]. This shift was further solidified by the European Regional/Spatial Planning Charter of 1983, which framed regional development as the projection of social, environmental, and economic policies to foster equitable development and organized spatial structures [13]. Consequently, this multidimensional planning approach became institutionalized in Europe, with notable success in rural areas, demonstrating its efficacy.

In contrast, the United States has adopted a functional approach to regional planning, focusing on infrastructure development to support production rather than addressing territorial imbalances [13]. Cultural differences also play a role: North Americans tend to exhibit less attachment to specific territories compared to populations in Europe and Latin America [13].

Schejtman and Berdegué [2] apply the concept of territorial development specifically to rural areas, describing it as a dual process of productive transformation and institutional change aimed at reducing poverty and inequality. Berdegué et al. [23] expand on this framework, defining territorial development as a process that drives changes in the social, economic, and institutional structures of territories, fostering growth, social inclusion, and environmental sustainability. They emphasize institutional change as a key element, involving coordinated actions between local and external actors and the reform of formal and informal rules that perpetuate the exclusion of marginalized populations from productive processes and their benefits.

The successful formulation and implementation of development strategies depend on these interactions and the specific characteristics of the territory. A thorough understanding of a territory requires the assessment of its social, historical, economic, institutional, physical, environmental, and material dimensions. Without accurate and comprehensive information on these aspects, development strategies risk being poorly adapted to the territory's unique conditions, hindering local participation, impeding process evaluation, and reducing resource allocation efficiency.

2.3. Dimensions and Variables for Analyzing the Territory

A comprehensive literature review reveals diverse approaches to identifying the dimensions essential for territorial analysis. Fuini [19], Sili [24], Rodríguez Miranda et al. [25], and Suárez-Roldán et al. [26] propose a framework based on four dimensions: economic, social, environmental, and governance. These dimensions, they argue, structure the territory and are indispensable for its characterization. Romo et al. [27], Mozzato et al. [28], and Labianca [29] focus on social, economic, and environmental dimensions, incorporating governance into the social dimension through mechanisms such as civic participation.

Regarding the variables that define these dimensions, scholars such as Fuini [19], Romao and Neuts [24], Mozzato et al. [28], Dunets et al. [30], Khomiiuk et al. [31], Rodríguez Miranda et al. [25], Romagnoli et al. [32], Labianca [29], and Suárez-Roldán et al. [26] identify key economic indicators, including predominant economic activities, new income opportunities, employment rates, and per capita family income. Several authors, including Fuini [19], Sili [24], and Ruperti Cañarte et al. [33], also highlight the role of social and productive infrastructure, such as transportation systems, communication networks, markets, and irrigation systems.

The social dimension encompasses variables related to quality of life, social cohesion, and cultural identity. Scholars like Pinzón and Echeverri [34], Romo et al. [27], Mozzato et al. [28], Rodríguez Miranda et al. [25], Ruperti et al. [33], Labianca [29], and Suárez-Roldán et al. [26] emphasize the availability of essential services such as education, healthcare, potable water, and sanitation. Historical factors are also considered integral by researchers such as Romao and Neuts [35], Dunets et al. [30], Millán et al. [36], and Fuini [19].

The environmental dimension focuses on natural resources and ecosystem dynamics. Studies by Pinzón and Echeverri [34], Romo et al. [27], Dunets et al. [30], and Ruperti et al. [33] underscore the importance of water resources: availability, quality, and vulnerability to pollution. Fuini [19], Sili [24], Costa and Lima [37], Dunets et al. [30], and Suárez-Roldán et al. [26] take into account biodiversity and ecosystems. Sili [24], Khomiuk et al. [31], and Labianca [29] add the edaphological analysis of the soils. Climate variables, such as precipitation and temperature, are also highlighted by Dunets et al. [30]. Few authors, such as Romo et al. [27], Pinzón and Echeverri [34], and Mozzato et al. [28], address the vulnerability of territories to socio-economic or natural risks.

Governance is another critical dimension, with scholars Romao and Neuts [35] and Romagnoli et al. [32] emphasizing the role of local institutions in public decision-making. Costa and Lima [37] highlight inter-institutional cooperation, while Pinzón and Echeverri [34] focus on mechanisms for civic participation.

2.4. Integration of the CFS-RAI Principles and the WWP Model

The Principles for Responsible Investment in Agriculture and Food Systems (RAI), endorsed by the Committee on World Food Security (CFS) in 2014, offer a comprehensive framework for promoting sustainable and responsible investments in agriculture and food systems. These principles aim to address the diverse contexts and needs of stakeholders, including states, small-scale producers, commercial enterprises, and civil society organizations [38].

The WWP model is a participatory and multidimensional approach to territorial development that emphasizes the integration of human values, expert knowledge, and local expertise. It fosters active collaboration among all stakeholders involved in a development project, including local communities, technical teams, and policymakers [14].

In 2016, the WWP framework was validated as an instrument to analyze rural development, as well as rural prosperity and resilience, in a world of growing demands and limited resources. This validation was done under the international RETHINK research program supported by the European Commission and funding bodies in 14 countries under the FP7 and the RURAGRI ERA-NET [39,40]. Over a maturation period of more than 25 years, the WWP model has been successfully applied in territorial development processes across Spain and Latin America [14,41–44].

This methodological framework transcends technical–economic perspectives and integrates social learning processes in the analysis and the building of prosperity from three dimensions: ethical–social, technical–entrepreneurial, and political–contextual. These dimensions work in an integrated manner to ensure that territorial development initiatives have a positive and lasting impact on communities. The integrating element of these dimensions is social learning, as highlighted by Friedmann [45]. By incorporating social learning processes, the WWP framework transcends a purely technical–economic perspective, facilitating the collaborative analysis and co-construction of local prosperity.

The integration of the CFS-RAI Principles and the WWP model offers a comprehensive and multidimensional framework to holistically address the complexities of rural development. This comprehensive and multidimensional framework was applied in the establishment of LAGs in Jauja, Peru, as well as in the design of family agriculture projects in the Dominican Republic. In the latter case, the framework enabled producers to improve their sustainable livelihoods, enhance the quality of their products, strengthen their organizational capacity, and access the international market [46]. By combining these approaches, a territorial development strategy can facilitate the following achievements:

1. Design interventions that are technically viable, socially equitable, and politically sustainable, ensuring long-term success and community acceptance.
2. Foster active collaboration among diverse territorial stakeholders, including local communities, institutions, local authorities, and external investors, thereby promoting shared responsibility and engagement.
3. Align decisions with community values and needs, while safeguarding and enhancing the territory's natural resources for sustainable development.

This integration is achieved by aligning the WWP model's dimensions with the key criteria of the CFS-RAI Principles, yielding positive outcomes in the following dimensions: economic viability and sustainability; adherence to standards and inclusive governance; and social inclusion and equity.

2.4.1. Dimension of Economic Viability and Sustainability

Aligned with the technical–business dimension of the WWP model, this aspect emphasizes the development and application of viable and effective locally adapted technical and productive solutions. Its objective is to enhance productivity, infrastructure, and services by tailoring scientific and technical knowledge to the specific needs of the territory. The key actions include the following:

1. Assessing the territory's current productive capacities, including its infrastructure, natural resources, and available technologies.
2. Identifying opportunities to enhance productivity and economic sustainability through the adoption of appropriate technologies and the implementation of efficient agricultural practices.

2.4.2. Dimension of Adherence to Standards and Inclusive Governance

This dimension corresponds to the political–contextual aspect of the WWP model, focusing on governance and institutional capacity to support sustainable long-term development projects. It emphasizes participatory and transparent governance practices. The proposed actions include:

1. Analyzing the existing governance structures within the territory, including institutional capacity and mechanisms for citizen participation.
2. Evaluating compliance with established standards and identifying legal frameworks that either facilitate or hinder development interventions.

2.4.3. Dimension of Social Inclusion and Equity

Aligned with the ethical–social aspect of the WWP model, this aspect prioritizes social justice, respect for human rights, and social cohesion. It seeks to strengthen social capital by fostering trust and collaboration among the various actors involved in the development process. The key actions include the following:

1. Assessing the social and cultural dynamics of the territory, identifying the main social actors, their values, and the power relations that shape interactions.
2. Analyzing the level of social inclusion, with a focus on identifying existing inequalities and exploring opportunities to enhance equity and social cohesion.

By addressing these dimensions, territorial development strategies become not only technically sound and economically viable but also socially just and politically sustainable, ensuring a positive and enduring impact on the territory.

2.5. Andean Highlands of Peru

Highland regions are fragile ecosystems where human intervention often leads to severe resource degradation, including erosion, landslides, river sedimentation, and loss of fertile soil [47]. Ecogeological studies reveal a distinct vertical pattern in environmental and human diversity in mountain regions. Interaction between altitude, climate, and soil fertility has defined the upper limits for agriculture and grazing, as well as the upper bounds for different crop types within the agricultural range [48].

The Andean Highlands of Peru is a region of high mountains located in the tropical latitudes and is home to some of the highest human settlements [47]. Altitudinal zonation governing environmental diversity in the Peruvian Andes, along with the interaction between altitude, climate, and soil fertility, has defined four ecological zones in this region. Each possesses a distinct vegetation, climate, and agricultural potential. These vertical life zones are as follows:

Quechua (2300–3500 m)—Characterized by a temperate and dry climate, this zone is suitable for cultivating crops such as corn, wheat, and other Andean staples.

Suni or Jalca (3500–4000 m)—With a cold and dry climate, this zone is ideal for growing potatoes, barley, and some vegetables. It is also well-suited for cattle grazing.

Puna (4000–4800 m)—Dominated by natural grasslands, this zone is ideal for raising llamas, alpacas, and other South American camelids.

Janca or Cordillera (above 4800 m)—This zone experiences an extremely cold climate and is snow-covered for most of the year. Vegetation is sparse, and the area is primarily used for extensive grazing.

The environmental characteristics of the Andean Highlands and the curiosity to understand how ancient populations adapted to this mountainous environment led to the development of the concept of "vertical zonation". Based on studies of pre-Columbian cultures, Murra proposed the concept of vertical control: "Each ethnic group strived to control a maximum number of vertical lifes zones or ecological niches to access resources that, in Andean conditions, were available only there" [49] (p. 94). Years later, some authors suggested that this vertical control strategy for exploiting vertical zonation could be replaced or complemented by a specialization strategy. This implied that ethnic groups controlled only one vertical life zone or ecological niche through production specialization, exchanging their products with groups from other vertical life zones [50]. Both strategies highlight the vertical interdependence between Andean populations.

Studies on social dynamics in the Andean Highlands reveal distinctive patterns of social relationships. Notably, there is minimal division of labor by age or gender, accompanied by a high degree of flexibility in social agreements at the community, family, and individual levels [47].

In terms of production and labor activity, both are significantly lower in the Andean Highlands than in lowland areas [51]. This inverse relationship between altitude and productivity is particularly stark. For instance, potato production in a lowland Peruvian province is 60 times higher for 1000 workdays than in a highland province under similar conditions [47].

According to the 2017 demographic census, the Andean Highlands are home to 8,268,000 people, representing 28.1% of Peru's population. Another feature of this region is that population density decreases with altitude [51]. Moreover, as population density increases with urbanization and the development of non-agricultural economic activities [52], the highlands become dependent on lowland populations for goods and services [53].

Efforts to implement effective territorial planning in the Andean Highlands face significant challenges due to a lack of comprehensive studies. Official data from 2017 indicate that of the 1874 districs in Peru, 815 (43.5%) have fewer than 500 inhabitants and consequently lack their own territorial planning instruments. According to Peru's territorial planning and urban development regulations, districts with populations below 5000 must be grouped in the territorial planning process (Supreme Decree No. 022-2016-Housing). However, this approach often fails to account for the unique contextual characteristics of smaller highland communities.

3. Methodology

This study employed a case study methodology in Peru, a country where the Andean Highlands constitute 29% of the national territory. To ensure both internal and external validity, the research incorporated official data and field observations. Each step of the analysis process was meticulously documented to guarantee reliability and reproducibility [54].

The primary aim of the research was to develop a methodology that facilitates a comprehensive understanding of the physical space's characteristics, the relationships among the population, and the interactions between the population and the physical environment. This approach was designed to support the effective territorial development program.

The study focused on evaluating the physical and demographic characteristics of Paucar del Sara Sara province, where a local NGO initiated a territorial development program. This program was conceptually guided by the WWP model, a framework that has been successfully implemented in various regions of Peru [10,11].

To evaluate the physical characteristics and social relationships in the study area, the integration of the CFS-RAI Principles with the WWP model was employed, applying the CFS-RAI Principles to the rural development process.

Paucar del Sara Sara province is situated in Peru's high Andean region and is one of 11 provinces in the Ayacucho department. Ayacucho covers an area of 43,000 km^2, with a population of 616,176 (INEI, 2017) [55]. The department has a Human Development Index (HDI) of 0.4 and is governed by a coordinated regional development plan for 2024–2034. The region's diagnostic framework was made based on the priority variables of six dimensions: socio-demographics, environment, infrastructure services, economy, disaster risk management, and governance. At the local level, however, only two of Ayacucho's eleven provinces have adopted a coordinated local development plan, and just six of its one hundred nineteen districts have implemented such plans.

Paucar del Sara Sara province spans 2097 km^2 and has a population of 9609. Despite its relatively small population, it faces significant challenges: 13.9% of residents lack access to water services, and 46.8% lack sewage services (INEI, 2017) [55]. The poverty rate in the province stands at 31.9% (INEI, 2017) [55], with an HDI of 0.4212 (2019) [56]. Nevertheless, Paucar del Sara Sara holds substantial development potential due to its capacity for organic agriculture and the presence of partnerships with mining companies operating in the area (see Figure 1).

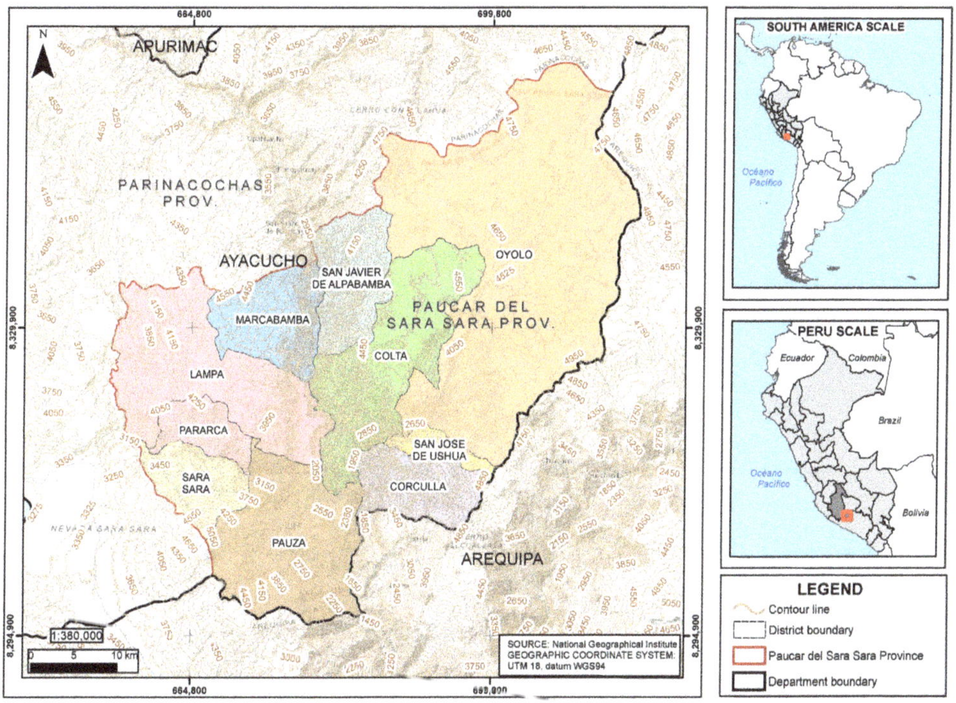

Figure 1. Geographical location of the province of Paucar del Sara Sara. Source: own elaboration.

3.1. Case Study: Territorial Analysis of Paucar del Sara Sara

The province of Paucar del Sara Sara—derived from the Quechua pawqar ("flower garden" and Sara Sara ("cornfield", referencing the corncob-shaped rocks of the eponymous volcano)—is one of the eleven provinces comprising the Ayacucho department in the high Andean region of Peru [57]. It is the youngest province in Ayacucho, officially established on 2 January 1985 [57]. Administratively, the province is divided into ten districts: Pauza, Colta, Corculla, Lampa, Marcabamba, Oyolo, Pararca, San Javier de Alpabamba, San José de Ushua, and Sara Sara (see Figure 2).

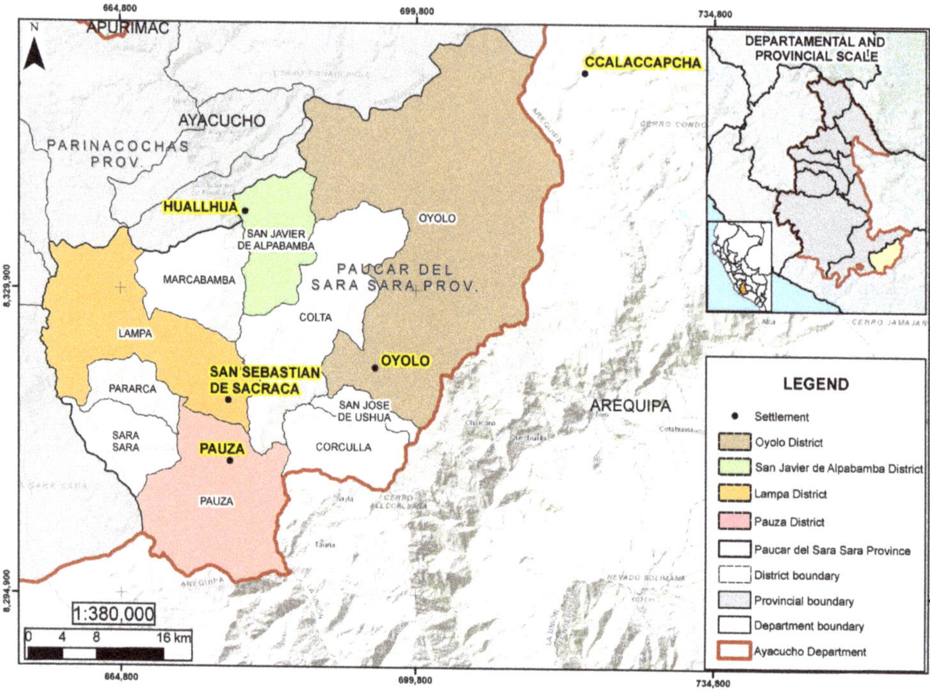

Figure 2. Geographical distribution of the districts in the Province of Paucar del Sara Sara. Source: own elaboration.

Paucar del Sara Sara lies within the Ocoña River basin, characterized by its rugged terrain and the presence of potentially active volcanoes, including Coropuna, Solimana, and Sara Sara. The basin encompasses ten sub-basins, covering a total area of 16,045 km^2. During the rainy season (January–March), high-water flows significantly activate the region's hydrography, contributing to some of the largest river flows in southern Peru.

The province is characterized by vertical life zones, sub-territories determined by altitude and shaped by the branching Andes Mountain range. As mentioned above, these vertical life zones critically influence the province's agricultural and livestock activities, offering diverse opportunities and challenges based on their unique characteristics.

The distribution of the population across these zones has given rise to distinct customs, cultural expressions, and social networks, which not only enhance local diversity but also facilitate communication and economic exchange between communities.

The general objective of the study was to thoroughly understand the study area to implement a territorial development program in Paucar del Sara Sara province. Guided by the WWP model, the research emphasized the importance of fostering collaboration between the local population and the research team to analyze the social and physical characteristics of the territory. Accordingly, the fieldwork was designed to be highly participatory: In total, 88 in-depth interviews were conducted with individuals representing various social roles, including homemakers, community leaders, leaders of productive associations, merchants, school administrators and teachers, public officials, and health or sports promoters. In addition, 15 workshops were held, engaging 192 community members. Furthermore, contact was established with an additional 102 individuals through informational sessions and visits to reservoirs, canals, and agricultural plots.

3.1.1. Field Work

Three field visits were conducted to gather extensive information about the territory. The geographic delineation of the study area was informed by an analysis of basins and sub-basins, supported by cartographic material using Geographic Information Systems (GIS) (Software ArcGis 10.8) and an in-depth examination of the region's water resources. The study also drew upon official publications from entities such as municipal authorities, the Geological, Mining, and Metallurgical Institute (INGEMMET), the Ministry of Transport and Communications, the National Institute of Statistics and Information Technology (INEI), the National Water Authority (ANA), and the National Institute of Civil Defense (INDECI).

The research extended beyond geographic and hydrological aspects to examine social, political, cultural, and religious phenomena. Researchers prioritized participatory engagement with the local population through semi-structured interviews and on-site workshops. To ensure data coherence, assess quality, and refine work guides, the research team maintained constant communication and collaboration with the community. Thematic maps were created to document the aspirations, challenges, and needs of residents.

The first field visit (4–17 May 2023) focused on identifying localities with the potential to serve as development hubs. The research team visited populated centers including Pauza, Oyolo, San Javier de Alpabamba, Huallhua, Belén, Santa Rosa de Cascara, Patarí, and Huancute.

The second field visit (7–11 June 2023) aimed to cover the remainig localities in the province, including Cotahuasi, Cochapampa, Pampamarca, Mungui, Cachana, Chapi, and Oyolo. During these visits, the team assessed the social, economic, personal, and political realities of the province, along with its development potential. Ethnographic methods included in-depth interviews and home visits to observe daily life, such as cooking, eating, knitting, and recreational activities.

The third field visit (7 September–15 October 2023) tested and refined hypotheses generated during earlier visits, particularly regarding the dynamics of vertical life zones. Localities revisited included Pauza, Sacraca, Huallhua, Oyolo, and Ccalaccapcha. This iterative approach ensured a comprehensive and grounded understanding of the territory's potential for development.

This visit involved participatory workshops with productive associations, irrigation boards, prominent families, local leaders, and key stakeholders. Researchers also conducted in-depth interviews in homes and agricultural plots, supplemented by informal conversations in public spaces such as town squares and shops. Visits to irrigation canals, reservoirs, and agricultural areas provided additional insights.

3.1.2. Processing Field Information

The collected data was systematized using a range of analytical tools, including livelihoods analysis, value chains assessments, empathy maps, dream maps, journey maps, social network analysis, and social cartography. The team notably adapted the Livelihoods Model developed by the Tropical Agricultural Research and Higher Education Center (CATIE) and the FAO to suit the territorial context. This model analyzed five key forms of capital—environmental, physical, financial, human, and social—to identify the territory's strengths and weaknesses. This comprehensive approach provided a multidimensional understanding of the territory, enabling a more effective analysis and strategy development.

4. Results: The Analysis of the Case Study Under the Integration of CFS-RAI Principles and WWP Model

To identify the key variables necessary for understanding the territory, this case study was analyzed using the three dimensions derived from the integration of the CFS-RAI Principles and the WWP framework: economic viability and sustainability dimension; adherence to the standards and inclusive governance; and social inclusion and equity. This analytical approach not only allowed for a systematic organization of these variables but also facilitated the potential replication of the model in other similar regions. Each variable was aligned with the relevant CFS-RAI principles and WWP dimensions for greater applicability and coherence.

This study also defined territorial units for planning purposes. In this research, these units were designated as Hatun Wasi (meaning "big house" in Quechua). Specifically, two distinct territorial units were identified: Hatun Wasi 1 and Hatun Wasi 2.

4.1. Territorial Analysis of Paucar Del Sara Sara

This section details the data collection processes and key findings across the evaluated dimensions.

4.1.1. Dimension Economic Viability and Sustainability

The study identified the primary economic activities of the population, along with the distribution of monetary resources at both family and community levels. High-potential local products, including both raw materials and processed goods, were identified, and value chains within the area were analyzed. Social cartography was utilized to illustrate resource flows and land use within the territories.

Particular attention was given to irrigation canals, roads, and areas vulnerable to climate-related risks. Using climate scenarios, the study assessed the potential impacts of climate change on agricultural zones and densely populated areas, identifying the months most affected by frosts. Existing agricultural zones and areas with agricultural potential were also mapped.

Agriculture remains the primary economic activity in the province. In Hatun Wasi 1, organic farming predominates, with key products including avocados, alfalfa, and fruits such as prickly pear, apples, and peaches. Farmers in this area use modern tools more widely, and some producers, through intermediaries, have successfully exported organic avocado products to international markets. Conversely, in Hatun Wasi 2, the cultivation of potatoes, alfalfa, and quinoa prevails, primarily using ancestral tools. Production levels in this region remain low, with much of the harvest used for subsistence. Climatic conditions, challenging terrain, and poor soil quality severely limit agricultural activity in the highest-altitude communities of Hatun Wasi 2.

Access to credit was found to be more prevalent in Hatun Wasi 1 due to its proximity to the provincial capital. However, a lack of knowledge regarding efficient investment and the absence of a savings culture diminishes the positive impact of these credits.

Water resource management also differs between the two territories. In Hatun Wasi 1, water is regulated by the Local Water Agency, with water-user committees managing agricultural access. However, limited availability results in significant constraints, as water is provided only once every 20 days. The use of inefficient irrigation methods, such as gravity and sprinkling, leads to substantial water loss through flooding and filtration. In Hatun Wasi 2, water for agriculture is abundant but poorly distributed due to damaged infrastructure, with gravity irrigation remaining the dominant method.

The commercial culture in Hatun Wasi 1 is primarily based on monetary exchange. In Hatun Wasi 2, barter systems are common, and commercial relationships are often restricted to family networks. In such cases, sponsorship rituals, where outsiders are integrated through familial ties, facilitate trade.

Commercial interactions further illustrate these dynamics. For example, in Hatun Wasi 1, commercial interactions are observed between Pauza, Sacraca, and Huallhua; Pauza acts as a hub, receiving potatoes from Huallhua and fruits from Sacraca, while exporting horticultural products to Huallhua and goods from Lima to other areas. In Hatun Wasi 2, Oyolo serves as the main commercial hub, where potatoes and tubers are traded. Oyolo also supplies products from Pauza to other localities. These interactions confirm the persistence of interconnections between the vertical life zones across the territory.

Additionally, the study identified 68 small businesses or associations registered for tax purposes, with 11 operating in agriculture. The remaining entities focus on transportation, engineering, housing construction, and the trade of food, household goods, construction materials, hardware, machinery spare parts, and motorcycles.

4.1.2. Dimension of Adherence to Standards and Inclusive Governance

Governance in the study area was categorized based on three criteria: system type, decision-making processes, and individual–group relations.

System type: governance structure can be modern, led by district mayor; or traditional, governed by communal assembly and resident associations.

Decision-making: processes can be participatory, requiring majority approval from the population; or representative, where decisions are made by elected representatives.

Individual-group relations: these can be liberal, with individuals or organizations pursuing their own interests; corporate by interests, where people join associations to achieve common goals; or corporate by kinship, where people organize within families.

Political culture also varies across territories. In Hatun Wasi 1, political attitudes are generally neutral, though some young professionals exhibit reformist tendencies. However, some people are influenced by those who offer them some form of support, while others openly oppose private entrepreneurial activity. In Hatun Wasi 2, most residents are non-confrontational but have been politically manipulated through money or favors. In high-altitude communities, decisions are made by community-wide consensus, prioritizing collective development.

In the area of study, local authority is determined by factors such as longevity in the community, family size, the prestige of kinship networks, the scale of organized festivals, and participation in cultural and religious activities. In Hatun Wasi 1, success in business also contributes to an individual's authority.

The political boundaries of Hatun Wasi 1 and Hatun Wasi 2 do not align with official district divisions. These territories extend beyond these borders and are defined by a unique political and social history.

However, central public administration allocates resources based on district boundaries rather than the territorial logic of historically connected localities like Hatun Wasi 1 and Hatun Wasi 2. As a result, despite the implementation of five public projects, most residents remain unaware of these initiatives.

4.1.3. Dimension of Social Inclusion and Equity

Using social network analysis, the study identified the dynamics of local group interactions, assessing the strength of individual connections and the population's capacity for collaboration toward shared goals.

The findings revealed the coexistence of two distinct associative logics within the study area. The first, a modern associative logic, is characterized by individuals freely choosing their associations. The second, a traditional associative logic, is grounded in relationships confined to family networks.

The work culture across the region reflects two predominant approaches: entrepreneurial, which emphasizes innovation and risk-taking; and inclusive, which prioritizes collective success and equitable benefit distribution. However, conflicts over resource allocation among families were noted in areas where the inclusive vision dominates.

The study identified social leaders with the capacity to drive development projects, along with individuals skilled in forging connections and engaging influential interest groups. Data collection also included education, health, security, and demographic profiles, alongside an assessment of social infrastructure such as health centers, schools, transport terminals, financial institutions, markets, and housing conditions. Access to water for human consumption varied, with minimal filtration losses but irregular and occasionally turbid supplies.

The study also highlighted the cultural identity, skills, and tangible and intangible cultural assets of the territories. Each region boasts unique dances, songs, and cultural expressions that enrich its identity and heritage.

Hatun Wasi 1 exhibits a modern associative logic where individuals voluntarily join associations. Pauza hosts an association of organic avocado producers, a livestock producers' association, and a guinea pig producers' group. Sacraca similarly features associations for prickly pear and guinea pig producers. Entrepreneurialism and individual risk-taking define the work culture here, with an emphasis on individual productive success.

Hatun Wasi 1 demonstrates a strong affinity for literature, theater, music, and dance. Notably, Pauza, recognized as the "Cervantine Capital of the Americas", holds historical significance as the first location in the Americas to stage a performance of Don Quixote de la Mancha in 1607, just two years after its publication in Spain. This tradition continues through annual theatrical performances organized by local youth, inviting participants from across Hatun Wasi 1.

Despite its cultural richness, Hatun Wasi 1 faces challenges such as the lack of hospital-level healthcare facilities, inefficient water distribution, an aging population, and labor shortages. In Huallhua, the highest-altitude community in this vertical life zone, social tensions arise from a division between groups supporting and opposing a nearby mining operation.

In contrast, Hatun Wasi 2 retains traditional associative practices, with interactions primarily confined to family networks. This dynamic is particularly pronounced in Ccalaccapcha, where economic activities are limited to familial clans. Outsiders involved in economic activities are often required to assume the role of godparents to a family member's child as a form of social integration. In Oyolo, associative practices revolve around neighborhood leadership, but overall levels of associativity remain low—a legacy of the region's history of internal conflict during the terrorism era (1980–1992), which left deep scars on families.

Oyolo serves as the social and economic hub for Hatun Wasi 2, with families from 82 dispersed small groups relying on the town for supplies, trade, education, and public administration. Ccalaccapcha maintains familial and commercial ties with Oyolo and the southern highlands of the country.

Despite the region's abundant natural resources, Hatun Wasi 2 has a notably low work culture. Many residents, particularly in Oyolo, exhibits an assistentialist mindset, relying on public authorities or NGO initiatives to address their problems.

Spanish is the predominant language in the study area, except in Ccalaccapcha, located in Hatun Wasi 2, where Quechua is the mother tongue. Religious affiliation also varies, with Ccalaccapcha being predominantly Evangelical, while the rest of the region is deeply Catholic.

Cultural expressions in Hatun Wasi 2 center around religious festivities. Key local concerns include the overgrowth of grass in fields and, in the community of Ccalacchapcha, the degradation of wetlands and deteriorating pastures.

4.2. Defining the Territorial Units for the Territorial Development Program in Paucar del Sara Sara

After visiting the localities and analyzing the area of study under the three CFS-RAI principles and WWP model dimensions, the study identified territorial units with unique physical and social characteristics. These units, termed Hatun Wasi 1 and Hatun Wasi 2, share several features: a locality that serves as a focal point for activities and interactions; a specific level of openness to new members; a market system connecting different vertical life zones within the unit; and a network of bridle paths that define and facilitate mobility within the territorial unit. Figure 3 illustrates the two territorial units identified in the province of Paucar del Sara Sara, highlighting their spatial dynamics and interactions across vertical life zones.

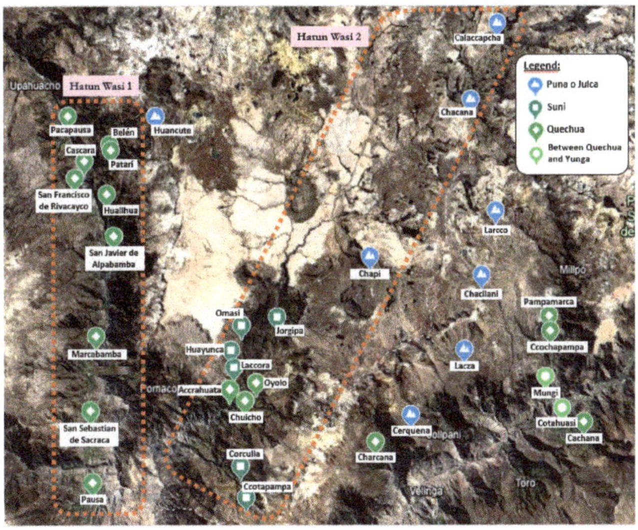

Figure 3. Definition of territorial units for the territorial development program showing the settlements and their location across different levels of vertical life zones. Source: own elaboration.

4.2.1. Hatun Wasi 1

This territorial unit comprises the localities of Pauza, Sacraca, Marcabamba, San Javier de Alpabamba, Huallhua, San Francisco de Ravacayco, Cascara, Patarí Pacapausa, and Huancute. While some of these localities belong administratively to the province of Paucar del Sara Sara and others to Parinacochas, they share familial, commercial, and cultural ties, as detailed in the territorial analysis.

The main center of gravity is Pauza, a historically significant location. It is also acknowledged by other populated centers within the territorial unit as a hub of culture and commerce. Currently, Pauza serves as a gathering place for young people from surrounding areas who participate in various cultural activities, including running, football matches between communities, cooking events, fairs, carnivals, traditional TikTok video recordings, raffles, parades, and concerts featuring guitars, harps, and violins.

In addition to its cultural significance, Pauza plays a crucial economic role. It is the primary location where surrounding communities source agricultural inputs, and it provides essential services, such as hotels, restaurants, and hardware stores, for those involved in artisanal mining (see Table 1).

Table 1. Settlements included in Hatun Wasi 1 by vertical life zone and altitude.

Settlement-Town	District	Province	Vertical Life Zone	Altitude (masl)
Pauza	Pauza	Paucar del Sara Sara	Quechua	2536
Marcabamba	Marcabamba	Paucar del Sara Sara	Quechua	2615
San Javier de Alpabamba	San Javier de Alpabamba	Paucar del Sara Sara	Quechua	2628
San Sebastian de Sacraca	Lampa	Paucar del Sara Sara	Quechua	2695
Pacapausa	Pacapausa	Parinacochas	Quechua	2821
Cascara	San Francisco de Ravacayco	Parinacochas	Quechua	2531
San Francisco de Ravacayco	San Francisco de Ravacayco	Parinacochas	Quechua	2833
Huallhua	San Javier de Alpabamba	Paucar del Sara Sara	Quechua	2915
Belén	Pacapausa	Parinacochas	Quechua	3160
Patarí	San Francisco de Ravacayco	Parinacochas	Quechua	3390
Huancute	San Francisco de Ravacayco	Parinacochas	Puna	4154

Source: own elaboration.

4.2.2. Hatun Wasi 2

This territorial unit includes the localities of Oyolo, Accrahuata, Accopampa, Mallco, Chuicho, Ccotapampa, Corculla, Laccora, Jorgipa, Omasi, Huayunca, Chacana, Ccalaccapcha, and Chapi. All of these localities belong to the province of Paucar del Sara Sara. Table 2 presents the localities encompassed in this territorial unit.

Table 2. Settlements included in *Hatun Wasi* 2 by vertical life zone and altitude.

Settlement-Town	District	Province	Vertical Life Zone	Altitude (masl)
Accrahuata	Oyolo	Paucar del Sara Sara	Quechua	2793
Accopampa	Oyolo	Paucar del Sara Sara	Quechua	2874
Mallco	Oyolo	Paucar del Sara Sara	Quechua	2935
Chuicho	Oyolo	Paucar del Sara Sara	Quechua	3155
Oyolo	Oyolo	Paucar del Sara Sara	Quechua	3410
Ccotapampa	Corculla	Paucar del Sara Sara	Suni	3503
Corculla	Corculla	Paucar del Sara Sara	Suni	3503
Laccora	Oyolo	Paucar del Sara Sara	Suni	3679
Jorgipa	Oyolo	Paucar del Sara Sara	Suni	3862
Omasi	Oyolo	Paucar del Sara Sara	Suni	3873
Huayunca	Oyolo	Paucar del Sara Sara	Suni	3899
Chacana	Oyolo	Paucar del Sara Sara	Puna	4512
Ccalaccapcha	Oyolo	Paucar del Sara Sara	Puna	4523
Chapi	Oyolo	Paucar del Sara Sara	Puna	4612

Source: own elaboration.

This territorial unit is characterized by two centers of gravity. The most prominent is Oyolo, followed by Ccalaccapcha.

Oyolo holds a rich historical legacy dating back to the viceroyalty period when it served as a hub for social control and interaction across various vertical life zones. However, its prominence waned before the end of the viceroyal era. Oyolo is home to three of the four principal ethnic groups in this territorial unit—the Qollana, Chaupichaca, and Pomatambo.

These groups continue to inhabit the territory based on a pre-viceroyal logic of hanan (upper) and hurin (lower) divisions. Each ethnic group takes pride in demonstrating its identity to the others. Nevertheless, younger generations in Oyolo, who primarily speak Spanish and are less engaged in agricultural activities, often struggle to recognize these ethnic differences.

Ccalaccapcha, on the other hand, has historically played a pivotal role as a provider of protein through camelid breeding. The relationship between Oyolo and Ccalaccapcha remains strong, facilitating social and economic exchanges along this route. From an Andean perspective, which values control over multiple vertical life zones, Oyolo has historically sought to integrate Ccalaccapcha into its market and social administration. However, Ccalaccapcha's high population density, thriving camelid production, and strong economy have enabled it to maintain its autonomy. At present, the two centers maintain a cooperative and politically harmonious relationship.

5. Discussion

A review of the literature reveals that territorial characterization is generally approached from a multidimensional perspective. However, there is no consensus on the dimensions or variables required for comprehensive analysis.

The application of the CFS-RAI Principles and the WWP Model in Paucar del Sara Sara has allowed for an integrative, multidimensional analysis of the territory. This approach incorporates three primary dimensions: economic viability and sustainability; adherence to standards and inclusive governance; and social inclusion and equity.

In terms of the economic viability and sustainability dimension, the case study emphasizes the importance of understanding the commercial culture of the territory and the mechanisms of product exchanges, including monetary transactions, barter, sponsorships, or exchanges across ecological zones. This critical aspect of economic interaction is absent in the literature [17,28,34].

Additionally, the literature overlooks the work culture of the territory [17,32–34]. In contrast, the case study identifies three distinct approaches: an entrepreneurial vision; a model where the value generated by individual businesses is expected to be distributed equitably among all community members; and a third scenario, where social conflicts with companies are used as a means of subsistence.

Regarding the primary sector, existing research [25,26,28,35,37] largely neglects opportunities for developing local or regional markets based on native products [24,30–32,34]. In Peru's Andean localities, challenging land travel and high freight costs render many external goods unaffordable. Furthermore, the literature neglects to address the ease or difficulty of accessing credit or the financial culture present in the territory. Additionally, the literature fails to examine the availability of labor and its educational level within the territory [2,16,29,33,35].

Although climate is acknowledged as a significant factor in the literature [30,37], the spatial and temporal vulnerabilities affecting populations and the environment are underexplored. Furthermore, georeferencing productive and social infrastructure—a critical tool for effective development planning—is largely absent.

The adherence to standards and inclusive governance dimension underscores the need to enhance the existing body of research on roles, responsibilities [17,29], and institutional descriptions [17,24,25,29,32,34,41] with an analysis of the local political culture. Key variables include the governance system of the territory (modern or traditional); the model of political decision-making; and the nature of the relationship between the individual and the social group.

Additionally, this approach seeks to evaluate the level of coordination among companies, public administration, and the local population; the projects promoted by the public administration; and the extent of the population's awareness and understanding of these initiatives.

Existing literature on vulnerability and risk often fails to address two critical aspects: the resources available to local institutions for managing risks and the cultural responses to these challenges, such as community awareness and the use of signage [17,28,34].

The social inclusion and equity dimension of the case study reveals significant gaps in the literature, particularly in its treatment of social organization and relationships within communities [29,30,32–34]. The study highlights diverse forms of social organization in high Andean areas of Peru, including kinship-based structures, sponsorship networks, and community-driven frameworks.

Furthermore, while existing literature discusses local leadership, it rarely evaluates the capacity of leaders to effectively drive development initiatives [2,8,15]. The WWP methodology, which emphasizes the active participation of the local population, underscores the importance of identifying and assessing leaders with the potential to guide development efforts successfully.

The integrated approach applied in this study has enabled the identification of territorial units for local planning, respecting the interconnections among vertical life zones. In the case of Paucar del Sara Sara, two distinct territories—Hatun Wasi 1 and Hatun Wasi 2—were identified, each with unique characteristics and relationships among their localities.

The findings suggest that territorial planning in Peru's Andean regions differs significantly from frameworks based on natural basins or sub-basins, as proposed by some authors [17,24,34], or the political boundaries established by the national planning system. Instead, effective planning requires an understanding an understanding an understanding of the social, cultural, and commercial relationships within local communities.

The interactions between vertical life zones play a defining role in shaping territorial units. The case study's comprehensive approach has redefined territorial characterization by incorporating variables often overlooked by a strictly functional approach.

Drawing on the functional dimensions discussed in the academic literature and the integrated approach of the CFS-RAI Principles and the WWP model, a matrix was developed to define intervention territories in Peru's Andean regions. This matrix serves to delineate the territory and to provide a structured framework for understanding its opportunities, risks, and social, political, and cultural dynamics.

Table 3 outlines this matrix, detailing the variables associated with each dimension according to territorial and functional approaches. To enhance usability, the territorial dimensions are labeled in alignment with the WWP model.

Table 3. Proposed Matrix for Territorial Analysis in Andean Regions.

Functional Approach	Sub-Dimensions	Key Dimensions of the Territorial Approach: Political-Contextual	Ethical-Social	Technical-Entrepreneurial
Economic	Economic dynamics	-	Demographical variables Socio-economic variables (labor availability, education level)	Main economic activities Financial culture Commercial culture Work culture Vertical life zone relationships
	Infrastructure	Community centers	Social infrastructure	Productive infrastructure Geospatial Mapping of Infrastructure
	Market Size	-	-	Local products Opportunities for innovation Advancement of Regional Market Dynamics
Governance-Institutional	Government Entities	Political boundaries Local political system Social decision-making processes Relationship individual-community	-	-
	Local Organization	Social associations	-	Productive associations
	Social governance	Relationship state-private sector-civil society	-	-
Socio-cultural	Innovation and Technology	-	-	Technological change culture
	Historiography	Authority traditions Historical political relationships	Historical community development Cultural identity	-
	Cultural Expressions	-	Cultural assets: tangible and intangible	Economic potential of cultural heritage
	Basic Services	-	Living conditions	-
Physical-environmental	Water resources	Basin boundaries	Domestic water use	Water use—productive activity
	Biodiversity	-	-	Economic potential of natural heritage
	Soils	-	Domestic land use	Land use—productive activity
	Climate	-	-	Climate scenarios
	Vulnerabilities and risks	On roads and telecommunication systems	On social infrastructure	On productive infrastructure

Source: own elaboration.

6. Conclusions

The high Andean regions of Peru present a unique geographical and social complexity shaped by rich cultural diversity, exceptional biodiversity, and fragile ecosystems. The CFS-RAI Principles and the WWP model provide an integrated framework analyzing these territories through three key dimensions: economic viability and sustainability, compliance with standards and inclusive governance, and social inclusion and equity.

The integrative nature of these dimensions enables the identification of interrelated functionalities, facilitating the delineation of the intervention space, referred to as the territory [16]. In the case study of Paucar del Sara Sara, this concept is reflected in the territorial units for planning identified as Hatun Wasi 1 and Hatun Wasi 2.

The technical and entrepreneurial dimension highlights the need to go beyond identifying the main economic activities and productivity opportunities. It emphasizes the importance of evaluating access to credit, knowledge of financial concepts, and the local culture of savings. Additionally, understanding the commercial culture, work culture, and the quality of water and natural heritage management is crucial for comprehensive territorial analysis.

The political and contextual dimension underscores the significance of the governance systems operating within the territory, as well as the relationship between local authorities and community participation. In the case study, the political–institutional framework revealed the coexistence of modern and traditional governance systems. This duality presents significant challenges for local development and the successful implementation of long-term projects.

The ethical and social dimension of the case study revealed that, while cultural diversity has the potential to foster cohesion and a sense of belonging, it can be weakened by the conflict between modern and traditional associative logics. The coexistence of these differing logics within the territory creates tensions that can impede development if not managed with cultural and political sensitivity. In the case of Hatun Wasi 1 and Hatun Wasi 2, this perspective highlighted low social cohesion among the groups of inhabitants.

This research underscores that analyzing a territory with social and cultural complexities requires an approach that integrates the relationships between various functional dimensions. This is particularly critical in the Andean territories of Peru, where social, ethical, economic, and environmental factors are deeply interconnected and where comprehensive studies of local realities are scarce.

In the case of Paucar del Sara Sara, this multidimensional analysis facilitates the design of a local development program, complete with strategic objectives and tailored projects for each territorial unit. Furthermore, it allows for the identification of local leaders who will play a pivotal role in the program's execution.

The findings of this research, combined with a thorough literature review, informed the creation of a methodological matrix for analyzing Andean territories in Peru (see Table 3). This matrix provides a systematic framework for diagnosing challenges and opportunities, offering a practical tool for territorial analysis. While the matrix includes a preliminary set of variables, it allows for adjustments based on its application in local planning processes.

In Peru, local planning traditionally relies on demographic, economic, and social indicators, often using political boundaries as the primary territorial units. The proposed methodology, which aligns planning efforts more closely with local realities and emphasizes active community participation, introduces higher initial costs. However, future analysis should evaluate whether these initial expenses are justified by the long-term benefits of more effective and inclusive planning.

This methodology holds relevance for mining companies operating in the Andean regions of Peru. For such companies, establishing strong, sustainable relationships with local communities is critical to their operational success. Effective engagement requires a nuanced understanding of the complexities of the local context, which this approach provides.

The findings challenge the current methodology used to define areas of direct and indirect influence for mining operations in Peru's high Andean regions. The interconnections between vertical life zones—shaped by complex social, economic, historical, and cultural factors—demand a reconsideration of the equidistance criterion often used to delineate areas of influence.

Mining activities impact not only the populations immediately surrounding their operations but also those connected through vertical life zones. As such, accurately defining these areas of influence is crucial for mitigating social conflicts that can disrupt mining operations.

A well-informed approach to defining influence areas would enhance the ability to address the needs and concerns of all affected populations, fostering more constructive and sustainable relationships between industries and local communities.

Author Contributions: Conceptualization, A.F. and A.V.-F.; methodology, A.F. and A.V.-F.; validation, A.F. and A.V.-F.; formal analysis, A.F. and A.V.-F.; investigation, A.F., A.V.-F., M.I.R.-V. and G.C.-G.; writing—original draft preparation, A.F. and G.C.-G.; writing—review and editing, A.F., A.V.-F. and G.C.-G.; supervision, A.F. and A.V.-F.; project administration, A.V.-F. All authors have read and agreed to the published version of the manuscript.

Funding: This research was funded by Fontana, Velasquez, Pazos, Perez Consultores. The APC was funded by PAD Escuela de Direccion, Universidad de Piura.

Institutional Review Board Statement: The study was conducted in accordance with the Declaration of Helsinki, and approved by the Ethics Committee of PAD Escuela de Direccion, Universidad de Piura.

Informed Consent Statement: Informed consent was obtained from all subjects involved in the study.

Data Availability Statement: Main Official Data Resources: municipal authorities—Paucar del Sara Sara Municipality: https://www.gob.pe/munipaucardelsarasara and https://www.transparencia.gob.pe/enlaces/pte_transparencia_enlaces.aspx?id_entidad=11877; Lampa Municipality: https://www.transparencia.gob.pe/enlaces/pte_transparencia_enlaces.aspx?id_entidad=10769; Oyolo Municipality: https://www.munioyolo.gob.pe/; the Geological, Mining, and Metallurgical Institute (INGEMMET): https://www.gob.pe/ingemmet; the Ministry of Transport and Communications: https://www.gob.pe/mtc; the National Institute of Statistics and Information Technology (INEI): https://www.gob.pe/inei/; the National Water Authority (ANA): https://www.gob.pe/ana; the National Institute of Civil Defense (INDECI): https://www.gob.pe/indeci; and Friendly inquiry—MEF: https://apps5.mineco.gob.pe/transparencia/Navegador/default.aspx; all accessed on accessed on 4 February 2025. Specific data generated in this study is unavailable due to privacy.

Conflicts of Interest: The authors declare no conflicts of interest.

References

1. Fernández, J.; Fernández, M.I.; Soloaga, I. *Enfoque Territorial y Análisis Dinámico de la Ruralidad: Alcances y Límites para el Diseño de Políticas de Desarrollo Rural Innovadoras en América Latina y el Caribe*; CEPAL: Washington, DC, USA, 2019.
2. Schejtman, A.; Berdegué, J. *Desarrollo Territorial Rural*; RIMISP: Santiago, Chile, 2004.
3. Torre, A. Contribution to the theory of territorial development: A territorial innovations approach. *Reg. Stud.* **2023**, 1–16. [CrossRef]
4. Wanner, A.; Seier, G.; Pröbstl-Haider, U. Policies related to sustainable tourism—An assessment and comparison of European policies, frameworks and plans. *J. Outdoor Recreat. Tour.* **2020**, *29*, 100275. [CrossRef]
5. Bonilla-Mejía, L.; Londoño-Ortega, E. Geographic isolation and learning in rural schools. *Borradores Econ.* **2021**, *1169*, 1–32.
6. Angeon, V.; Saffache, P. Les petites economies insulaires et le développement durable: Des réalités local résilientes? *Stud. Caribb.* **2008**, *11*. [CrossRef]

7. De Arruda Mauro, R.; Cintrão, J.F.F.; Melhen, J.E.; de Oliveira, E.A. Dimensões do desenvolvimento territorial e políticas públicas: Perspectivas e desafios a partir da Constituição Federal de 1988. *Rev. Gest. Ambient. Sustentabilidade* **2018**, *7*, 489–506.
8. Favareto, A.; Lotta, G. Inovações institucionais nas políticas para o desenvolvimento territorial em três estados brasileiros. *REDES Rev. Desenvolv. Reg.* **2017**, *22*, 11–38. [CrossRef]
9. Bock, B.B. Rural marginalization and the role of social innovation; a turn towards nexogenous development and rural reconnection. *Sociol. Rural.* **2016**, *56*, 552–573. [CrossRef]
10. Jiménez Aliaga, R.; De los Ríos-Carmenado, I.; San Martín Howard, F.; Calle Espinoza, S.; Huamán Cristóbal, A. Integration of the Principles of Responsible Investment in Agriculture and Food Systems CFS-RAI from the Local Action Groups: Towards a Model of Sustainable Rural Development in Jauja, Peru. *Sustainability* **2022**, *14*, 9663. [CrossRef]
11. Sastre-Merino, S.; Negrillo, X.; Hernández-Castellano, D. Sustainability of rural development projects within the working with people model: Application to Aymara Women Communities in the Puno Region, Peru. *Cuad. Desarro. Rural* **2013**, *10*, 219–243.
12. Cazorla-Montero, A.; De los Ríos-Carmenado, I. From "Putting the Last First" to "Working with People" in Rural Development Planning: A Bibliometric Analysis of 50 Years of Research. *Sustainability* **2023**, *15*, 10117. [CrossRef]
13. Cazorla, A.; De Los Ríos, I.; Salvo, M. Working with People (WWP) in rural development projects: A proposal from social learning. *Cuad. Desarro. Rural* **2013**, *10*, 131–157.
14. Ríos-Carmenado, I.D.L.; Becerril-Hernandez, H.; Rivera, M. La agricultura ecológica y su influencia en la prosperidad rural: Visión desde una sociedad agraria (Murcia, España). *Agrociencia* **2016**, *50*, 375–389.
15. Espitia, L.F.; Pérez, C.P. El proceso de planeación para el desarrollo territorial "Caso municipio de Lejanías, Meta-Construcción de un modelo de desarrollo sostenible para la vigencia 2016–2019". *Lebret* **2017**, *9*, 121–147. [CrossRef]
16. Raynaut, C. A gênese da abordagem territorial e participativa do desenvolvimento rural: Raízes conceituais e experiências internacionais. *Territ. E Cid.* **2014**, *56*, 56–78.
17. Renault Adib, A. *Guía para la Formulación y Gestión de Planes de Desarrollo Rural Sostenible: Un Abordaje Participativo con Enfoque Territorial*; IICA: Asunción, Paraguay, 2010.
18. Torre, A. Nuevas propuestas para analizar el desarrollo territorial. *Eutopía. J. Terr. Econ. Dev.* **2020**, *17*, 11–24. [CrossRef]
19. Fuini, L.L. A territorialização do desenvolvimento: Construindo uma proposta metodológica. *Interações* **2014**, *15*, 21–34. [CrossRef]
20. Estrada-Meléndez, O.H.; Olivares-González, A. Implicaciones del desarrollo territorial en los territorios dependientes: Evaluación del caso de Puerto Rico. *Bitácora Urbano Territ.* **2017**, *27*, 29–42. [CrossRef]
21. Economic Commission for Latin America and the Caribbean (ECLAC). Towards an ecosystem of territorial development policies and instruments: Legal frameworks and development plans and policies at the national level. In *Planning for Sustainable Territorial Development in Latin America and the Caribbean*; United Nations Publication: Santiago, Chile, 2019.
22. Henriques, J.M. On becoming healthier communities: Poverty, territorial development and planning. *J. Public Health* **2013**, *31*, 58–73. [CrossRef]
23. Berdegué, J.A.; Bebbington, A.; Escobal, J. Conceptualizing spatial diversity in Latin American rural development: Structures, institutions, and coalitions. *World Dev.* **2015**, *73*, 1–10. [CrossRef]
24. Sili, M. Un modelo para comprender la dinámica de los territorios rurales: El caso de la Argentina. *Mundo Agrar.* **2016**, *17*, e003.
25. Rodríguez Miranda, A.; Vial Cossani, C.; Parrao, A. Índice compuesto y multidimensional de desarrollo regional: Una propuesta para América Latina. *Rev. Iberoam. Estud. Munic.* **2021**, *23*, 1–33. [CrossRef]
26. Suárez-Roldán, C.; Hernández-Atencia, Y.; Méndez-Giraldo, G. Method of Selection of Rural Territory in the Development of a Territorial Diagnosis. *J. Eng.* **2024**, *1*, 8795216. [CrossRef]
27. Romo, V.Y.; Parra, C.M.; Rüchardt, M.D. Vulnerabilidad socio-territorial: Posibilidad de determinarla a partir de una construcción metodológica. *Rev. Geog. Venez.* **2017**, *58*, 148–163.
28. Mozzato, D.; Gatto, P.; Defrancesco, E.; Bortolini, L.; Pirotti, F.; Pisani, E.; Sartori, L. The role of factors affecting the adoption of environmentally friendly farming practices: Can geographical context and time explain the differences emerging from literature? *Sustainability* **2018**, *10*, 3101. [CrossRef]
29. Labianca, M. Proposal of a Method for Identifying Socio-Economic Spatial Concentrations for the Development of Rural Areas: An Application to the Apulia Region (Southern Italy). *Sustainability* **2023**, *15*, 3180. [CrossRef]
30. Dunets, A.N.; Gerasymchuk, N.A.; Kurikov, V.M.; Noeva, E.E.; Kuznetsova, M.Y.; Shichiyakh, R.A. Tourism management in border destinations: Regional aspects of sustainable development of protected natural areas. *Entrep. Sustain. Issues* **2020**, *7*, 3253. [CrossRef] [PubMed]
31. Khomiiuk, N.; Bochko, O.; Pavlikha, N.; Demchuk, A.; Stashchuk, O.; Shmatkovska, T.; Naumenko, N. Economic modeling of sustainable rural development under the conditions of decentralization: A case study of Ukraine. *Manag. Econ. Eng. Agric. Rural Dev.* **2020**, *20*, 317–332.
32. Romagnoli, F.; Masiero, M.; Secco, L. Windstorm Impacts on Forest-Related Socio-Ecological Systems: An Analysis from a Socio-Economic and Institutional Perspective. *Forests* **2022**, *13*, 939. [CrossRef]

33. Ruperti Cañarte, J.S.; Mendoza García, J.G.; Lucas Intriago, M.A.; Franco Moreira, J.A. El desarrollo territorial y el pensamiento económico. *Rev. Soc. Tecnol.* **2021**, *4*, 399–415. [CrossRef]
34. Pinzón Botero, M.V.; Echeverri Álvarez, I.C. La sostenibilidad ambiental regional: Una propuesta metodológica para su estudio más allá de la ecología urbana. *Luna Azul* **2012**, *34*, 131–147.
35. Romao, J.; Neuts, B. Territorial capital, smart tourism specialization and sustainable regional development: Experiences from Europe. *Habitat Int.* **2017**, *68*, 64–74. [CrossRef]
36. Millán, S.Q.; Arias-Ortega, K.; Troncoso, G.M.; Torres, H.; Morales, S.; Peña-Cortés, F. Conocimientos geográficos y territoriales con base epistémica en la memoria social mapuche. *Rev. Bras. Ciências Sociais* **2021**, *36*, e3610603. [CrossRef]
37. Costa, T.; Lima, M.J. Cooperation in tourism and regional development. *Tour. Manag. Stud.* **2018**, *14*, 50–62. [CrossRef]
38. Food and Agriculture Organization of the United Nations (FAO). *Principios para la Inversión Responsable en la Agricultura y los Sistemas Alimentarios*; FAO: Rome, Italy; IFAD: Rome, Italy; WFP: Rome, Italy, 2014.
39. De los Ríos, I.; Rivera, M.; García, C. Redefining Rural Prosperity through Social Learning in the Cooperative Sector: 25 Years of Experience from Organic Agriculture. *Land Use Policy* **2016**, *54*, 85–94. [CrossRef]
40. Rivera, M.; Knickel, K.; De Los Rios, I.; Ashkenazy, A.; Pears, D.Q.; Chebach, T.; Šūmane, S. Rethinking the Connections between Agricultural Change and Rural Prosperity: A Discussion of Insights Derived from Case Studies in Seven Countries. *J. Rural Stud* **2018**, *59*, 242–251. [CrossRef]
41. Cazorla-Montero, A.; Ríos-Carmenado, I.D.L.; Díaz-Puente, J.M. The LEADER community initiative as rural development model: Application in the capital region of Spain. *Agrociencia* **2005**, *39*, 697–708.
42. Ríos-Carmenado, I.D.L.; Díaz-Puente, J.M.; Cadena-Iñiguez, J. The initiative LEADER as a model for rural development: Implementation to some territories of México. *Agrociencia* **2011**, *45*, 609–624.
43. Ceron, C.A.A.; De los Rios-Carmenado, I.; Fernández, S.M. Illicit crops substitution and rural prosperity in armed conflict areas: A conceptual proposal based on the Working with People model in Colombia. *Land Use Policy* **2018**, *72*, 201–214. [CrossRef]
44. Ortuño, M.; De los Ríos, I.; Sastre-Merino, S. The Development of Skills as a Key Factor of the Cooperative System: Analysis of the Cooperative of Artisan Women Tejemujeres-Gualaceo-Ecuador from the WWP Model. *Sustainability* **2022**, *14*, 16233. [CrossRef]
45. Friedmann, J. *Planning in the Public Domain: From Knowledge to Action*; Princeton University Press: Princeton, NJ, USA, 1987.
46. Ávila Cerón, C.A.; Castañeda, R.; De Los Ríos, I. Desarrollo de un Modelo de Agricultura Familiar Alineado con los Principios de Inversión Responsable: Integración Cadena Turística en República Dominicana. In Proceedings of the 28th International Congress on Project Management and Engineering, Jaén, Spain, 3–4 July 2024.
47. Guillet, D. Toward a cultural ecology of mountains: The Central Andes and the Himalayas compared. *Curr. Anthropol.* **1983**, *24*, 561–574. [CrossRef]
48. Troll, C. Geo-ecology of the mountainous regions of the Tropical Americas. In Proceedings of the Unesco Mexico Symposium, Mexico City, Mexico, 1–3 August 1966.
49. Murra, J. Los Límites y las Limitaciones del Archipiélago Vertical Andino. 1974. Available online: https://repositorio.unal.edu.co/bitstream/handle/unal/26857/13631-47851-1-PB.pdf?sequence=1 (accessed on 20 January 2025).
50. Recasens, A.V. La cara oculta de los Andes: Nota para una redefinición de la relación histórica entre la sierra y la selva. *Boletín Am.* **1992**, *42*, 7–22.
51. Caballero, J.M. *Economía Agraria de la Sierra Peruana: Antes de la Reforma Agraria de 1969*, 1st ed.; IEP Ediciones: Lima, Perú, 1981.
52. Boserup, E. *Population and Technological Change: A Study of Long-Term Trends*; University of Chicago Press: Chicago, IL, USA, 1981.
53. Thomas, R.B. Human Adaptation to a High Andean Energy Flow System. Ph.D. Thesis, Pennsylvania State University, University Park, PA, USA, 1973.
54. Yin, R.K. *Case Study Research: Design and Methods*, 4th ed.; SAGE: Thousand Oaks, CA, USA, 2009; Volume 5.
55. Census. 2017. Available online: https://www.inei.gob.pe/media/MenuRecursivo/publicaciones_digitales/Est/Lib1568/ (accessed on 20 January 2025).
56. IDH. 2017. Available online: https://www.undp.org/sites/g/files/zskgke326/files/migration/pe/8624fa5f0070c47aea699a2be6adfab7552b9a45c2446a267fec91a116c3150a.xlsx (accessed on 21 March 2024).
57. Vivanco, C. Paucar del Sara Sara en Ayacucho. Historia Milenaria de Llaqtakuna: Evolución, cultura y tradición. *Alteritas. Rev. Estud. Sociocul. Andin.-Amaz.* **2017**, *7*, 271–304.

sustainability

MDPI

Article

University–Business Link for Sustainable Territorial Development Through the Principles for Responsible Investment in Agriculture and Food Systems (CSA-IRA): Working with People in the Dominican Republic

Milagros del Pilar Panta Monteza [1,*], Ubaldo Eberth Dedios Espinoza [2], Gustavo Armando Gandini [3] and Jorge Luis Carbajal Arroyo [1]

1 Faculty of Economics and Business Administration, Professional School of Administration and Management, Ricardo Palma University (URP), Av. Alfredo Benavides 5440, Santiago de Surco 15039, Peru; jorge.carbajal@urp.edu.pe

2 SMI, San Miguel Industry, Autopista Duarte Km. 18 Santo. Domingo Oeste, Santo Domingo 11512, Dominican Republic; ubaldo.dedios@ism.global

3 BANELINO, Organic Bananas of the Northwest Line. Av. Miguel Crespo, Santa Cruz de Mao 61000, Dominican Republic; ggandini@banelino.com.do

* Correspondence: milagros.panta@urp.edu.pe; Tel.: +51-986740874

Abstract

There is little evidence of the implementation of the Principles for Responsible Investment in Agriculture and Food Systems between universities and businesses, and there is even less research that prioritizes people and implements sustainable development with a territorial focus. In this article, we address a form of collaborative work that integrates academia with business, where the Principles for Responsible Investment in Agriculture and Food Systems (CFS-RIA) are seen as an opportunity to promote and strengthen the management of a business in the communities where it operates, and determine a new way of working from its links with the university. The experience is developed in the provinces of Santiago Rodríguez, Valverde (Mao), and Dajabón in the Dominican Republic, with the aim of contributing, using this new approach, to economic, social, environmental, and governance development in the territory. The conceptual and methodological basis for the university–business link is Working With People, a model that integrates key elements of planning such as social learning, collaborative participation, and project management models. The main catalysts of the experience are the business values and the stakeholders who insert the principles into their programs and projects. Among these is an innovative Family Social Responsibility Program with female entrepreneurs and organic banana production. It is concluded that the implementation of the CFS-RIA Principles has a significant impact on the sustainable development of the region and that the university–business link reinforces the social responsibility of companies, providing an opportunity for the entry of new actors.

Keywords: university–business; territorial sustainable development; principles for responsible investment in agriculture and food systems; collaborative participation; working with people; female entrepreneurship

1. Introduction

Bringing universities closer to the business world and the needs of society remains a challenge in Latin America and the Caribbean. There is an urgent need to develop responsible proposals and sustainable mechanisms that strengthen this connection. In this context, the university is seen as an ideal system for problem-solving through research [1], relevant teaching, and effective engagement with society [2]. Fostering this connection requires creative, high-level faculty who lead research projects with social and technical value, capable of providing innovative solutions to businesses both locally and internationally. This connection between teaching and research not only transmits knowledge but also develops the ability to understand and therefore explain it [3]. Universities contribute to progressing society and are not isolated, imparting more or less relevant knowledge and preparing future professionals. For its part, society, with its companies, associations, federal or state government, turns to universities when faced with problems, opportunities, or needs. Moreover, all of this is achieved directly and with few bureaucratic complications typical of European or Latin American environments [4]. This reinforces the need for integrative initiatives between universities and companies that promote sustainable development of territories [5].

In this context, our experience of a university–business link for sustainable regional development is presented in this paper, based on the Principles for Responsible Investment in Agriculture and Food Systems (CFS-RIA). This initiative integrates academia and the business sector to promote entrepreneurship focused on training people, developing skills, and internalizing values, thus improving living conditions in the intervention territories [6].

The voluntary CFS-RIA Principles were approved in October 2014 by the United Nations Committee on World Food Security. These principles address different needs and realities of actors such as states, small producers, commercial companies, and civil society organizations. In 2016, FAO (the United Nations Agency for the Food and Agriculture Organization that leads the international effort to end hunger) signed an agreement with GESPLAN, a research group of the Polytechnic University of Madrid (UPM) that works in the field of project planning, management, and evaluation to promote teaching, research, and social engagement in Latin America, the Caribbean, and Spain, taking these principles as a reference. This agreement facilitated the research team's access to useful knowledge and experiences of promoting cooperative actions in agriculture, food, the environment, and climate change.

This experience is methodologically based on the Working with People (WWP) approach, a participatory and multidimensional model of territorial development that articulates human values, expert knowledge, and local knowledge. This approach encourages collaboration between communities, technical teams, and policymakers [7], organized around three interrelated dimensions: ethical–social, technical–business, and political–contextual. The WWP metamodel has been successfully applied in different rural contexts [8–18], contributing to improving governance and promoting private entrepreneurship in depopulated areas of Spain [19], as well as public–private partnership initiatives [20].

This experience employs a territorial approach that conceives sustainable development [21] as focusing on training people, especially women and young people, and is aligned with the vision and mission of two universities (one local and one international) and two socially responsible companies, key players in the Network of Latin American Universities, established in 2018 and committed to the dissemination of the CFS-RIA Principles. This network has promoted innovative projects with high economic, social, environmental, and governance impacts [22].

The project was implemented in the provinces of Santiago Rodríguez, Valverde, and Dajabon, in the border area of the Dominican Republic. This region is strategic due to its proximity to Haiti and its inclusion in the Special Zone for Comprehensive Border Development, which promotes investment through economic incentives [23].

https://doi.org/10.3390/su18031179

Since the implementation of the CFS-RIA Principles, new opportunities have emerged to coordinate territorial development and define sustainable perspectives for the future of these regions [24]. Both universities and businesses have proven to be key players in raising awareness among public and private stakeholders and in the implementation of sustainable development projects, with the former being appropriate systems for addressing the challenges of sustainable development [25], and the latter being catalysts for strengthening corporate social responsibility.

2. Case Study

2.1. *Country Covered by the Proposal*

The Dominican Republic, located in the insular Caribbean, is part of the Central American region Figure 1. It is surrounded by the Atlantic Ocean to the north and the Caribbean Sea to the south, bordering the Mona Channel to the east—which separates it from Puerto Rico—and the Republic of Haiti to the west. The proposed project is located within a border territory of approximately 276 km, specifically in an area bordering Haiti. This region possesses a strategic uniqueness as it is included in the framework of Law Nos. 12-21 [23], which establish a special regime for border development. This regulation creates a Special Zone for Comprehensive Border Development, which offers tax and tariff incentives aimed at encouraging private investment as an engine of economic and social development. The purpose is to stimulate the local economy and generate decent employment in historically disadvantaged territories. This special zone covers the provinces of Pedernales, Independencia, Elías Piña, Dajabón, Montecristi, Santiago Rodríquez, and Bahoruco, all with high potential to become centers of sustainable territorial development.

Figure 1. Map of the Dominican Republic, providing context for the presented experience.

2.2. *Geographic Coverage*

These provinces are located on the border with Haiti and cover a very interesting area of agricultural production and tourism, featuring a university presence that makes it especially interesting. The geographic area covers 3 of the 32 provinces of the Dominican Republic—Santiago Rodríguez, Valverde, and Dajabón (see Table 1). Santiago Rodríguez

https://doi.org/10.3390/su18031179

(with the municipalities of San Ignacio de Sabaneta, Monción, and Villa Los Almácigos)—and is located in the northwest of the country, in the Cibao subregion known as the Northwest Line. Its capital is the city of San Ignacio de Sabaneta. It borders the province of Valverde to the north (with three municipalities—Mao, Esperanza, and Laguna Saladas—and an area of 823 km^2 and 183,738 inhabitants) and the province of Dajabón to the west (with five municipalities: Dajabón, El Pino, Loma de Cabrera, Partido, and Restauración).

Table 1. Provinces covered by the proposal.

Province	Area (km^2)	Population (Inhab)	Density (Inhab/km^2)	Number of Municipalities
Santiago Rodríguez	1147.5	58,044.0	50.6	3
Valverde	823.4	183,738.0	263.6	3
Dajabón	1020.7	87,274.0	85.5	5
Total	2.991.6	362,344.0	121.1	11

2.2.1. Mechanisms for Participant Identification

Participants with potential for responsible entrepreneurship and territorial innovation can only be identified through institutional selection mechanisms. They can be immediate family members of workers (mothers, sisters, brothers, husbands, or wives) who are unemployed, have lost their job, had an initial business idea, or had an operational business but need to develop an entrepreneurial mindset to grow. Participation is voluntary.

2.2.2. Participation in the Program

Participation in the program consisted of the following phases: (i) a call within the organization, inviting participation in the program; (ii) an informative meeting to explain what the program consists of, who it is aimed at, its duration, and the team that will develop it; (iii) the completion of a survey gathering information about their interest in the initiative (or lack thereof) and asking what type of activity they would feel motivated to carry out; and (iv) the signing of a commitment letter if they decide to participate.

2.2.3. Selection Metrics

Selection metrics included ratio of accepted applicants to total applicants; gender composition; social approach to entrepreneurship; dropout rate during the training phase.

2.2.4. Implications for Internal and External Validity

Two aspects are considered to assess the internal validity of entrepreneurship projects in relation to sustainable development: completion rates and validated projects. Both of them strengthen internal validity. Other aspects that have been considered involve the person and the project, as an integral approach that allows the evaluation of both personal capabilities and the technical and strategic viability of the enterprise. With regard to external validity, the results of this study may be relevant to other groups of people or contexts if there are (i) key institutions that ensure continuity, (ii) a clear methodological framework, (iii) a supportive policy environment, and (iv) access to differentiated markets.

2.3. *Key Actors Involved*

For the university–business link to effectively contribute to sustainable territorial development, local actors must meet certain key characteristics. First, they must have a strong presence in the territory, which implies a deep understanding of the local context and an active presence in community processes. They must also play a central role within the framework of local development, among others [26], acting as catalysts for initiatives

https://doi.org/10.3390/su18031179

that integrate economic, social, environmental, and governance interests to contribute to sustainability [27] and develop their institutional capacity [28].

2.3.1. San Miguel Industry

(SMI) is a Peruvian company with over 36 years of experience in the market. In 2004, it began the construction and installation of a production plant in the province of Santiago Rodríguez, Dominican Republic, which has established itself as one of the most modern in the Caribbean region. With an initial investment of approximately USD 11 million, SMI has experienced sustained growth, characterized by high market acceptance and a strategy of continuous expansion through new product lines, infrastructure improvements, and social investments through its corporate social responsibility (CSR) program.

SMI is distinguished by its strong social sensitivity and by considering people as the core of its management model. Within this framework, it actively promotes training and the development of competencies and skills among its stakeholders. In its official website 69, SMI describes its Mision, Vision and Values (with Leadership among them) as follows: "People are the most valuable to us. We value our employees, clients, and community equally".

One of the most notable initiatives was the development of the course "IPMA International Competencies in Project Management: Foundations for Entrepreneurship", based on international intellectual property standards from the GESPLAN research group [29], and adapted to the context of the SMI CSR program. The main objective of the course was to strengthen people's capacities to formulate and direct entrepreneurial projects, with a focus on personal competencies (ethical, emotional, and social), as well as to contribute to improving the quality of life of families linked to the company.

The course was aimed at the immediate family members of SMI employees and incorporated elements that balance and make decision-making more assertive in the management of entrepreneurial projects [30]. In addition, it offered training aligned with the international standards of the International Project Management Association (IPMA), thus preparing participants to formulate projects and access the international certification system in project management promoted by IPMA [31].

2.3.2. Organic Bananas from the Northwest Line

(BANELINO) is an organization located in the provinces of Montecristi and Valverde, in the border area with Haiti. It was established in 1996 to promote the economic, social, and environmental progress of small organic banana producers and their communities. It currently brings together 200 producers, each with an average of 3 hectares, who develop their activities with a focus on competitive, sustainable, healthy, and high-quality production. BANELINO is dedicated not only to the economic strengthening of its producers but also to the development of marginalized communities in its area of influence, through participatory processes that promote collective action and integrate organized civil society [32,33]. Its intervention model is structured around three fundamental axes: (1) producer competitiveness (infrastructure, technical assistance, and training); (2) social development (education, health, housing, sports, and culture); and (3) environment.

BANELINO faces key challenges affecting its members, such as food insecurity, limited empowerment of women and youth, market instability, intense competition from new entrants in the organic banana market, and fluctuating prices. In the face of these challenges, BANELINO has managed to position itself as a leader in the international organic banana market thanks to a strategy based on internal cohesion, technical support, and comprehensive support for its producers.

https://doi.org/10.3390/su18031179

In total, 95% of its production is organic and 80% is marketed under Fairtrade conditions, through its certification by Fairtrade Labelling Organizations, FLO International [34–37]. In addition, it has certifications such as organic and Global GAP, and it is in the process of obtaining DEMETER certification, which endorses biodynamic agricultural practices that promote biodiversity and soil health.

Among the services offered by BANELINO are technical assistance, certification management, and collective marketing, to improve the quality of life of small producers, their families, and their communities. It is worth highlighting the growing role of women producers, who now constitute 31% of the total membership [38].

The environmental commitment is also evident: producers apply sustainable practices in soil management, water use, and the conservation of natural resources. Guaranteed sales at fair prices have not only improved the living conditions of producer families but also reinvested in community development programs in areas such as education, health, housing, food security, culture, and the environment.

BANELINO represents a successful model for how rural organizations can effectively connect with global markets, ensuring that agricultural investment is carried out under principles of social equity and environmental sustainability. Like SMI, its experience contributes significantly to sustainable territorial development through capacity building, social innovation, and improved livelihoods, fully aligned with the CFS-RIA Principles and the Working with People methodological approach.

The activities of BANELINO and SMI promote sustainable territorial development by encouraging education, capacity building, participant competencies, and innovation to improve production and entrepreneurship conditions. These actions align with the needs of rural communities and transform them while generating positive and lasting impacts such as those mentioned throughout this paper.

2.3.3. The Technological University of Santiago

UTESA has two campuses in the provinces of Mao and Dajabón, located in the northwest region of the Dominican Republic. The Faculty of Agricultural Sciences at the UTESA-Mao campus has a strategy aimed at strengthening the agricultural sector, promoting sustainable rural development, and improving the quality of life of small and medium sized producers.

The institution bases its mission on the development of strong leaders through teaching, applied research, and engagement with the community, within the framework of a corporate model with a global reach that promotes entrepreneurship, innovation, and sustainability. From a humanistic perspective, UTESA is committed to providing solutions to the social and productive challenges of the region through its strategic use of science and technology.

Its institutional vision is to consolidate itself as a benchmark in the Dominican university ecosystem for its academic excellence, capacity for continuous innovation, and commitment to sustainable development, social responsibility, and territorial transformation.

Within this framework, UTESA has established strategic alliances with companies such as San Miguel Industry (SMI) and organizations such as BANELINO, through which it has developed training programs that promote skills for productive entrepreneurship and sustainable project management.

2.3.4. The GESPLAN Research Group at the Polytechnic University of Madrid

UPM has established itself as an international benchmark in the field of project planning, management, and evaluation. Its organizational model is based on project-based work, integrating strategic elements that represent the essence of a research university:

https://doi.org/10.3390/su18031179

high-level teaching, especially in postgraduate programs, strong ties to the community, and innovative research aimed at solving real-world problems.

GESPLAN stands out for its applied approach and its ability to transfer technical and methodological knowledge to academic institutions as well as to businesses and social organizations. In the Dominican Republic, the group provides collaborative technical advice to universities, such as UTESA, and to local companies, such as San Miguel Industry and BANELINO, strengthening the university–business–community bond through participatory strategic planning.

One of GESPLAN's most significant contributions is the development of training programs that promote skills for productive entrepreneurship and sustainable project management. A key example is the course "IPMA International Competencies in Project Management: Foundations for Entrepreneurship," adapted to rural contexts based on international standards. This course is aimed at strengthening the capacities of local stakeholders, especially women and youth, to formulate and lead sustainable projects that improve their living conditions and promote local development.

Another contribution has been the promotion and integration of the CSA-IRA Principles, aligning them with the Sustainable Development Goals (SDGs). This support has enabled the implementation of sustainable territorial development projects with an inclusive and participatory approach, under the Working with People methodological model, which promotes collaborative processes among key stakeholders in the territory. Thanks to its transdisciplinary vision and international experience, GESPLAN enhances local capacities, contributing to the strengthening of rural innovation ecosystems and the empowerment of communities from an ethical, environmental, and socially responsible perspective.

3. Materials and Methods

3.1. Methodological Approach

The conceptual basis of this research is the Working with People (WWP) metamodel, understood as a professional practice of cooperation that seeks to connect knowledge and action through joint projects, based on active participation, social learning, and the integration of ethical values by the actors involved in the development processes [7].

The WWP conceptual framework is structured around three interrelated components:

- Ethical–social;
- Technical–entrepreneurial;
- Political–contextual.

These components interact and mutually reinforce each other through social learning processes, as has been demonstrated in experiences of the European Union's LEADER Community Initiative [39,40] and in projects carried out in Ecuador [41], Peru [14,42], and Colombia [43]. These components cover the four areas of the system of social relations, namely, the political field, public administration, the private and business sector, and civil society, which allows for a systemic understanding of society [44]. Figure 2 illustrates these components and their interactions, as well as the overlaps generated through social learning processes [45].

The interaction of these elements promotes social learning as a driver for the sustainable development of ventures focused on the training of people, the development of their skills, and the internalization of values within a given context. This process seeks to transform knowledge into concrete and effective actions, as exemplified in the entrepreneurial projects addressed in this research.

The university–business link operationalizes the CFS-RAI Principles through activities and observable indicators through the three dimensions of the Working with People metamodel (ethical–social, technical–entrepreneurial, and political–contextual). The methodol-

https://doi.org/10.3390/su18031179

ogy follows a logical model and indicators that describe the causal logic and measurement procedures. The political–contextual dimension facilitates agreements and resources; the ethical–social dimension builds social capital and capacities; the technical–entrepreneurial dimension transforms knowledge into projects and productive practices. For each relevant CFS-RAI principle (P1–P10), the proposed indicator(s) and the results are shown in Table 2.

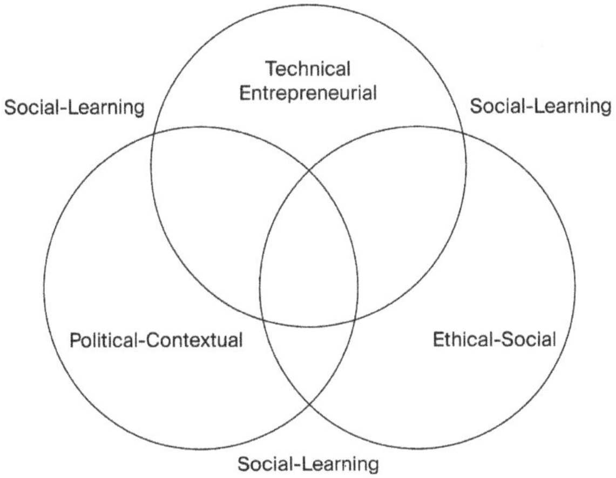

Figure 2. Logical model of the WWP metamodel (dimensions: ethical–social, technical–entrepreneurial, and political–contextual). Source: Created by the authors based on the conceptual representation by Cazorla et al. [7].

Table 2. Corresponding WWP dimensions, CFS-RIA Principles, indicators and results.

WWP Dimensions	CFS-RAI Principles		Observable Indicators (Quantitative)	Results
Ethical–Social	P1	Contribute to food security and nutrition	% of organic production % of production sold in Fairtrade	100% 80%
	P3	Promote gender equality and the empowerment of women	Number of women trained in entrepreneurship programs % of women organic banana producers	41 31%
	P4	Strengthen youth participation and empowerment	Number of university students trained in CFS-RAI Principles	180
Technical–Business	P2	Contribute to sustainable and inclusive economic development and poverty eradication	Number of women running business projects with very high and high feasibility Average income increase for banana producers % of products with Fairtrade certification	28 18% 100%
Political–Contextual	P9	Incorporate inclusive and transparent governance structures, processes and grievance mechanisms	Numbers and actors participating in governance structures Inclusion of public sector representatives	3 30%

Source: final project report, second edition, 2024. Unpublished institutional document [46].

The WWP metamodel emphasizes the need to take planning from theory to practice [7], promoting sustainable transformations in the territories by solving real and specific problems of each region.

The participating companies represent two complementary integration models that have contributed to territorial development, understood as a process that involves productive transformations and institutional changes aimed at generating new capabilities and opportunities for local actors [47]. According to Torre [48], this approach promotes social well-being and prosperity by balancing competition, cooperation, initiatives, and tensions between different actors.

3.2. Data Collection Instruments, Analysis, and Quality Control

Data collection was carried out through participatory workshops, meetings, competency assessments, technical reports, final reports, business plan evaluations, courses, and seminars, which allowed for the capture of multiple perspectives and practical experiences.

https://doi.org/10.3390/su18031179

The analysis was developed under the instrumental and multiple case study approach, considering the articulation of concrete experiences led by SMI and BANELINO, with the academic accompaniment of UTESA, and the technical–scientific support of the GESPLAN research group of the Polytechnic University of Madrid.

To do this, a triangulation of sources and techniques was applied, which included

- Document analysis of business plans, technical reports, institutional reports, training schedules, impact assessment documents, and entrepreneurship projects developed by the participants;
- Semi-structured interviews with participating entrepreneurs and local community members, conducted by staff SMI and BANELINO;
- Participant observation of training sessions, technical support processes (in-person and online), and on-site meetings to record social learning dynamics, inter-institutional relationships, and skill acquisition;
- Evaluative follow-up sessions that validated preliminary findings, gathered feedback from those involved, and strengthened the analysis from a dialogical and inclusive perspective.

The analysis focused on identifying social learning processes, the interactions between the three components of the WWP model (ethical–social, technical–entrepreneurial, and political–contextual), and the facilitating elements and obstacles in building the university–business–community link. Based on this systematization, common patterns, trajectories of change, differentiated impacts on the territories, and contributions to sustainability were recognized.

This approach allowed us to understand not only the tangible results of the projects, but also the processes of institutional transformation, capacity building, and coordination of actors that emerge in a complex context such as that of the Dominican Republic, especially in the provinces of Santiago Rodríguez, Valverde, and Dajabón.

This case study does not incorporate a control group before/after the experiment. The design applied was of an instrumental nature, featuring a case study with a predominance of qualitative techniques (documentary triangulation, semi-structured interviews, and participant observation). To ensure reproducibility and methodological transparency, all qualitative and mixed-methods instruments were standardized across both program editions (2023–2024). The data collection protocol included four components: semi-structured interview guides, participant observation sheets, project evaluation rubrics, and post-training questionnaires.

(1) The interview guide focused on (i) the perceived effectiveness of the training process; (ii) institutional coordination mechanisms; (iii) barriers to implementation; and (iv) perceived impact on projects. Each interview lasted 30–45 min, conducted either in person or via Zoom, depending on participant availability. All interviews were audio-recorded with the participant's consent and transcribed verbatim in Spanish. Transcriptions were reviewed independently by the project coordinator and processed posteriorly.

(2) Observation sheets: Participant observation was carried out during the synchronous sessions and mentoring activities. Standardized observation sheets were used to record (i) engagement with training, (ii) team interaction, (iii) the application of the WWP model, (iv) the application of the CFS RIA Principles, and (v) issues during project mentoring. Each session was logged using a standard Excel template developed by the GESPLAN-UPM coordination team.

(3) Scoring rubrics: The entrepreneurial projects were evaluated considering the following criteria in relation to sustainable development: (i) business innovation, (ii) territorial impact, (iii) social improvement, and (iv) profitability. Each criterion was scored a percentage between 20 and 30%.

https://doi.org/10.3390/su18031179

(4) Post-training questionnaires: After the training phase, participants completed an online self-assessment form that collected data on satisfaction, perceived learning, and future intentions to continue their entrepreneurship. The average completion time was 15 min, and the response rate was 84%.

Quality control procedures: All instruments were pre-tested during the pilot sessions of the first edition and revised for clarity and internal consistency. The complete instruments can be found in Supplementary Material File S1.

4. Results

To increase internal validity, Table 3 presents a comparison between the first (2023–2024) and second (2024–2025) cohorts as a non-concurrent reference group, allowing a quasi-experimental descriptive analysis of inter-cohort differences.

Table 3. Historical reference cohort.

Indicator	First Edition	Second Edition	Total
Selected entrepreneurs	28	51	79
Number of entrepreneurs who completed the training and were certified	14	27	41
Percentage of entrepreneurs who were certified	50%	53%	
Gender of entrepreneurs	14	27	41
Women	14	26	40
Men	0	1	1
Average annual income (USD)	4500–6000		
Average investment in training per participant (USD)	400	400	
Projects with high and very high feasibility generated	9	19	28
Projects with low viability	5	8	13
Projects launched with seed capital	4	9	13
Amount requested by the female entrepreneurs as seed capital (USD)	49,122.81	55,337,223.00	5,386,345.81
Amount delivered by IMS (USD)	28,508.77	2,089,671.00	2,118,179.77
% financing ISM	58%	39%	
% financing by women	52%	61%	
Number of tutors responsible for monitoring the entrepreneurs	2	3	5

Source: Authors' own elaboration based on the final project report, second edition, 2024. Unpublished institutional document [46].

A university–business link can contribute to and facilitate effective planning and project design processes for sustainable territorial development, which, according to this experience, has the following results.

4.1. Training Programs Aligned with the CFS-RIA Principles for Building People's Capacities

BANELINO and SMI have developed training programs aimed at strengthening the capacities of producers and their families, explicitly incorporating the CFS-RIA Principles, specifically 3 and 4, into their institutional work.

BANELINO maintains a Quality Management System (QMS) tailored to small-scale producers and maintains an ongoing training program for field technicians, producers, and their workers. This program focuses primarily on topics such as food hygiene, health, safety, and well-being at work, pesticide use and management, first aid, good agricultural practices, and the pillars of the certification standards applied. Additionally, it inspects 100% of producers annually to ensure compliance with certifications. This not only allows BANELINO to strengthen local human capital but also illustrates how the relational co-operation pillar of the WWP approach is realized in an organizational structure based on trust, transparency, and long-term commitment.

SMI, with its Family Social Responsibility Program, has incorporated a course called "IPMA International Competencies in Project Management: Foundations for Entrepreneurship." Its specific objective is to provide training and support for formalizing a self-sustaining business in the community. The course was the basis for formulating an en-

https://doi.org/10.3390/su18031179

trepreneurial project based on IPMA competencies. Working under the WWP approach meant understanding the entrepreneurial project as the primary means of ensuring a connection with society [7,31], which is necessary to solve real-life problems. SMI demonstrates that when social responsibility programs are aligned with participatory methodologies such as the WWP and the CFS-RIA Principles, an effective, replicable intervention model with high territorial impact is produced. The commitment to project management training based on IPMA standards allows for a strategic vision that transcends simple financial support, generating autonomy and leadership capabilities.

The course was structured using Project-Based Learning (PBL) [49], which promotes the development of practical skills. Participants worked on real-life personal projects, analyzing situations, tasks, and outcomes, which facilitated a meaningful and transformative learning experience. This learning strategy strengthens the connection with society by generating an anticipated professional experience, articulating cooperative learning, and promoting a service-based approach [50,51].

The course was taught virtually during 2023 and 2024, with synchronous and asynchronous sessions through the Moodle platform of the Polytechnic University of Madrid (UPM), and included periodic face-to-face sessions. In the 2023 edition, 28 people participated, of whom 14 (50%) completed the course. In 2024, 51 people enrolled and 27 completed it, achieving a completion rate of 53%, as shown in Table 3.

The first edition had 28 participants, and 14 projects led by women were considered suitable, particularly in the service sector (bakery, pastry, and restaurants). In the second phase, 51 participants were trained. Of these, 27 successfully completed their initiatives in more diverse sectors than the first edition, such as decoration, event services, food, and education.

The success rates of the second edition were slightly higher than those of the first edition (Table 3), with 53% of participants certified as suitable and 47% not suitable. The expression "certified as suitable" refers to participants who successfully demonstrated integrated competencies in the People, Practice, and Perspective dimensions established by the International Project Management Association (IPMA) framework. Certification was awarded not merely for course completion but for fulfilling the competence criteria assessed through project performance, self-evaluation, and the final project presentation. This improvement can be attributed to the potential of entrepreneurship as a tool for job creation, increasing family income, reducing inequality, and improving living conditions, especially for women, by fostering their economic independence and achievement orientation [52,53].

An analysis of Table 3 reveals that 41 individuals (40 women and 1 man) have developed entrepreneurial skills [31] and are ready to launch their own businesses, although many may not do so in the short term. It is noteworthy that 16% of the entrepreneurs received direct sponsorship from SMI in the form of seed capital to implement their business plans and apply the knowledge acquired during the course. This shows that SMI has understood that training, although essential, is not sufficient to ensure the implementation of productive ventures in their early stages [26]. Training must be complemented with other key resources, such as initial financing, strategic alliances with universities, the activation of networks such as the Network of Latin American Universities Committed to the CFS-RIA Principles, and specific programs to support family entrepreneurship.

The growth in the number of participants certified as suitable—from 14 in the first edition to 27 in the second (Table 3)—confirms the effectiveness of the Working With People methodological model (Figures 2 and 3), as well as its capacity to generate sustainable impact in the territory, regardless of the economic activities developed. It is worth noting that in the second edition, greater diversity was observed in the business proposals presented (Table 4), which can be attributed to the increase in the number of participants (Table 3). Of the plans evaluated, 13 were classified as having "very high", 7 as "high", and the rest as "medium" viability, reflecting not only the scope of the program but also the progressive strengthening of entrepreneurial skills in the territory.

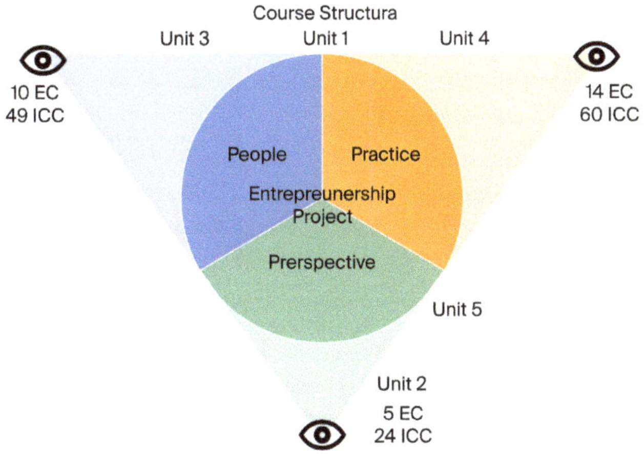

Figure 3. Themes connected to the three dimensions of IPMA competencies.

Table 4. Quantitative indicators for female projects.

Indicator	Notes/Assumptions	Ratio
Predominant economic sector	Commerce	41.9
	Agriculture, livestock, forestry, and fishing	13.9
	Other service activities	11.2
	Accommodation and food services	11.1
	Manufacturing industries	6.9
Number of employees	The entrepreneur is the sole employee	67.2
Business location	Shared with the dwelling	46.2
	Located in a commercial plaza store	15
	Street vendor	4.9
	Located in a marketplace	2
Business operation time	Between 1 and 3 years of operation	30.5
	Less than 1 year of operation	12.8
Formalization	Do not have a National Taxpayer Registry (RUC)	90.8
Owner's age	Entrepreneurs aged 26 to 35	23.7
	Entrepreneurs aged 36 to 45	18.9
	Entrepreneurs under 25 years old	12.3
Number of participants (n)	Participants in both editions	79
Completion rate	Assumed completion rate	53.0%
Number of graduates	Participants who completed the training	41
Number of validated projects	Projects validated across editions	41
Number of advanced/certified projects	Projects advanced to certification/incubation	27
Estimated annual income gain per participant (USD)	Baseline USD 5000; +18% increase	900.00
Cost per participant (USD)	Unit cost assumed	400.00
Cost per graduate (USD)	Total cost/number of graduates	75238
Cost per validated project (USD)	Total cost/validated projects	770.73
Cost per advanced/certified project (USD)	Total cost/advanced projects	1170.37
Individual cost/benefit ratio (cost per participant/inc)	400/900	0.444
Benefit–cost ratio (B/C) per participant	Inverse of cost/benefit	225
Total estimated annual benefit (USD) for graduates	n_graduates × income gain per participant	36,900.00
Aggregated B/C (total benefits/total cost)	Aggregated ratio across graduates	1.196
Cost per job (0.5 job/project) (USD)	Total cost/total jobs (20.5)	1541.46
Cost per job (1.0 job/project) (USD)	Total cost/total jobs (41)	770.73
Cost per job (1.5 jobs/project) (USD)	Total cost/total jobs (61.5)	513.82
Cost per job (3.0 jobs/project) (USD)	Total cost/total jobs (123)	256.91

Source: Authors' own elaboration based on the final project report, second edition, 2024. Unpublished institutional document [46] and the national statistics.

https://doi.org/10.3390/su18031179

These results not only reflect the specific success of a training program, but also demonstrate the effectiveness of the Working with People (WWP) methodology as a tool for generating sustainable and lasting impacts in the region. By integrating technical knowledge with community participation and inter-institutional links, this methodology has made it possible to identify real needs and design contextualized responses, generating significant learning and strengthening both individual and collective capacities.

Of the 79 entrepreneurs who participated in the call, 78 were women and only 1 man registered (Table 3), which may indicate that women choose entrepreneurship because they have greater schedule flexibility that allows them to combine work, housework, and raising children, as well as gain greater independence [54].

Women's preference for entrepreneurship in the trade and services sector (Table 4) may be influenced by the limitations imposed by other sectors and the lower initial capital and infrastructure requirements for this branch of activity. Furthermore, as organization in these sectors is more flexible (activities are often carried out within the home and require little financial investment), it allows them to balance paid work with family responsibilities and tasks [55,56]. Table 4 showns that, concerning the number of employees, 67.2% of the entrepreneurships were operated solely by the entrepreneur, indicating a predominance of individual microenterprises. In relation to the business location, 46.2% of the ventures were run from the entrepreneur's home, 15% operated from commercial plaza stores, 4.9% were street vendor businesses, and 2% were located in traditional market places. With respect to entrepreneurship operation time, 30.5% of the enterprises had been operating between one and three years, while 12.8% had less than one year of activity, reflecting an early stage of entrepreneurship consolidation. Regarding formalization, 90.8% of the entrepreneurs did not have a National Taxpayer Registry (RNC), suggesting a prevailing level of informality. In terms of the age of business owners, the majority were between 26 and 35 years old (23.7%), followed by those aged 36 to 45 (18.9%) and those under 25 years old (12.3%). Finally, the incorporation of technologies in the entrepreneurship represents a turning point for organizations, as it significantly contributes to the modernization and streamlining of processes, which increases their productivity and competitiveness, thereby generating competitive advantages and a higher level of innovation.

Most entrepreneurs use mobile phones (37.1%), followed by the internet, with 23.6% reporting that they have it. Meanwhile, only 21.1% incorporate the use of social networks into their business and 8% use email.

These data are important to analyze because, according to the report, border areas have the lowest proportion of Micro, small and medium sized enterprises (MIPYMES). Therefore, it is crucial to promote sustainable and human-centered territorial development.

Skills training requires a significant investment, especially in terms of time. The course lasted 100 h of theoretical and practical training, with a further 15 h taken up by the defense of entrepreneurship projects, as indicated in Table 5, and was conducted over 12 weeks.

Table 5. Academic load in hours.

Workload	N of Theory Hours	N of Hours of Practice in the Entrepreneurship Project	Entrepreneurship Project Presentation	Total Hours
Total	45	55	15	115

Source: Guide for the female entrepreneur: IPMA International Project Management Competencies 2021 [57].

Investing 115 h of a resource they do not have at their disposal demonstrates the importance that the participants placed on skills training and their high motivation to develop project management capabilities under international standards. These skills were

https://doi.org/10.3390/su18031179

applied to their entrepreneurial projects, considering three key dimensions: practical skills, perspective skills, and personal skills [31].

Figure 3 presents the thematic structure of the course and its connection with the three dimensions of the international standard of competencies for the management of entrepreneurship projects (NCB-IPMA).

The content of the training program was divided into five units, each of which included theoretical aspects and practical applications through case studies. Table 6 connects the academic units that address the areas of people competency, perspective, and practice in entrepreneurial project management with the course content.

Table 6. Academic units and course content.

Units	Topics	Workload (Hours)	Duration
Unit 1: Basis for the certification of competencies in Project Management: the entrepreneurship project	IPMA—International Project Management Competencies: ✓ Global Competency Certification System. ✓ Basis for individual competence in project management.	5	Week 1
	✓ The entrepreneurship project ✓ Main dimensions of the project ✓ The entrepreneurship project cycle ✓ The project agents	5	Week 2
Unit 2. Perspective skills for project management	Strategy and business: ✓ Governance, structures, processes ✓ Compliance, standards, and regulations ✓ Power and interest ✓ Culture and values	5	Week 3
	✓ Entrepreneurship project formulation		Week 4
Unit 3: Personal skills for project management	✓ Self-reflection and self-management ✓ Personal integrity and reliability ✓ Personal communication	5	Week 5
	✓ Relationships and participation ✓ Leadership ✓ Teamwork ✓ Conflicts and crises ✓ Resourcefulness	5	Week 6
	✓ Results—orientation ✓ Entrepreneurial project development	5	Week 7
Unit 4. Practical skills for project management	✓ Project design ✓ Requirements and objectives ✓ Scope	5	Week 8
	✓ Time ✓ Organization and information ✓ Quality	5	Week 9
	✓ Finance ✓ Resources ✓ Procurement	5	Week 10
	✓ Planning and control ✓ Risk and opportunity ✓ Stakeholders ✓ Change and transformation ✓ Entrepreneurship project training	5	Week 11
Unit 5: Preparation for the IPMA exam (optional)	Practical case studies for exam preparation (IPMA) Entrepreneurship project preparation	5	Weeks 12
Unit 6: Entrepreneurship project	Entrepreneurship project formulation Personal Entrepreneurship project challenges Oral presentation	5	Weeks 1–12 3 h/week

Source: Guide for the female entrepreneur: IPMA International Project Management Competencies. Unpublished institutional document [57].

https://doi.org/10.3390/su18031179

The course was continuously assessed using a comprehensive approach that considered both the individual and the entrepreneurship project, allowing for the simultaneous evaluation of personal, practical, and perspective competencies. The weighting was distributed as follows:

- Fifty percent of the entrepreneur's competencies, based on three dimensions of competencies: personal, perspective, and practical.
- Fifty percent of the formulated entrepreneurial project.

The evaluation criteria for the activities to obtain the IPMA International Certificate in Project Management: Bases for Competence issue by the Polytechnic University of Madrid, were

- Concept evaluation questionnaires (30%);
- Participation, analysis, and discussion in forums (20%);
- Final report on the entrepreneurship project (20%);
- Presentation and final defense of the entrepreneurship project (30%).

The participants' oral defense of the project before a mixed panel, composed of professors from the GESPLAN Group of the Polytechnic University of Madrid and representatives of SMI, was one of the key activities to assess the skills acquired in project management applied to a real case: their business venture.

A fundamental component of the training process was personalized support by tutors (Table 4) who carried out continuous monitoring to ensure the understanding of the contents of each module and the progressive development of the final project.

The allocation of seed capital was based on a comprehensive and transparent evaluation framework that equally weighted was structured around four key criteria aligned with sustainable development objectives: innovation (20%), territorial development (30%), social impact (30%), and profitability and governance (20%). Each criterion comprised specific indicators—such as product or service differentiation, local job creation, empowerment of vulnerable groups, financial sustainability, and risk analysis—and was scored using a standardized 1–5 scale. This multi-criteria approach enabled the prioritization of initiatives with higher economic viability, social impact, and contribution to sustainable territorial development, ensuring a strategical and consistent use of seed capital.

As a result of the training process developed in collaboration between the university and the two participating companies in the Dominican Republic, a circle of trust has been generated among participants, where the CFS-RIA Principles are perceived as an opportunity to design projects aligned with the real needs of the territory. This is evident in the 27 ventures from the second edition (Table 3) and the 41 individuals certified as qualified (Table 3), who acquired skills in accordance with IPMA standards. In the case of BANELINO, this impact is reflected in the continued strengthening of the capacities of the 200 organized small producers.

This university–business link has contributed to the strengthening of "living laboratories," conceived as real learning and experimentation environments, in which development projects not only generate income for the families involved but also promote the socioeconomic development of their communities.

4.2. Integration of the CFS-RIA Principles into Institutional Work to Promote Sustainable Territorial Development

Companies effectively incorporate some of the CFS-RIA Principles into their business activities, adapting them to the specific context in which each operates, and placing responsible practices and sustainability objectives at the heart of their strategies. BANELINO integrates the CFS-RIA Principles by promoting sustainable agricultural practices, social

https://doi.org/10.3390/su18031179

equity, and economic development, and SMI by fostering the entrepreneurial project management skills of its beneficiaries. Table 7 details the integration into the activities of the two companies.

Table 7. Integration of the CFS-RIA Principles into the institutional work of two companies.

Principles	Dimension	BANELINO	SMI
P1 Contribute to food security and nutrition	Economic	Promoting sustainable organic banana production, ensuring healthy food for consumers, and improving food safety.	
P2 Contribute to sustainable and inclusive economic development and poverty eradication.	Economic	Representing 200 small-scale producers, each with an average of 3 hectares of land. Access to international markets, obtaining fair prices, and improving their income, contributing to local economic development and poverty reduction.	Community participation and inclusion through the family entrepreneurship program fosters community integration in the development of sustainable initiatives.
P3 Promoting gender equality and women's empowerment	Social	Inclusion and empowerment of women in the agricultural sector, constituting 31% of the total membership	Promotion and economic empowerment of families through capacity building and economic autonomy through the Family Social Responsibility Program based on training in entrepreneurship and project management.
P4 Promoting youth participation and empowerment	Social	Establishment of agreements with UTESA and UPM, allowing for the transfer of knowledge, technical assistance, and the generation of new territorial development projects.	Facilitating the creation of new university extensions in the area in collaboration with UTESA, allowing young people to access careers aligned with the agro-industrial sector.
P5 Respect land tenure, fisheries and forests, and access to water	Environmental	Developing operations within the Dominican legal framework and promoting the sustainable use of natural resources.	
P6 Conserve and sustainably manage natural resources, increase resilience, and reduce disaster risk	Environmental	Promoting organic production; agricultural practices are implemented that conserve the soil, protect biodiversity, and promote environmental sustainability.	Implementing strategies for sustainable business production and management.
P7 Respect cultural heritage and traditional knowledge, and support university and innovation	Economic		Working on cultural integration programs, strengthening relationships with the local community, and promoting entrepreneurship based on traditional knowledge.
P8 Promote safe and healthy agricultural and food systems	Environmental	Organic banana production ensures that food is free of chemical residues, offering healthier products to consumers	
P9 Incorporate inclusive and transparent governance structures, processes, and grievance mechanisms.	Governance	Operating as an association that promotes the active participation of its members in decision-making, ensuring transparency and accountability in its operations.	Signing agreements with UTESA and UPM for the development of local talent and the strengthening of corporate governance in the region. It also collaborates with the International Network of Universities and Businesses promoted by FAO and UPM.
P10 Assess and address impacts and promote accountability	Governance	Evaluating the impact of its programs on regional development.	

Source: Authors' own elaboration on based Principles for Responsible Investment in Agriculture and food systems [6].

There are few documented experiences of implementing the CFS-RIA Principles through university–business links, and even less research with a territorial focus that prioritizes people and drives sustainable development. Those that do exist focus on

https://doi.org/10.3390/su18031179

technology transfer and human capital development [58], but few have analyzed their impact on territorial development from the perspective of the CFS-RIA Principles [6].

4.3. Impact of CFS-RIA Principles on the Economic, Social, Environmental, and Governance Development of the Territory

SMI's Family Social Responsibility Program and BANELINO's production of organic and Fairtrade bananas constitute two models of university–business links based on the CFS-RIA Principles that have generated economic, social, environmental, and governance impacts, as well as economic autonomy for BANELINO producers and SMI entrepreneurs. To measure the impact of the experience, it is necessary to emphasize the key actors committed to development in the territory and the training of its stakeholders. Table 8 shows the economic, social, environmental, and governance impacts of the experience. Since the SMI project is in the consolidation phase and does not yet have a comprehensive accounting record of costs and benefits, a partial estimate was applied based on three assumptions: average annual profit, average total cost per participant, and the cost–benefit ratio. According to the available data, the cost–benefit ratio (C/B) would be 2.25 in the first year, indicating that for every dollar invested, the estimated annual return is 2.25 dollars, without considering social or environmental benefits, which is a conservative but favorable estimate.

Table 8. Impact of the integration of the CFS-RIA Principles on territorial development.

Impact of the CFS-RIA Principles	San Miguel Industry (SMI)	BANELINO
Economic impact	41 viable family businesses in total. 13 projects benefited from seed capital (4 projects in 1st edition and 9 in (2nd Edition). Generation of direct and indirect employment for at least 50 people. Promotion of family economic autonomy.	Improved competitiveness of 200 small producers. Its focus on certified organic production and fair trade has generated economic opportunities for small producers, increasing their competitiveness in international markets. Strengthened agricultural and family businesses. Improved producer income.
Social impact	41 IPMA competency certificates awarded. Training, education, and technical assistance for their businesses. Strengthening IPMA competencies in entrepreneurial project management. Improvement in project management, marketing, and digital technology skills. Increased female participation in economic and social activities. 90% of the ventures are led by women, highlighting their contribution to strengthening the social fabric. Support networks between entrepreneurs and mentors are consolidated.	Improving the quality of life of producers and their families through programs in education, health, first aid, good agricultural practices, and workplace safety. Empowering small producers, including 31% of women producers
Environmental impact	Implementation of sustainable good practices in production and business management.	100% organic production; Fairtrade- and Global GAP-certified. Ecological focus has maintained continuous organic and Fairtrade certification since 2005, renewed annually under FLO-CERT registration (FLO ID 1515). Soil and water indicators: the Dominican Institute for Agricultural and Forestry Research (IDIAF) reports stable or improving soil organic carbon (2.7–3.1%) and the absence of residual synthetic fertilizers in water samples (IDIAF, 2022–2023). Each export batch is verified in compliance with the European Union organic production and certification framework, together with the USDA Organic Certification Program. These schemes include annual third-party audits covering soil fertility, agrochemical use, and waste management. The most recent verification report (2023) confirms that 100% of farms comply with the prohibition of synthetic pesticides and maintain soil organic matter levels above 2.5%. Production is currently being diversified with crops such as coconut and cacao. For the sustainable banana cycle, the carbon and water footprints are being measured. Use of alternative solar energy for irrigation systems. A factory of bio-inputs for nutrition and pest and disease control. Use of digital monitoring technologies for plantation health. Monitoring of soil microbiota. Measurement of soil organic carbon. Applications using drone technology.

https://doi.org/10.3390/su18031179

Table 8. *Cont.*

Impact of the CFS-RIA Principles	San Miguel Industry (SMI)	BANELINO
Governance impact	Consolidation of a university–business–community cooperation model based on the CFS-RIA Principles, establishing accountability, inclusion, and transparency mechanisms that strengthen territorial sustainability.	
Sustainability	Training in entrepreneurship programs and projects for the economic autonomy of families.	Agro-ecological production and sustainable certifications.
Replicability	Possibility of expanding the model to other companies in similar sectors.	Model applicable to other agricultural and productive cooperatives

Source: own elaboration on based Principles for Responsible Investment in Agriculture and Food Systems [6].

This experience in the Dominican Republic indicates that the strategic integration of key stakeholders and the implementation of responsible investments can transform rural communities and generate lasting positive impacts. Environmental evidence (some indicators and good agricultural practices) has been incorporated into the logic model under the "Technical–Business" and "Political–Contextual" dimensions of the WWP framework (Figure 2). This evidence demonstrates that the program's outcomes are aligned with verifiable environmental performance standards and contributes to the ecological sustainability of the banana value chain in the Dominican Republic.

By strengthening the university–business link, the CFS-RIA Principles have stimulated the creation of productive ventures that diversify family income sources. The strengthening of local capacities, skills development, access to productive resources, and technical training have increased the productivity and financial sustainability of small rural economic units. In the case of SMI, there is evidence of an improvement in the purchasing power and economic autonomy of families, especially women, through activities such as beauty, bakery and pastry services, event celebrations and decorations, cafeterias, and education.

Along this line, the research on competencies and capacities for the management of sustainable rural development projects in the value chain in Jauja-Peru [18] also shows that the CFS-RIA Principles promote entrepreneurship with endogenous resources, as well as the participation of women with communication, leadership, and collaboration skills (leading nine organizations), achieving a progressive improvement in organizational management and technical competencies, which, although still incipient, show progress towards economic sustainability.

The ethical–social approach of the WWP model enhances collective learning, community cohesion, and respect for shared values.

4.4. Association Between Participation Intensity and Project Performance

To assess whether program outcomes were influenced by the intensity of engagement rather than mere participation, a correlation analysis was conducted using the full sample of 79 participants. Three independent variables—attendance rate, mentoring hours, and seed-fund amount—and one composite performance index, combining project validation, certification status, and self-reported income gains six months after completion, were considered in the analysis.

Table 9 summarizes the Pearson correlation coefficients among these variables. The results indicate that performance outcomes are significantly associated with both mentoring intensity ($r = 0.52$; $p < 0.001$) and attendance rate ($r = 0.41$; $p < 0.01$). The correlation with the seed-fund amount is weaker but still positive and significant ($r = 0.28$; $p < 0.05$).

https://doi.org/10.3390/su18031179

Table 9. Associations between program intensity and performance indicators (n = 79). (Pearson correlation coefficients with significance levels).

Variable	Attendance	Mentoring Hours	Seed-Fund (USD)	Performance Index	Interpretation
Attendance (%)	-	0.38 **	0.21 ns	0.41 **	Moderate positive association
Mentoring Hours	0.38 **	-	0.25 *	0.52 ***	Strongest predictor
Seed-Fund (USD)	0.21 ns	0.25 *	-	0.28 *	Weak–moderate positive effect
Performance Index	0.41	0.52 ***	0.28 *	-	

Source: final project report second edition 2024. Unpublished institutional document [46]. * $p < 0.05$; ** $p < 0.01$; *** $p < 0.001$; ns not significant.

These findings suggest that participants who were more consistently engaged—by attending a greater share of training sessions and receiving more hours of mentoring—achieved higher levels of project validation, certification, and income improvement. The moderate association between attendance and mentoring (r = 0.38; $p < 0.01$) confirms that mentoring complemented, rather than replacing, formal training.

A simple multiple regression ($R^2 = 0.43$; $p < 0.001$) further indicated that mentoring hours were the strongest predictor of post-program performance, followed by attendance rate and, to a lesser extent, seed-fund support. This reinforces the idea that program quality and intensity, not participation alone, are key determinants of impact.

From an evaluative perspective, this pattern provides evidence of construct validity for the program's design: higher exposure to learning and personalized guidance is directly linked to measurable entrepreneurial outcomes. Moreover, the consistent direction of these associations across variables supports the internal coherence of the methodology.

The above results only concern empirical findings derived from the implementation and evaluation of the case study (enrolments, completion rates, average investment in training per participant, number of ventures, certifications, and observed effects). Comparative cases and broader interpretation of these findings are presented in Section 5 (Discussion).

5. Discussion

In this section, we interpret the study results in light of the literature and comparative experiences. We reflect on causal relationships between the WWP/CFS-RIA implementation and observed impacts and compare our findings with relevant cases (Villa Poppy, Kayambi, Jauja, National University of La Plata, and Aymara Women's Coordinator) to identify transferable lessons and limits to generalization.

5.1. Interpretation of Key Findings

Positive and statistically significant associations between mentoring, attendance, and performance.

As shown in Table 9, mentoring hours exhibited the strongest correlation with the composite performance index (r = 0.52; $p < 0.001$). Attendance rate also showed moderate significance (r = 0.41; $p < 0.01$), while seed-fund amount had a weaker yet positive relationship (r = 0.28; $p < 0.05$).

Performance was not solely determined by funding access but also by the intensity and quality of participant engagement. This reinforces the ethical–social dimension of the WWP, where relational learning, peer trust, and mentoring continuity outweigh short-term financial incentives. The causal pathway is therefore behavioral and institutional rather than purely economic.

Verifiable environmental and sustainability outcomes consistent with CFS-RIA Principles.

Table 8 presents empirical evidence of ecological impact: full (100%) compliance with organic certification, a stable 18–22% price premium over conventional banana exports, soil organic matter maintained between 2.7 and 3.1%, and absence of agrochemical residues.

https://doi.org/10.3390/su18031179

These results demonstrate that the political–contextual dimension of WWP—through institutional coordination among SMI, Banelino, and UPM—enabled long-term ecological commitments. The certification and monitoring system provides credible verification consistent with the CFS-RIA Principles of responsible investment and measurable impact.

The implementation of the WWP model in this case study reveals the empirical correspondence between CFS-RIA dimensions and the CFS-RIA Principles (Table 2): the political–contextual phase allowed for the formalization of alliances and the provision of resources (agreements and institutional support); the ethical–social phase strengthened capacities and empowered women and youth (27 certified in the second edition); and the technical–entrepreneurial phase enabled the formulation and validation of viable ventures. These results confirm that without the prior consolidation of governance and trust, the technical transfer and sustainability of the projects would have been limited.

5.2. Comparative Cases and Lessons Learned

Experiences such as the case of Villa Poppy in the Dominican Republic [43] show, as does BANELINO, how producers are managing to improve their livelihoods, the quality of their products, organizational strengthening, and their connection to new markets in the tourism and export sectors in the municipality of Constanza. The CFS-RIA Principles, which are presented as a guide for the implementation of productive projects, have not only allowed farmers to access new markets but have also encouraged them to improve their production standards and establish inter-institutional alliances, facilitating their marketing with hotels and the public sector.

This experience, which brings together small-scale producers (between 2 and 10 hectares on average), reinforces the importance of adopting practices and policies that promote environmental sustainability, respect the rights of agricultural workers, foster gender equality and the empowerment of women, and support the economic development of rural communities.

The inter-institutional articulation model implemented in the Villa Poppy project includes key actors with a backbone role in economic, social, environmental, and governance development [26], such as BANELINO and SMI, which strengthen their relationships with other actors, improving the living conditions of the families involved.

Both the experience of Villa Poppy and that of BANELINO and SMI are configured as successful cases of inter-institutional collaboration that contribute to the sustainability of the territory [27] through the work of the actors themselves, the provision of technical and financial assistance through seed capital (Table 3), the development of skills and capacities, and the granting of access to demanding markets, thereby strengthening community organization and promoting sustainable practices that align with the CFS-RIA Principles.

Family farming production in the Puebla Valley (Mexico): Linking, transfer, and dissemination of innovation characterize another experience with a territorial approach based on the CFS-RIA Principles to strengthen the university–business link [59]. The experience shows how the application of the three dimensions of the WWP model and the CFS-RIA Principles improve the economic and social stability of small producers and entrepreneurs, similarly to what was observed in BANELINO and SMI, by focusing on the family as a production unit and promoting equal opportunities, the empowerment of women in a transversal manner, and the participation and empowerment of young people by granting them access to management techniques to ensure high productivity. Furthermore, inclusive governance structures are incorporated into the planning model as social learning [33,39,40], giving prominence to farmers. The association of producers is strengthened, and in the process, they learn how to collaborate and defend their interests.

In Ecuador, the case of the Kayambi people illustrates how rural communities can transform their realities through strategies that integrate diverse actors and resources,

https://doi.org/10.3390/su18031179

promoting development models that merge local needs with global commitments and, therefore, fostering innovation in sustainable rural development [60,61]. In innovation processes, the participation of beneficiaries, academia, decentralized local governments, and companies [5] is essential for the promotion of training, financing, the formation of agro-ecological producer associations, biomarkets, food baskets, etc., and the creation of sustainable value chains and inclusive community business models.

As in BANELINO and SMI, collaboration between key stakeholders improves the living conditions of producers, students, and local communities. It is observed that the associations formed have achieved a high percentage of women members, as in the case of BANELINO. An important contribution has been the marketing of products through agro-ecological fairs, which have promoted the production, consumption, and marketing of products through fair prices and weights, improving incomes. The involvement and participation of the population in all stages of the project ensures the continuity and sustainability of activities. The associations of agro-ecological producers have managed to raise awareness about the use of natural fertilizers in a way that does not harm families and protects the environment.

5.3. Comparisons with Other Universities and Community Initiatives

The study carried out in Jauja, Peru [14], implemented by the National University of San Marcos, focused on the innovative figure of the Local Action Groups (LAGs) as a key element for the articulation and formulation of projects that integrate the CFS-RIA Principles in its Regional Development Centers, promoting university decentralization and strengthening relationships with local stakeholders. It shows that the link with the university strengthens territorial governance and generates learning spaces for students and teachers with research that solves local livestock problems and technical assistance to producers, a finding aligned with the case of BANELINO.

The participation of research groups in solving societal problems, supporting the development of organizational skills at the local level for endogenous and sustainable development [15], creating living laboratories, etc., demonstrates a real link between universities and organized society. By integrating the CFS-RIA Principles into their business model, companies respond to the needs of the territory, as is the case with SMI in entrepreneurship projects, capacity building, and skills development that meet the needs of the population (Tables 3, 5 and 6).

Another study in Jauja, titled "Competencies and Capabilities for Managing Sustainable Rural Development Projects in the Value Chain: Perceptions of Small and Medium-Sized Enterprise Agents in Jauja, Peru" [18], has generated confidence and proactivity among the actors in the chain (as well as BANELINO and SMI) that are interested in continuing with the implementation of the CFS-RIA Principles and achieving sustainability with projects that show commercial innovation. The authors of this study designed an innovative program for the development of competencies, which balances the three dimensions of project management competencies (perspective, people, and practice) as defined by the International Project Management Association (IPMA), concerning the CFS-RIA Principles, as carried out in the "IPMA International Competencies in Project Management: Foundations for Entrepreneurship" course (Table 6) of SMI.

Research aimed at strengthening people's capabilities has been developed by the National University of La Plata [62], increasing the university's involvement with society and its environment. As a result, the university's contribution facilitated work that benefits the community and, in addition, reinforced the academic training of undergraduate students through the work experiences they carried out in different communities, leading to the creation of cooperation opportunities between the university and the Punta Indio commu-

https://doi.org/10.3390/su18031179

nity. This experience, as in the case of BANELINO and SMI, provides new development strategies to strengthen the socioeconomic base. A process of competencies is developed with the local action group for the management of agricultural projects (in the case of BANELINO) and capacity building through training, business management advice, and the formalization of companies (in the case of SMI), favoring development in rural areas. The Local Action Group (LAG)'s activities have led to 18 projects, focusing on the following areas: valorization of local production, training, promotion of tourism heritage, environmental protection and conservation, and the maintenance of sites of interest in Punta del Indio. The individuals involved apply this new knowledge in responsible investment projects based on sustainability and contribute to economic development.

Another research project where a societal problem is investigated by the university is the case of the Aymara Women's Coordinator (Coordinadora de Mujeres Aymaras), an association that invests in training for women's leadership and the sustainable use of alpaca wool in the Peruvian highlands through crafts. The association promotes capacity building [63] through links with the university, in this case, the Polytechnic University of Madrid through the GESPLAN research group. This experience emphasizes the social learning model [33,39,40] as the most effective among planning models, since it allows for continuous exchange between all stakeholders involved in a development project [64], thus ensuring sustainability.

It is considered that the actions carried out promote gender equality and the empowerment of women (CFS-RIA Principle 4); in addition, the satisfaction of the artisans and women producers of BANELINO is reflected in the growing female membership of the organizations.

Through various interventions, leaders have been trained, the organizational structure has improved, and beneficiaries' income has improved by 30%. Additionally, production has been diversified, focusing on women's, men's, children's, and baby clothing and household items, which are advertised on the company's website. Its efficiency indicators show that for every dollar invested, three dollars have been generated for the beneficiaries. Garments have been sold in nine countries, and outsourcing services are offered to third parties for the production of their designs. The income generated has allowed them to invest in their limited agricultural activity and food security.

Another research case is the one developed in Peru involving the Peruvian Poultry Association (APA) [65] which implemented a territorial strategy to prevent social conflicts and promote sustainable development in the poultry sector. This strategy included collaboration with the National Agrarian Health Service (SENASA) to eradicate avian diseases through vaccination programs, georeferencing of poultry farms, and health surveillance throughout the country. These actions strengthened the sector's competitiveness and contributed to national food security.

Companies that integrate the territorial dimension into their planning can contribute to sustainability, reduce social tensions, and enhance their legitimacy. Both BANELINO and SMI embrace a territorial perspective, working hand in hand with local stakeholders, governments, and communities to design interventions that generate shared value.

The adoption of the CFS-RIA Principles has contributed to strengthening the community fabric, gender equity, and social cohesion. The focus on the inclusion of youth and women in productive processes and decision-making strengthens the region's social capital and improves self-esteem and recognition among traditionally marginalized populations. The active participation of universities such as UTESA and UPM has been key to civic education, empowering local stakeholders through education and ongoing technical support.

https://doi.org/10.3390/su18031179

A research that strengthens social capital deals with Good Practices of Food Banks in Spain: Contribution to Sustainable Development [66], which analyzes 54 food banks based on four dimensions of sustainability: social, economic, environmental, and governance, in line with the CFS-RIA Principles, ensuring availability, access, and consumption of food in a sustainable, equitable and safe manner, highlighting awareness-raising activities against food loss and waste. The results identified, as with BANELINO and SMI, the strengthening of the social fabric through volunteering, inter-institutional collaboration, citizen participation, and the empowerment of social actors by integrating participatory governance mechanisms where the voice of multiple actors is recognized: civil society, the private sector, universities, public administration.

5.4. Transferability Conditions

The experience analyzed demonstrates potential transferability to other regions and sectors, provided that key enabling conditions are met. Transfer requires, first, the presence of a coordinating institution—such as a university, cooperative, or local development agency—capable of acting as a trusted intermediary among stakeholders. Second, access to differentiated or sustainable markets (e.g., Fairtrade, organic, or territorial brands) is essential to sustain economic incentives. Third, a supportive policy environment and multi-level governance mechanisms facilitate continuity and integration with local and national development strategies.

Conversely, in territories lacking institutional stability or where producer organizations are weakly articulated, the model would require preliminary investment in capacity building and social capital formation before replication.

Therefore, the WWP–CFS-RAI Principles approach is transferable but context-dependent, demanding adaptation to each region's organizational maturity, market opportunities, and policy coherence.

This case study validates the proposal of the Metauniversity—a "university of universities"—as an innovative instrument to scale knowledge transfer, research, and community engagement [67].

5.5. Limitations

The limitations of this study involve selection bias, a lack of experimental data, cost estimates and assumptions, and the limitations of direct environmental measurements. Future measurements (soil/water direct sampling) should be proposed. In the following paragraph, we describe each of these limitations, explaining how they affect the interpretation and scope of the results.

First, the study is subject to selection bias, which stems from the participant recruitment process. The beneficiaries are family members of employees of SMI who voluntarily responded to an open call and met specific eligibility criteria, such as prior interest in entrepreneurship and availability to complete the training program. As discussed in the Sections 4 and 5, this self-selection mechanism likely resulted in a sample with higher levels of motivation, commitment, and social capital than the general population. Consequently, while the observed outcomes—such as high completion rates and project viability—are consistent with the objectives of the program [66], they are most directly applicable to populations with similar levels of motivation, engagement, and contextual support.

Second, the study lacks an experimental or concurrent quasi-experimental design, such as a randomized control group or systematic pre- and post-intervention measurements for all participants. Although a historical reference cohort (first and second editions of the program) was incorporated to provide descriptive comparisons, the absence of comprehensive experimental data limits the ability to establish strong causal relationships

https://doi.org/10.3390/su18031179

between the training intervention and the economic, social, or environmental outcomes discussed. Therefore, the positive effects identified in the Discussion should be considered as associative and indicative rather than strictly causal.

Third, the cost and cost-effectiveness estimates presented in the study are based on partial financial information and explicit assumptions derived from available program documentation. As noted in the results, the lack of disaggregated data—such as detailed costs related to mentoring, cost related instructional hours, infrastructure, and follow-up activities—defines the level of detail at which the economic indicators can be reported in the present analysis. Accordingly, these estimates offer a valuable first approximation that provides initial insights into the program's economic performance and lays the groundwork for more detailed evaluations in future research.

Finally, the study addresses environmental sustainability primarily through verified certification standards rather than direct bio-physical measurements. The environmental performance is assessed through recognized certification frameworks, which provide a consistent and relevant for evaluating practices related to organic production, resource management, and sustainability. At the same time, this approach highlights opportunities for future research to integrate direct soil and water monitoring in order to further strengthen the environmental evidence base.

Taken together, these limitations do not diminish the relevance of the findings but rather define their analytical scope and applicability. At the same time, they reinforce the contribution of the study as an applied and exploratory case and point to clear directions for future research aimed at strengthening causal inference, economic precision, representativeness, and environmental measurement in university–business sustainability initiatives.

6. Conclusions

A university–business link, anchored in the CFS-RIA Principles and methodologies such as Working with People, has proven to be an effective way to activate sustainable territorial development dynamics, where stakeholders not only benefit but also become agents of change. The experiences of SMI and BANELINO reveal that when working with people—and not just for them—it is possible to consolidate learning, production, and cooperation ecosystems that transcend the local level to build sustainable, inclusive, and replicable models. This synergy between training, participation, and sustainability constitutes a solid base that invites us to reconsider the role of universities and businesses in the future of territories. The proposed approach positions itself as a key catalyst for social and economic innovation processes in vulnerable territories.

The WWP methodology has proven to be an effective strategy for fostering rootedness, the social appropriation of knowledge, and the sustainability of initiatives. By prioritizing collaboration, active participation, the empowerment of local stakeholders, and the building of trusting relationships, the university–business link becomes a platform for regional innovation and the co-creation of sustainable solutions.

The adoption of a previously planned evaluation system grounded in explicit sustainability and responsibility criteria enabled a coherent allocation of seed capital. The integrated assessment of entrepreneurial competencies and the technical, social, and economic quality of the projects supported the identification of initiatives with a greater capacity to generate sustainable, inclusive, and territorially embedded value. This approach, aligned with the Committee on World Food Security's Principles for Responsible Investment in Agriculture and Food Systems (CFS–IRA), ensured transparent funding decisions oriented toward women's economic empowerment, local employment creation, strengthening of productive systems, and long-term financial sustainability. Consequently, the seed capital supported both the launch of new ventures and the consolidation of existing businesses,

https://doi.org/10.3390/su18031179

confirming the achievement of the program's objectives and highlighting its potential as a replicable model of responsible investment for sustainable territorial development.

The findings indicate that this experience holds potential for transfer to other regions, provided that essential enabling conditions are in place, including a coordinating institution, access to differentiated markets, and a supportive policy framework. In territories with weak organizational structures, initial efforts must focus on capacity building and social capital formation. Thus, the WWP–CFS-RAI Principles approach is transferable, but its effectiveness depends on context-specific institutional maturity, market opportunities, and governance coherence.

The experience in the Dominican Republic constitutes a concrete and replicable way to promote sustainable territorial development when it focuses on the training of individuals and social capital formation, demonstrating that this link enables context-specific responses to social, economic, and environmental challenges by activating local capacities and strengthening the socio-productive fabric through a logic of shared responsibility. Companies that integrate the territorial dimension into their planning can contribute to sustainability, improve living conditions, strengthen organizations, reduce social tensions, and improve their legitimacy.

Both the ventures launched within the SMI course, with its Family Social Responsibility Program, and the continuous improvement programs implemented by BANELINO reflect a process of institutional transformation geared toward sustainability. Training based on international standards such as the IPMA project management model [67], not only enhances individual capabilities but also fosters a strategic vision of entrepreneurship, with an emphasis on autonomy, viability, and social commitment. This case study demonstrates that the performance index was not solely determined by funding access but also by the intensity and quality of participant engagement (Table 9). This reinforces the ethical–social dimension of the WWP, where relational learning, peer trust, and mentoring continuity outweigh short-term financial incentives.

These findings suggest that higher levels of exposure to training and personalized guidance are linked to better outcomes in project validation, certification, and income progression.

The moderate correlation between attendance and mentoring confirms that both components functioned complementarily. The multiple regression model ($R^2 = 0.43$; $p < 0.001$) further identifies mentoring intensity as the strongest predictor, followed by attendance and seed-capital support. Overall, the pattern strengthens evidence of construct validity and demonstrates that program intensity—not participation alone—is a determining factor in outcome achievement.

Implementing training and technical support programs, promoted by the university in collaboration with socially responsible companies such as SMI and BANELINO, has generated significant impacts in terms of economic empowerment, social innovation, and environmental sustainability. These processes have strengthened skills in leadership, project management, and entrepreneurship, especially among women, consolidating an inclusive development approach with a gender perspective.

In short, this experience provides valuable evidence for the design of policies and programs that seek to integrate sustainability, participation, and competitiveness. Its replication and adaptation to other territorial contexts is recommended based on the principle of working with the people, not just for the people. Only in this way will it be possible to move toward more just, resilient, and sustainable development models.

Supplementary Materials: The following supporting information can be downloaded at https://www.mdpi.com/article/10.3390/su18031179/s1, Supplementary Material S1—Data Collection Instruments.

https://doi.org/10.3390/su18031179

Author Contributions: Conceptualization: M.d.P.P.M., J.L.C.A., U.E.D.E. and G.A.G.; Methodology: M.d.P.P.M. and J.L.C.A.; Validation: M.d.P.P.M., U.E.D.E. and G.A.G.; Formal Analysis: M.d.P.P.M.; research: M.d.P.P.M. and J.L.C.A.; Resources: M.d.P.P.M.; Data Curation: M.d.P.P.M.; Writing—Original Draft Preparation: M.d.P.P.M.; Writing—Review and Editing: M.d.P.P.M., J.L.C.A. and U.E.D.E.; Visualization: M.d.P.P.M. and G.A.G.; Supervision: M.d.P.P.M.; Project Management: M.d.P.P.M.; Funding Acquisition: M.d.P.P.M., U.E.D.E., J.L.C.A. and G.A.G. All authors have read and agreed to the published version of the manuscript.

Funding: This research received no external funding.

Institutional Review Board Statement: Ethical review and approval were waived for this study by the Ethics Committee of the Polytechnic University of Madrid and the GESPLAN Research Group, in accordance with institutional regulations, as the research did not involve interventions, clinical trials, or sensitive data.

Informed Consent Statement: Informed consent was obtained from all subjects involved in this study.

Data Availability Statement: The original contributions presented in this study are included in the article/Supplementary Materials. Further inquiries can be directed to the corresponding author.

Acknowledgments: The authors sincerely thank San Miguel Industry (SMI) and BANELINO for granting access to their areas of direct influence and for supporting the fieldwork that underpins this study. Their collaboration made it possible to generate empirical evidence contributing to the analysis of sustainable territorial development and responsible university–business engagement.

Conflicts of Interest: The authors declare no conflicts of interest.

Abbreviations

The following abbreviations are used in this manuscript:

WWP	Working with People
CFS-RAI PRINCIPLES	Principles for Responsible Investment in Agriculture and Food Security
SMI	San Miguel Industry
UPM	Polytechnic University of Madrid
UTESA	Technological University of Santiago
IPMA	International Project Management Association
BANELINO	Organic bananas from the northwest line
GESPLAN	Research group in project planning
FAO	Agency from the Food and Agriculture Organization
QSM	Quality Management System
PBL	Project-Based learning
ONE	National Statistic Office Dominican Republic
RNC	National Taxpayer Registry
MIPYMES	Micro, small and medium sized enterprises
LAG	Local Action Group
CMA	Aymara Women's Coordinator

References

1. Cazorla, A.; De los Ríos-Carmenado, I.; Dávila, G.; Mínguez, E.; Fontana, A.; Navarro, J.A.; Galindo, R.; Montes, A.; Sommer, D.; García Fermín, F.; et al. *El Papel de Los Profesores En Las Universidades de Investigación: Relevancia En Los Rankings*; UNICARIBE: Santo Domingo, Dominican Republic, 2023; ISBN 978-9945-9208-3-3.
2. Lavalle Iriarte, C. Las Universidades de Investigación Nuevo Paradigma Mundial: Modelo de Gobernanza y Aplicación a Cinco Universidades. Ph.D. Thesis, E.T.S. de Ingeniería Agronómica, Alimentaria y de Biosistemas (UPM), Madrid, Spain, 2017.
3. Salgado Guerrero, J.P.; Herrán Gómez, J. Gobierno Para Una Universidad de Investigación: La Universidad Politécnica Salesiana. In *Buen Gobierno en las Universidades de Investigación en un Entorno Digital Global*; Fondo Editorial: Cuenca, Ecuador, 2021; pp. 55–68, ISBN 978-9972-46-693-9.

https://doi.org/10.3390/su18031179

4. Cazorla, A.; Fernández-Pello, C.; Fontana, A.; Garcia, A.; De Giusti, M.R.; Gómez, F.; Herrán, J.; Mínguez, E.; Muñoz, J.P.; Nicolas, V.; et al. La vinculación con la sociedad en las universidades de investigación. In *La Universidad Motor de Trasnformación de la Sociedad*; Editores Cazorla & Stratta: Madrid, Spain, 2017; Volume 1, pp. 142–161.
5. Garzón Vera, B.; Cárdenas Tapia, J.; Merchán Arízaga, X.; López Villavicencio, F.; Samaniego Sagbay, V.; Pillco Guamán, E.; Gordillo Gordillo, E.; Arce Cuesta, D.; Jumbo González, W.Í.; Paucar Paucar, J.E.; et al. *Incidencia de los proyectos de Vinculación con la Sociedad de la Universidad Politécnica Salesiana*; Editorial Abya-Yala: Quito, Ecuador, 2022; Volume 2, ISBN 978-9978-10-707-2.
6. Principles for Responsible Investment in Agriculture and Food Systems. Available online: https://openknowledge.fao.org/server/api/core/bitstreams/a5473764-5e32-4988-a6c4-4383f59a0292/content (accessed on 4 May 2025).
7. Cazorla, A.; De los Ríos-Carmenado, I.; Salvo, M. Working With People (WWP) in Rural Development Projects: A Proposal from Social Learning. *Cuad. De Desarro. Rural.* **2013**, *10*, 131–157.
8. Acosta Mereles, M.L.; De los Ríos Carmenado, I.; Ávila Cerón, C.A.; Castañeda Sepulveda, R. Responsible investment in the processes of substitution of illicit crops from the "working with people" model: Case study Guaviare (Colombia). In Proceedings of the 27th International Congress on Project Management and Engineering, Donostia-San Sebastian, Spain, 10–13 July 2023. [CrossRef]
9. Sastre-Merino, S.; Negrillo, X.; Hernández-Castellano, D. Sustainability of Rural Development Projects within the Working With People Model: Application to Aymara Women Communities in the Puno Region, Peru. *Cuad. Desarro. Rural.* **2013**, *10*, 219–243.
10. Cazorla, A.; Negrillo, X.; Montalvo, V.; De Nicolas, V.L. Institutional Structuralism as a Process to Achieve Social Development: Aymara Women's Community Project Based on the Working with People Model in Peru. *J. Soc. Soc. Welfare* **2018**, *45*, 55.
11. Ceron, C.A.A.; De los Rios-Carmenado, I.; Martín Fernández, S. Illicit Crops Substitution and Rural Prosperity in Armed Conflict Areas: A Conceptual Proposal Based on the *Working With People* Model in Colombia. *Land Use Policy* **2018**, *72*, 201–214. [CrossRef]
12. De los Ríos-Carmenado; Rivera, M.; García, C. Redefining Rural Prosperity through Social Learning in the Cooperative Sector: 25 Years of Experience from Organic Agriculture in Spain. *Land Use Policy* **2016**, *54*, 85–94. [CrossRef]
13. De Nicolás, V.L. Towards a Transformational Hydraulic Engineering Project for the Territory: A Focus on the Working With People (WWP) Model. *Land Use Policy* **2016**, *54*, 246–252. [CrossRef]
14. Jiménez Aliaga, R.; De los Ríos-Carmenado, I.; San Martín Howard, F.; Calle Espinoza, S.; Huamán Cristóbal, A. Integration of the Principles of Responsible Investment in Agriculture and Food Systems CFS-RAI from the Local Action Groups: Towards a Model of Sustainable Rural Development in Jauja, Peru. *Sustainability* **2022**, *14*, 9663. [CrossRef]
15. Jiménez Aliaga, R.; De los Ríos-Carmenado, I.; Huamán Cristóbal, A.E.; Aliaga Balbín, H.; Marroquín Heros, A.M. Competencies and Capabilities for the Management of Sustainable Rural Development Projects in the Value Chain: Perception from Small and Medium-Sized Business Agents in Jauja, Peru. *Sustainability* **2023**, *15*, 15580. [CrossRef]
16. Castañeda, R.; Avila Cerón, C.; De los Ríos-Carmenado, I.; Domínguez, L.; Gomez, S. Implementing the Voluntary Guidelines on the Responsible Governance of Tenure of Land, Fisheries and Forests from the Working with People Model: Lessons from Colombia and Guatemala. *J. Peasant. Stud.* **2023**, *50*, 1820–1851. [CrossRef]
17. De los Ríos-Carmenado, I.; Sastre-Merino, S.; Fernandez, C.; Nuñez, C.; Reyes, E.; Garcia-Arjona, N. Proposals for Improving Evaluation Systems in Higher Education: An Approach from the Model "Working with People". *J. Technol. Sci. Educ.* **2016**, *6*, 104–120. [CrossRef]
18. De los Ríos-Carmenado, I.; Mur, C. Inclusive and transparent governance structures from the IRA Principles. FESBAL Case—Food Bank. In Proceedings of the 27th International Congress on Project Management and Engineering, San Sebastian, Spain, 10 July 2023.
19. López, M.; Cazorla, A.; Panta, M.d.P. Rural Entrepreneurship Strategies: Empirical Experience in the Northern Sub-Plateau of Spain. *Sustainability* **2019**, *11*, 1243. [CrossRef]
20. De Los Ríos-Carmenado, I.; Ortuño, M.; Rivera, M. Private–Public Partnership as a Tool to Promote Entrepreneurship for Sustainable Development: WWP Torrearte Experience. *Sustainability* **2016**, *8*, 199. [CrossRef]
21. Klarin, T. The Concept of Sustainable Development: From Its Beginning to the Contemporary Issues. *Zagreb Int. Rev. Econ. Bus.* **2018**, *21*, 67–94. [CrossRef]
22. Cazorla, A.; De Los Rios-Carmenado, I. Hacia una Metauniversidad para un desarrollo sostenible: Inversión Responsable en agricultura y Sistemas Alimentarios. Towards a Meta-University for Sustainable Development: Responsible Investment in Agriculture and Food Systems. In Proceedings of the 28th International Congress on Project Management and Engineering, Jaén, Spain, 5–6 July 2024. [CrossRef]
23. Law No. 12-21. Creación de la Zona Especial de Desarrollo Integral Fronterizo y un Régimen de Incentivos en la Zona de Pedernales, Independencia, Elías Piña, Dajabón, Montecristi, Santiago Rodríguez, and Bahoruco. O. G. No. 11010 del 23 de Febrero 23, 2021. Available online: https://dgii.gov.do/legislacion/leyesTributarias/Documents/Leyes%20de%20Incentivos%20y%20Fomentos/12-21.pdf (accessed on 14 May 2025).
24. Torre, A. Nuevas propuestas para analizar el desarrollo territorial. *Eutopía. Rev. De Desarro. Económico Territ.* **2020**, *17*, 11–24. [CrossRef]

25. Cazorla-Montero, A.; de los Ríos-Carmenado, I.; Pasten, J.I. Sustainable Development Planning: Master's Based on a Project-Based Learning Approach. *Sustainability* **2019**, *11*, 6384. [CrossRef]
26. Panta Monteza, M.d.P. Propuesta Metodológica Para el Desarrollo de Emprendimientos Productivos Promovidos Por Mujeres de Escasos Recursos: Aplicación al Valle de Cañete—Perú. Ph.D. Thesis, E.T.S. de Ingeniería Agronómica, Alimentaria y de Biosistemas (UPM), Madrid, Spain, 2019.
27. Solano, D. Desarrollo, Sostenibilidad y Capacidades: Una Trilogía Indesligable. *J. Econ. Financ. Adm. Sci.* **2007**, *12*, 9–27. [CrossRef]
28. Huerta, A.R. Capacidad institucional: Revisión del concepto y ejes de análisis. *Doc. Y Aportes En Adm. Pública Y Gestión Estatal* **2019**, *19*, 81–107. [CrossRef]
29. Inicio. Available online: https://cursosipma.es/ (accessed on 4 May 2025).
30. De los Ríos-Carmenado, I.; Rahoveanu, A.T.; Gallegos, A.A. Project Management Competencies for Regional Development in Romania: Analysis from "Working with People" Model. *Procedia Econ. Financ.* **2014**, *8*, 614–621. [CrossRef]
31. IPMA-MEXICO. Bases Para la Competencia Individual en Dirección de Proyectos, Programas y Carteras de Proyectos. Available online: https://ipmamexico.com/wp-content/uploads/2025/02/Base-para-la-Competencia-Individual4-MX2.pdf (accessed on 24 May 2025).
32. Pérez, C.; Fajardo, M.; Granja, R.Y. Solidarity territories for capacity development and collective action from the local. *J. Rural. Community Dev.* **2022**, *17*, 102–127.
33. De los Ríos-Carmenado, I.; Cadena-Iñiguez, J.; Díaz-Puente, J.M. Creación de grupos de acción local para el desarrollo rural en México: Enfoque metodológico y lecciones de experiencia. *Agrociencia* **2011**, *45*, 815–829.
34. Calisto Friant, M. Comercio justo, seguridad alimentaria y globalización: Construyendo sistemas alimentarios alternativos. *Íconos Rev. De Cienc. Soc.* **2016**, *55*, 215–240.
35. Egas, F. Comercio justo o economía solidaria. *Rev. Afese* **2017**, *49*, 1–4.
36. García Chiang, A. El comercio justo: ¿una alternativa de desarrollo local? *Polis* **2011**, *7*, 105–140.
37. Coscione, M. América Latina y el sentido originario del comercio justo. *Eutopía. Rev. De Desarro. Económico Territ.* **2015**, *7*, 11–26. [CrossRef]
38. Galán, B. Reglamentaciones jurídicas sobre el acceso a la tierra de la mujer rural en países de América Central y el Caribe. *Land Reform* **2000**, *1*, 81.
39. De los Ríos-Carmenado, I.; Díaz-Puente, J.M.; Cadena-Iñiguez, J. La iniciativa LEADER como modelo de desarrollo rural: Aplicación a algunos territorios de México. *Agrociencia* **2011**, *45*, 609–624.
40. Cazorla-Montero, A.; De los Ríos-Carmenado, I.; Díaz-Puente, J.M. La iniciativa comunitaria leader como modelo de desarrollo rural: Aplicación a la región capital de españa. *Agrociencia* **2005**, *39*, 697–708.
41. Ortuño, M.; De los Ríos-Carmenado, I.; Sastre-Merino, S. The Development of Skills as a Key Factor of the Cooperative System: Analysis of the Cooperative of Artisan Women Tejemujeres-Gualaceo-Ecuador from the WWP Model. *Sustainability* **2022**, *14*, 16233. [CrossRef]
42. Fontana, A.; Velasquez-Fernandez, A.; Rodriguez-Vasquez, M.I.; Cuervo-Guerrero, G. Territorial Analysis Through the Integration of CFS-RAI Principles and the Working with People Model: An Application in the Andean Highlands of Peru. *Sustainability* **2025**, *17*, 1380. [CrossRef]
43. Ávila Cerón, C.A.; Castañeda, R.; De Los Ríos, I. Desarrollo de un Modelo de Agricultura Familiar Alineado con los Principios de Inversión Responsable: Integración Cadena Turística en República Dominicana. Development of a Family Agriculture Model Aligned with the Principles of Responsible Investment: Integration Tourism Chain in the Dominican Republic. In Proceedings of the 28th International Congress on Project Management and Engineering, Jaén, Spain, 3–4 July 2024. [CrossRef]
44. Friedmann, J. *Planning in the Public Domain: From Knowledge to Action*; Princeton University Press: Princeton, NJ, USA, 1987; ISBN 978-0-691-07743-7.
45. Cazorla-Montero, A.; De los Ríos-Carmenado, I. From "Putting the Last First" to "Working with People" in Rural Development Planning: A Bibliometric Analysis of 50 Years of Research. *Sustainability* **2023**, *15*, 10117. [CrossRef]
46. UPM-SMI. *Final Project Report Second Edition 2024—Unpublished Institutional Document*; UPM-SMI: Selangor, Malaysia, 2024.
47. Suárez, M.; Robaina, N.; Del Prado, L.; Noboa, A. Universidad y desarrollo local: Procesos innovadores en el tercer nivel de gobierno. *Integr. Y Conoc.* **2022**, *11*, 34–53. [CrossRef]
48. Torre, A. Contribution to the Theory of Territorial Development: A Territorial Innovations Approach. *Reg. Stud.* **2025**, *59*, 2193218. [CrossRef]
49. García Bacuilima, J.L. Modelo de Aprendizaje Basado en Proyectos Para el Desarrollo de Competencias de Contabilidad: Aplicación en la UPS. Ph.D. Thesis, E.T.S. de Ingeniería Agronómica, Alimentaria y de Biosistemas (UPM), Madrid, Spain, 2022.
50. De los Ríos-Carmenado, I.; Cazorla, A.; Díaz-Puente, J.M.; Yagüe, J.L. Project–Based Learning in Engineering Higher Education: Two Decades of Teaching Competences in Real Environments. *Procedia—Soc. Behav. Sci.* **2010**, *2*, 1368–1378. [CrossRef]

https://doi.org/10.3390/su18031179

51. De los Ríos-Carmenado, I.; Rodriguez, F.; Perez, C. Promoting Professional Project Management Skills in Engineering Higher Education: Project-Based Learning (PBL) Strategy. *Int. J. Eng. Educ.* **2015**, *31*, 184–198.
52. Mesa, F.A.H.; Morales, A.J.M.; Hurtado, P.D. Emprendimiento femenino en América Latina: Una aproximación teórica. *Económicas CUC* **2023**, *44*, 191–234. [CrossRef]
53. Saavedra García, M.L.; Briseño Aguirre, N.D.L.L.; Velázquez Rojas, K.G. Análisis y Evolución del Emprendimiento Femenino en Latinoamérica. *RECAI* **2022**, *11*, 1–29. [CrossRef]
54. Cardella, G.M.; Hernández-Sánchez, B.R.; Sánchez-García, J.C. Women Entrepreneurship: A Systematic Review to Outline the Boundaries of Scientific Literature. *Front. Psychol.* **2020**, *11*, 1557. [CrossRef]
55. Hablich, F.; Bermúdez, C.; Espinoza, E.; Hablich, F.; Bermúdez, C.; Espinoza, E. Determinantes de La Actividad Emprendedora En La Mujer de Hispanoamérica. *Rev. Univ. Y Soc.* **2018**, *10*, 106–112.
56. García Tapias, T. El papel de la mujer dentro del sector terciario en el medio rural de Castilla y León. Estudio de caso: GAL Zona Centro Valladolid. 2022. Available online: https://uvadoc.uva.es/bitstream/handle/10324/58459/TFG_F_2022_103.pdf?sequence= (accessed on 24 May 2025).
57. GESPLAN—UPM. *Guide for the Female Entrepreneur: IPMA International Project Management Competencies-2024*; GESPLAN—UPM: Madrid, Spain, 2024.
58. González Campo, C.H.; Hur tado Ayala, A. Transferencia tecnológica, capital humano y cooperación: Factores determinantes de los resultados innovadores en la industria manufacturera en Colombia 2007–2008. *Inf. Técnico* **2012**, *76*, 32–45. [CrossRef]
59. FAO. Inversión Responsable en la Agricultura y los Sistemas Alimentarios Guía Práctica Para Parlamentarios y Asesores. Available online: https://parlatino.org/wp-content/uploads/2017/09/guia-responsable.pdf (accessed on 29 May 2025).
60. Cachipuendo, C.; Sandoval, C.; Sandoval, J. Sentinel-2A imaging in mapping greenhouse rose production and rainwater harvesting for agricultural irrigation use. *Environ. Res. Commun.* **2024**, *6*, 111005. [CrossRef]
61. Cachipuendo, C.; Requelme, N.; Sandoval, C.; Afonso, A. Sustainable Rural Development Based on CFS-RAI Principles in the Production of Healthy Food: The Case of the Kayambi People (Ecuador). *Sustainability* **2025**, *17*, 2958. [CrossRef]
62. Fernández, R.S.; De los Ríos-Carmenado, I.; González, M.L. Developing Competencies for Rural Development Project Management through Local Action Groups: The Punta Indio (Argentina) Experience. In *International Development*; Appiah-Opoku, S., Ed.; InTech: Singapore, 2017; ISBN 978-953-51-3103-8.
63. Sastre Merino, S. Desarrollo de Capacidades Para el Liderazgo en Proyectos de Desarrollo Rural. Aplicación a Comunidades Aymaras. Ph.D. Thesis, E.T.S.I. Agrónomos (UPM), Madrid, Spain, 2014.
64. Negrillo Deza, X. Los Proyectos de Desarrollo Rural Como Proceso a Través de sus Protagonistas: El Caso de la Coordinadora de Mujeres Aymaras (Puno, Perú). Ph.D. Thesis, E.T.S. de Ingeniería Agronómica, Alimentaria y de Biosistemas (UPM), Madrid, Spain, 2018.
65. Fontana, A.; Sastre-Merino, S.; Baca, M. The Territorial Dimension: The Component of Business Strategy That Prevents the Generation of Social Conflicts. *J. Bus. Ethics* **2017**, *141*, 367–380. [CrossRef]
66. Mur, C.; De Los Ríos-Carmenado, I.; Mejía, J.; Nole, P. Buenas prácticas de gobernanza en los Bancos de Alimentos: Criterios ESG desde el modelo APS-WWP—[Good Governance Practices in Food Banks: ESG Criteria from the APS-WWPModel]. In Proceedings of the Innovación Educativa en los Tiempos de la Inteligencia Artificial, Madrid, Spain, 18–20 October 2023; Servicio de Publicaciones Universidad: Zaragoza, Spain, 2023.
67. Cazorla, A.; Cáceres, A.; Lavalle, C. Toward a Metauniversity for Sustainable Development: Responsible Agriculture Investment and Food Systems. *Sustainability* **2025**, *17*, 9698. [CrossRef]

https://doi.org/10.3390/su18031179

sustainability

MDPI

Article

Sustainable Rural Development Based on CFS-RAI Principles in the Production of Healthy Food: The Case of the Kayambi People (Ecuador)

Charles Cachipuendo [1,*], Narcisa Requelme [2], Catalina Sandoval [3] and Ana Afonso [4]

1 Grupo de Investigación de Ciencias Ambientales (GRICAM), Universidad Politécnica Salesiana, Cuenca 010105, Ecuador

2 Grupo de Investigación de la Leche (GILEC), Universidad Politécnica Salesiana, Cuenca 010105, Ecuador; nrequelme@ups.edu.ec

3 Escuela de Ingenieros Técnicos Superiores en Topografía, Geodesia y Cartografía, Universidad Politécnica de Madrid, Avda. Puerta de Hierro 2, 28040 Madrid, Spain; c.sandoval@alumnos.upm.es

4 Planificación y Gestión Sostenible del Desarrollo Rural-Local (GESPLAN), Universidad Politécnica de Madrid, Avda. Puerta de Hierro 2, 28040 Madrid, Spain; ana.afonso@upm.es

* Correspondence: ccachipuendo@ups.edu.ec

Abstract: Food production faces significant economic, social, and environmental challenges. The Principles for Responsible Investment in Agriculture (CFS-RAI), aligned with the Sustainable Development Goals (SDG), provide a framework for implementing actions and investments in sustainable food production. Farmers are the primary investors in their agricultural activities, playing a fundamental role in the sustainability of rural territories. This study aims to analyze the rural development innovation system implemented by the Fundación Casa Campesina Cayambe and its alignment with the CFS-RAI principles and SDG. Focusing on 129 women producing healthy food within the Kayambi people and adopting a mixed-methods approach, including participant observation, interviews, and focus groups, the research examines the integration of public institutions, academia, and civil society in promoting sustainable rural development. The findings indicate that key innovation processes include a participatory, community-based, and gender-sensitive microcredit system implemented by the Foundation. Additionally, the technical and scientific support provided by the Universidad Politécnica Salesiana enhances investment capacity and knowledge acquisition, strengthening both individual and collective capacities for food production and commercialization. The study concludes that a decentralized and coordinated cooperation system, involving multiple territorial stakeholders, fosters the economic autonomy of female producers while adhering to the principles of CFS-RAI and the SDG.

Keywords: CFS-RAI principles; healthy food production; working with people; agroecology

1. Introduction

The production of healthy food in rural territories faces a series of interrelated challenges across economic, sociocultural, political, technological, and environmental dimensions [1]. These challenges directly or indirectly involve various territorial actors, including community-based producer organizations, the state, academia, and both national and international civil society organizations (NGOs). In many cases, the latter end up replacing the state, which, due to institutional weaknesses, struggles to enforce public policies and secure funding [2]. This situation is particularly evident in low-income countries, where the absence of state support negatively impacts rural development processes [3].

https://doi.org/10.3390/su17072958

In rural territories, agriculture constitutes the primary economic activity for the population. However, state policies often prioritize incentives for agro-export, marginalizing local food production essential for ensuring food sovereignty [4]. This phenomenon exacerbates the vulnerability of rural areas to global market dynamics, leading to competition for productive resources and negatively impacting not only food sovereignty but also economic and social resilience [5].

Historically, rural populations have created spaces of resistance to secure fundamental rights such as access to land and water and to be recognized as active development stakeholders [6]. In response to the challenges of globalization and social, economic, and environmental changes, these populations have demonstrated the ability to adapt and establish alliances with public, private, and academic institutions [7]. These collaborations have facilitated participatory planning for rural development projects focused on sustainability and healthy food production, aligned with the Sustainable Development Goals (SDG) and the Principles for Responsible Investment in Agriculture (CFS-RAI) [8]. An example of this is the work carried out by female food producers in Cayambe and Pedro Moncayo, in northern Ecuador.

International cooperation for development (ICD) has been studied from various perspectives; however, few studies have specifically analyzed financing mechanisms and how they could be optimized to ensure efficient agricultural investments. ICD-funded projects' evaluations are generally financial, without thorough examination of the sustainability of their impacts or the structural changes they may generate. Similarly, projects promote different types of innovations, yet it is essential to identify these innovations, the success factors that drive their adoption by beneficiaries, and how they strengthen synergies among local actors.

There are various studies on the application of CFS-RAI principles in different projects. In Spain, their connection with the rural food industry has been analyzed, showing that they can drive sustainable economic growth and contribute to the revitalization of rural areas [9]. In Colombia, these principles have been reported as a key support tool for actors involved in transforming regions affected by illicit crops, where land legalization, food security, and economic development are considered fundamental pillars for territorial transformation [10]. Similarly, in Ecuador, studies have examined sustainable livestock policies and their alignment with these principles [11].

However, few studies specifically analyze their implementation in indigenous communities. A notable example is a study in Peru that examines the integration of local actions with the CFS-RAI principles in designing a rural development strategy [12]. Likewise, in Ecuador, another study has explored the importance of skills development in strengthening the cooperative system of indigenous artisan women [13].

This research not only contributes to the study of food systems and rural development in indigenous communities but also aims to establish practical recommendations to improve the efficiency of investments and the sustainability of projects focused on sustainable rural development. In this context, the objective of this study is to analyze the innovation mechanisms for sustainable rural development, funded by international cooperation and implemented through investment projects led by the Foundation Casa Campesina Cayambe (FCCC), and their alignment with the CFS-RAI principles and the SDG. The specific objectives were (a) to assess the efficiency of ICD financing mechanisms for responsible investment in sustainable rural development; (b) to identify the innovations in sustainable rural development implemented by the FCCC; and (c) to examine the alignment of the FCCC's actions with the CFS-RAI principles in healthy food production.

This study focuses on food production in the Kayambi people's territory in the Ecuadorian Andes, where communities struggle over land and water access for food production [14]. Since the 1980s, various NGOs have promoted and strengthened agroecology in the region [15]. Through sustained technical support for communities, the FCCC has implemented multiple projects aimed at improving access to and control over productive resources. It has also provided training for producers and established local commercialization spaces (bio-markets). These projects have been funded through ICD, and the two most recent phases (Phase 1: May 2018–October 2021 and Phase 2: April 2022–September 2025) have benefited from the technical and scientific contributions of Universidad Politécnica Salesiana (UPS) as a member of the Network of Universities for the application of the CFS-RAI principles [16]. This local cooperation system presents an innovative proposal for sustainable rural development, aligned with the CFS-RAI principles and the SDG. This case illustrates how rural communities can transform their realities through strategies that integrate diverse actors and resources, promoting development models that merge local needs with global commitments and thereby fostering innovation in sustainable rural development.

The following Section 2 presents a review of the state of the art on the topics related to the research: sustainable cooperation and innovation; governance of food systems; Principles for Responsible Investment in Agriculture; and international cooperation for development. The Section 3 on materials and methods describes the study area, the approach, techniques, and tools used in each phase. The results and discussion on responsible investment in ICD, innovation in sustainable local development, and the CFS-RAI principles and SDG in healthy food production are presented in Section 4. Finally, key findings are presented in the conclusions Section 5.

2. State-of-the-Art Review

This section presents the state of the art on sustainable rural development as a key objective for transforming food systems into rural territories. It also examines the governance of food systems, guided by the CFS-RAI principles and the SDG, as essential frameworks for global governance. These principles ensure that public and private investments are relevant, efficient, and transparent, contributing to the realization of fundamental rights, particularly the right to food. Additionally, the role of international cooperation (ICD) is analyzed in relation to local organizational processes, participation, and the empowerment of key actors.

Sustainable rural development and innovation. Sustainable rural development seeks to improve the quality of life in rural communities through innovative approaches that integrate economic, social, and environmental dimensions. Social innovation has emerged as a crucial strategy for addressing rural development challenges by promoting creative and sustainable solutions [17]. This process incorporates diverse knowledge systems and fosters empowerment through the co-creation of socially innovative initiatives.

Innovation has become a key driver in tackling issues such as food security, climate change, and social inclusion. The integration of advanced technologies, participatory governance models, and inclusive financing mechanisms has been essential in transforming rural production systems and optimizing financial resource management. Key components of innovations such as collaborative economies, participatory decision-making, ethical banking, and technology for social purposes are adaptable to various rural contexts, enhancing their replicability and scalability [18]. Social innovation is based on empathy, systemic thinking, and adaptability, ensuring that solutions are both effective and long-lasting. Rural development strategies based on territorial integration, which consider the interrelationship between agricultural, environmental, and social sectors, have proven to be more successful in promoting sustainability as part of social innovation [19]. Models such as smart territories, which integrate technology and community participation, enable a more effective response to rural challenges by leveraging digitalization and strengthening local governance [20]. In this way, social innovation becomes a powerful tool for enhancing resilience and sustainability in rural environments.

Governance of Food Systems. The governance of food systems remains a subject of ongoing debate. Perspectives such as food sovereignty—defined as the right of people to determine their agricultural and food production policies in accordance with their cultural and nutritional needs—differ from the food security approach, which focuses on ensuring food availability and access. Given that nutrition is fundamental to public health, the need to produce healthy food becomes imperative.

Healthy eating is a polysemic and multidimensional concept, encompassing not only nutritional and biomedical aspects but also sociocultural, economic, political-legal, and environmental dimensions. It is characterized by the following: 1. A variety and complexity of fresh or minimally processed foods; 2. Safe and age-appropriate foods; 3. Culturally accepted foods; 4. Available and accessible food (food security); 5. Respect for community rights (food sovereignty) and compliance with existing laws; and 6. Environmental sustainability. Thus, adopting a holistic approach to healthy eating is a strategic way to ensure food security and sovereignty, environmental sustainability, and agro-food system resilience. Simultaneously, it significantly enhances public health through disease prevention and health promotion [21].

Beyond individual producers, food system governance involves networks of organizations of varying sizes and types. These networks, deeply rooted in the historical legacy of Andean communities, play a crucial role in promoting food sovereignty, environmental sustainability, and local economic development. They have facilitated women's access to and retention in markets, enabled support from public and private institutions, and contributed to rural territorial development.

These networks advocate for sustainable food production practices, strengthen solidarity economies by connecting producers and consumers, and ensure fair pricing through equitable commercialization mechanisms that eliminate intermediaries. Through these networks, women have maintained social cohesion around agroecological production, influencing the development of local public policies that address community-specific realities [15]. Strengthening local governance and developing cooperatives have proven to be effective strategies for enhancing the economic resilience of communities, allowing farmers to increase their bargaining power and achieve long-term financial stability.

Principles for Responsible Investment in Agriculture. As part of global food governance, FAO, in its role of promoting the right to food, introduced the Principles for Responsible Investment in Agriculture and Food Systems (CFS-RAI) in 2014. These principles provide a framework for guiding agricultural and food-related investments that foster food security and sustainable development. The CFS-RAI principles emerged in response to concerns over the negative impacts of certain investments, such as land grabbing and environmental degradation.

Their primary objective is to ensure that investments not only increase productivity but also contribute to inclusive and sustainable rural development, strengthening the rights of smallholder farmers, women, and indigenous communities. The ten CFS-RAI principles are as follows: 1. Contribute to food security and nutrition. 2. Promote sustainable and inclusive economic development and poverty eradication. 3. Foster gender equality and women's empowerment. 4. Enhance youth participation and empowerment. 5. Respect land tenure, fisheries, forests, and access to water. 6. Sustainably manage natural resources, increase resilience, and reduce disaster risks. 7. Respect cultural heritage and traditional knowledge, and support diversity and innovation; 8. Respect cultural heritage and traditional knowledge while supporting diversity and innovation. 9. Promote safe and healthy agricultural and food systems. 10. Establish inclusive and transparent governance structures, processes, and grievance mechanisms. The CFS-RAI principles are strongly aligned with the Sustainable Development Goals [22].

Responsible investment, according to the CFS-RAI principles, is based on three fundamental pillars: social inclusion, ensuring that vulnerable groups, such as women and smallholder farmers, actively participate in and benefit from investment processes; environmental sustainability, focusing on practices that preserve natural resources and biodiversity, promoting resilience to climate change; and capacity building, empowering rural communities through knowledge transfer, access to technology, and technical support, enabling autonomous and sustainable development.

International Cooperation for Development (ICD). ICD plays a crucial role in ensuring that agricultural investments are both efficient and responsible. It is important to recognize that ICD has undergone significant evolution. Initially, it was structured as economic assistance under the Marshall Plan, later shifting toward cooperation focused on industrialization and economic modernization in developing regions, particularly in Latin America [23].

Over time, ICD has transitioned from a dependency-based, aid-driven approach to more participatory and sustainable models, driven by initiatives such as the Millennium Development Goals and the Sustainable Development Goals (SDG) within the 2030 Agenda for Sustainable Development. However, structural challenges persist, including financial dependency, aid conditionality, and the limited autonomy of recipient countries in managing resources [24].

ICD has shifted towards decentralized models that involve multiple stakeholders, including Indigenous communities. The participation of these groups in development cooperation has been strengthened by international legal frameworks, such as the United Nations Declaration on the Rights of Indigenous Peoples, which emphasizes their right to self-determination and sustainable development [25]. Cooperation strategies in indigenous contexts focus on strengthening community self-management through integrated programs that address socioeconomic and cultural aspects.

Sustainable rural development requires ICD and multisectoral alliances that integrate producers, academic institutions, the private sector, and civil society. These collaborative networks facilitate information exchange, the formulation of policies tailored to territorial needs, and social innovation, reflected in a planning system based on the theory of liberating action [26]. This theory serves as a fundamental tool for understanding human behavior in various contexts, including rural development. Through the knowledge-action-communication framework, stakeholders identify and propose solutions to their challenges [27]. In many cases, rural development processes lack direct support from national governments, underscoring the crucial role of ICD, NGOs, and academia in driving sustainable initiatives.

3. Materials and Methods

3.1. Study Area

The research was conducted in the territory of the Kayambi people, which encompasses the Cayambe and Pedro Moncayo cantons in the Pichincha province, Ecuador. Hydrologically, this region is part of the Pisque River Basin. The territory is located at an altitude of 2500 to 3300 m above sea level, with an average temperature of 16 °C and annual precipitation of 700 mm. Three main production systems have been identified in the area: (i) greenhouse rose production for export, covering 5473.8 ha, which is the region's primary economic activity and main source of employment [28]; (ii) bovine milk production, mainly located in the higher-altitude areas and indigenous territories; and (iii) family farming, primarily focused on the production of vegetables, cereals, and potatoes [29].

This territory has been under the influence of the FCCC since 1994. With financial support from ICD originating from civil society, the FCCC implements development projects focused on production, health, education, and community organization. Over the past six years, the foundation has executed two projects with a total investment of $943,385.80, primarily aimed at healthy food production through agroecological practices and the traditional Andean chakra system, benefiting 129 female producers organized into 10 groups. Historically, the region has experienced struggles over access to land and water, where community organization has played a fundamental role in the process (Figure 1).

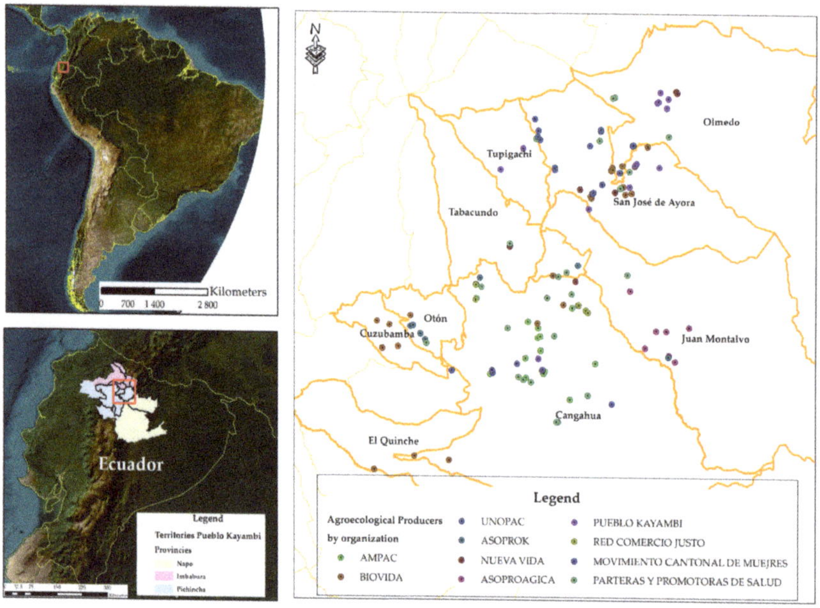

Figure 1. Location of healthy food producers in the territory of the Kayambi people.

3.2. *Methodological Approach*

This research employs a mixed-methods approach at a descriptive and correlational level, allowing for an analysis of the relationships between studied variables while providing an in-depth description. The conceptual framework is based on the Working with People (WWP) metamodel [30], which promotes participation, social learning, and the integration of ethical values in stakeholder engagement in development projects. This model connects knowledge and action toward a common objective, integrating technical results, the production of goods and services, and the human value contributed by project participants [31]. The WWP model consists of three key components: (i) Political-contextual: facilitates relationships among stakeholders and manages common objectives. (ii) Ethical-social: promotes responsible values and practices to build trust. (iii) Technical business: mobilizes resources, ensures quality, negotiates among stakeholders, and manages project risks.

The interaction of these components fosters social learning, connecting them through collective processes that transform knowledge into concrete and effective actions. Cazorla et al. [30] emphasize the need to shift planning from theory to practice, promoting sustainable transformations that address real contextual needs in project implementation.

3.3. Data Collection, Analysis, and Interpretation

The implemented methodology ensures a robust and multidimensional approach. To analyze the interventions by combining qualitative and quantitative methods, three stages of data collection and analysis were established (Figure 2). For each stage, techniques such as document review and institutional database analysis were used while maintaining confidentiality. Participant observation, interviews, and focus groups were conducted with the informed consent of the participants. Qualitative data were processed using NVivo (14) software, while quantitative data were analyzed using SPSS software, version 30.0.0.0 (172).

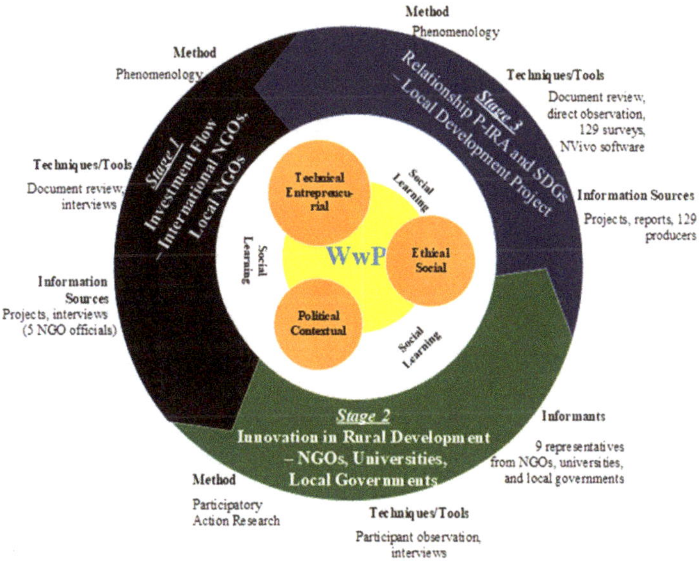

Figure 2. Stages, techniques, and tools for data collection and analysis using the WWP approach.

In the first stage, investment flows from civil society through NGOs were identified. Semi-structured interviews were conducted with officials from three NGOs, local government representatives, and academics from UPS, totaling nine key informants. The collected data were analyzed by categorizing common terms. Additionally, a documentary review of NGO-led projects was carried out to determine investment amounts, disaggregated by specific budget categories.

The second stage focused on identifying innovations in rural development. A participatory action research approach was applied, using participant observation and focus groups. Interviews were conducted with officials from decentralized autonomous governments (DAG), NGO technicians, UPS academics, and female leaders of beneficiary groups.

In the third stage, project actions were linked to the CFS-RAI principles and SDG. A phenomenological method was used, beginning with a literature review to define the concept of healthy food production. This process narrowed down the analysis to three key aspects: nutrition/health, social well-being, environmental sustainability, and food system governance.

Thus, categories, subcategories, and keywords were identified using NVivo (14) software to process document reviews and interviews. The alignment with the CFS-RAI principles was established based on information from projects, institutional reports, and data collected from all 129 project beneficiaries. These data were analyzed using SPSS software version 30.0.0.0 (172), applying statistical analyses based on frequencies, percentages, and means.

4. Results and Discussion

For our analysis, healthy food production is defined as a system that preserves the nutritional benefits of food and ensures equitable access to nutritious food. It also minimizes environmental impact, considering ecosystem sustainability, as well as the health and well-being of both producers and consumers, within an effective governance framework for local food systems. The following section presents the research findings.

4.1. Responsible ICD Investment in Healthy Food Production and Sustainable Rural Development

In the territory of the Kayambi people, rural development projects are implemented with funding from the ICD. This cooperation provides financial resources directed toward impoverished or emerging sectors, which are channeled through civil society organizations, businesses, clubs, and state entities. Funds are transferred to beneficiaries through state institutions in the recipient country, working in partnership with local foundations.

In the case of the FCCC, an innovative approach has been proposed for project implementation, with a particular focus on healthy food production (Figure 3).

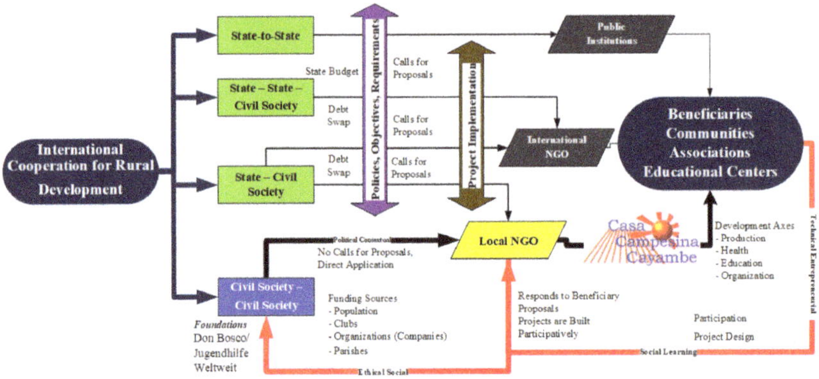

Figure 3. Structure, financing mechanisms, and project implementation by the FCCC with ICD.

As shown in Figure 3, there are four mechanisms for channeling financial resources: (i) State-to-State: Funds from the donor state are transferred to recipient states, which are responsible for implementing rural development projects. (ii) State-to-State and Civil Society (NGOs): Donor state funds are combined with those of the local recipient state, and project implementation is delegated to a local NGO. (iii) State-to-Civil Society: Funds are transferred from the donor state directly to a local NGO. (iv) Civil Society-to-Civil Society: Funds from private donors, businesses, community organizations, parishes, and clubs are collected by an NGO in the donor country and then transferred to a local NGO in the recipient country, as is the case with the FCCC. The civil society-to-civil society cooperation model developed by the FCCC operates without pre-established agendas or rigid conditions, allowing for the design of local development proposals with direct participation from beneficiary groups. This approach incorporates community wisdom, based on liberatory action theory [32]. As a result, the project is positively received, fostering stakeholder empowerment through a joint planning process aligned with common objectives [33] (Figure 4).

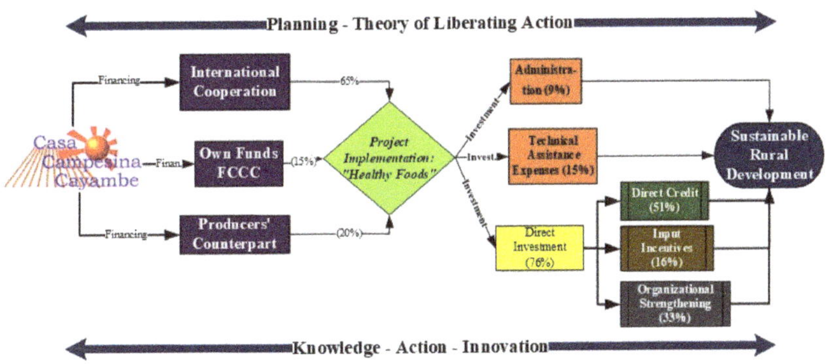

Figure 4. Financing, resource flow, and allocation in the FCCC "Healthy Foods" project.

Another fundamental element is the flexibility in project execution, which does not imply disregarding implementation regulations, schedules, and budgets but rather allows for adaptability to unforeseen changes. As shown in Figure 4, of the total funds raised, 10% remain with the donor organization for administration, 18% are allocated to the local NGO to cover administrative expenses and technical assistance, and 72% of the funds are directly invested in the beneficiary group. This cooperation model enhances budget execution efficiency due to the fundraising system, minimal bureaucracy within the cooperating NGO, the counterpart contributions from the local NGO, and the ability to form partnerships with other local actors. In essence, two cooperation systems coexist: international cooperation and local cooperation, both operating under a decentralized cooperation framework [34].

The direct investment of cooperation funds is focused on the following: generating individual and collective/community knowledge, strengthening organizational structures; providing incentives through productive inputs, such as seedling and seed distribution, which serve as seed capital; and investing in productive infrastructure through microcredit, which has proven to be an efficiently utilized production tool in rural sectors [35].

International cooperation has played a fundamental role in the development of Indigenous communities by financing various projects. The approaches, models, and conditions of these interventions vary depending on the context and the characteristics of the beneficiaries. In the specific case of indigenous communities, the application of a participatory approach throughout the entire planning cycle stands out. This approach recognizes communities as equal partners, allowing them to actively participate in decision-making regarding their own development [36]. Moreover, it entails the recognition of indigenous peoples' and nationalities' rights to benefit from their resources and to define their development priorities, thereby ensuring a more equitable and sustainable model [37].

4.2. Innovations in Sustainable Rural Development Implemented by FCCC

The current trend in International Development Cooperation focuses on innovation linked to sustainability dimensions. In the implementation of healthy food projects, four types of innovation have been identified: social, economic, technological, and food governance innovation. In these innovation processes, the participation of diverse territorial actors is fundamental, including beneficiary groups, academia (such as UPS), local decentralized autonomous governments (DAGs), and civil society, represented by foundations, businesses, and second-tier organizations. Based on action theory [38], these actors engage in planning processes through knowledge dialogue and practical implementation, leading to learning transformations that drive innovation [39]. This process embodies the knowledge-action-innovation approach.

Innovation in development also requires a shared vision based on common objectives, ensuring respect for the specific roles of each institution. These roles include research, training, financing through credit systems, technology transfer, and public policy development [33]. This interinstitutional coordination facilitates the creation of enabling spaces for promoting the production and distribution of healthy food while also fostering inclusive economic opportunities within the framework of the popular solidarity economy [40]. This is particularly beneficial for women in the territory, contributing to the preservation of food sovereignty. The implemented actions and innovation components align with a global governance process, contributing to the achievement of the SDG under the CFS-RAI principles (Figure 5).

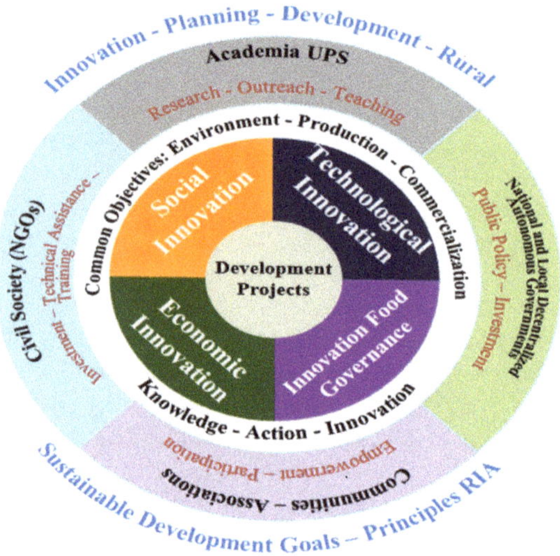

Figure 5. Synergies between innovation, planning, and development actors.

4.2.1. Social Innovation

Social innovation consists of creating producer networks through groups and/or associations to improve access to production resources and markets. It also reinforces local governance, citizen participation, participatory budgeting, and accountability, facilitating the enforcement of rights, food sovereignty, and project empowerment. Through organization, women are able to implement healthy food distribution systems for potential consumers, using innovative approaches such as bio-markets and food basket sales, thereby creating sustainable value chains and inclusive community business models [41]. The use of information technologies to improve commercialization is a key innovation today. Eighty percent of beneficiaries use WhatsApp, and UPS provided a WhatsApp Business course to develop alternative sales strategies, leveraging new communication and marketing systems. This strengthens rural associativity through digital platforms [42].

4.2.2. Economic Innovation

In project development, economic innovation is observed from various perspectives. During the production phase, the reuse of inputs such as solid and liquid bionutrients—through a combination of ancestral and contemporary practices—aligns with the concept of the circular bioeconomy [43]. This approach, based on waste reduction and the efficient use of resources, helps lower production costs and enhance sustainability [44]. In the commercialization of healthy foods, a phenomenon of collaborative economy emerges, where producers apply fundamental principles of social and solidarity economy: solidarity, unity, reciprocity, and complementarity [45]. These principles are also reflected in market spaces, where consumers demonstrate increased awareness of pesticide-free food consumption, thereby promoting fair trade practices that strengthen local markets and ensure more equitable prices.

One of the key economic innovations in sustainable rural development is the microcredit system implemented by the FCCC. Producers perceive this financing model as a production tool that expands their opportunities to increase food production. The loans are aimed at acquiring equipment, implementing irrigation systems, and constructing greenhouses and barns for small livestock farming.

For the microcredit system to function effectively and remain sustainable over time, several key factors must be considered. These include securing financial resources, shared management between beneficiaries and the FCCC Foundation, and regulatory oversight by participating in intercommunity organizations represented in credit committees. Additionally, decision-making regarding loan eligibility, loan monitoring, and ensuring repayment accountability are essential aspects for guaranteeing the system's long-term sustainability (Figure 6).

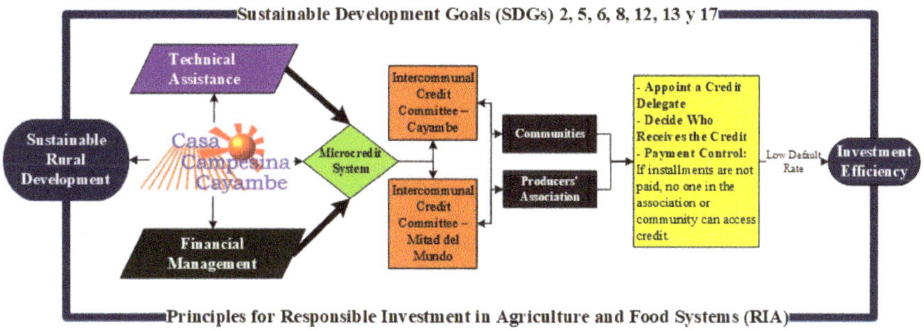

Figure 6. The financial management model of the FCCC microcredit system.

There are two credit committees involved in the formulation of regulations and oversight of invested resources, while the FCCC is responsible for the financial management of these funds. Each committee is composed of various communities within the territory, and each community elects a representative, known as the credit delegate.

This system differs from other microcredit models due to its socially managed structure, where the community plays a central role in governance. This approach ensures the long-term sustainability of resources from ICD while enhancing investment efficiency and loan recovery [46].

Access to credit is determined by the collective and community-based decision-making process of community members or producer groups. Loan approval depends on the applicant's behavior within the community and/or group, requiring compliance with community life regulations. In cases of non-payment, no other member can access financing until the outstanding debt is settled, fostering co-responsibility in credit management. This community-based control mechanism has resulted in low delinquency and default rates, reaching just 1% among the 129 producers. Furthermore, this financial governance model strengthens producers' autonomy and optimizes the use of resources from ICD, aligning with inclusive finance approaches and territorial development strategies [35].

4.2.3. Technological Innovation

Technological innovation in healthy food production is dynamic and constantly evolving due to scientific advancements and their interaction with ancestral knowledge. The adaptation of technologies is carried out in collaboration with project beneficiaries and with the support of UPS. Each stakeholder plays a specific role, facilitating interaction and knowledge exchange to achieve common goals. However, these innovations progress slowly due to various factors, such as uncertainty about potential reductions in production. Nevertheless, these perceptions can be managed by applying the diffusion of innovations theory through training, technical assistance, and experience exchange [27].

Among the main innovations in the production stage is the use of bionutrients and biostimulants, which are formulated from resources available within the chakra. Additionally, the promotion of crop association and the recovery of local products contribute to the conservation of agrobiodiversity—fundamental principles of agroecology, regenerative agriculture, and the Andean chakra system [45]. The integration of small animal husbandry is key to closing the healthy food production cycle. The combination of plant and animal biodiversity not only diversifies the range of products but also improves profitability. Another innovative process is agricultural production planning based on the lunar calendar phases. For example, research has documented that sowing in September during the "new moon" phase positively influences crop yield and development, which are technological practices applied in biodynamic agriculture and the Andean chakra system [47].

To optimize water use and enhance soil sustainability, drip and sprinkler irrigation systems have been implemented, increasing efficiency to 96% and 65%, respectively [48]. In these systems, pressurization is achieved through gravitational energy, avoiding the use of external energy sources. Additionally, rainwater harvesting has been integrated, covering up to 36% of water needs for greenhouse crops [28]. Through its research, UPS has developed strategies that reduce pressure on surface and groundwater sources, increasing resilience to the effects of climate change and water stress. This highlights how communities sustainably manage shared resources such as water and soil [49]. It has been shown that social appropriation of knowledge goes beyond scientific dissemination—it requires community participation to develop concrete solutions that directly benefit rural populations [50]. Technological innovation has transformed rural agriculture, optimizing production efficiency and strengthening agri-food systems against climate change and growing food demand [51].

In Latin America, improvements in traditional maize and potato varieties have increased yields and adaptability, while seed banks and in situ conservation have preserved genetic diversity [52]. In Africa, sustainable practices such as agroecology and biofertilizer use have revitalized traditional crops, improving food security and climate resilience [45]. Processing innovations have added value to crops like quinoa and amaranth, enhancing marketability and preservation [53]. Precision agriculture, adapted to traditional crops, has optimized irrigation and fertilization through sensors and drones, increasing productivity without compromising cultural characteristics [54]. These innovations highlight technology's potential to sustainably strengthen rural agriculture.

A fundamental aspect of innovation in rural development is knowledge transfer and capacity building. The implementation of agricultural extension programs and training enables farmers to adopt appropriate technologies and improve production practices sustainably. The collaboration between local actors has been crucial in developing training programs adapted to local needs, promoting a participatory approach that integrates traditional knowledge with technological innovation.

4.2.4. Innovation in Local Food Governance

The management of public policies related to territorial planning, production promotion, and the creation of commercialization spaces that integrate diverse stakeholders can be considered an innovation in food governance. This approach allows for an analysis of equity in resource access, empowerment of community association, and their relationship with public policies. In this regard, a multi-level governance process has been observed [55], facilitating an understanding of the interactions between different levels of government (national, provincial, and municipal) and producer organizations and academia in the management of bio-inputs, training, and regulatory frameworks for the use of public spaces.

In this context, provincial and municipal decentralized autonomous governments (DAGs), within their jurisdiction, provide bio-inputs and promote training processes in coordination with academia and the FCCC.

Thanks to the efforts of female producers and their demands to local authorities, public policies have been established through ordinances that allow for the use of physical spaces in various locations. In these spaces, bio-markets have been implemented, enabling direct consumer access to healthy food with quality assurance [15]. However, it is crucial to ensure continuous improvements in infrastructure and management to optimize long-term functionality and sustainability.

Thus, food governance encompasses processes, regulations, and institutions that govern food systems, involving government actors, civil society, the private sector, and academia in decision-making [11]. This approach enables the implementation of public policies related to territorial planning, equitable access to productive resources, and community participation in food commercialization.

From the perspective of food sovereignty theory, emphasis is placed on the importance of producers having control over their productive resources and markets [56]. The struggle of women producers to obtain commercialization spaces can be analyzed within this framework, highlighting their autonomy from large agro-industrial markets. In this regard, it is possible to consider common goods and the community-based management of natural resources.

One of the main challenges for the sustainability of healthy food production is the aging of producers. To address this issue, a training program in sustainable agriculture has been implemented for young family members of project beneficiaries. This initiative not only facilitates knowledge transfer but also creates economic opportunities for younger generations, supporting generational transition and youth entrepreneurship.

Finally, equitable resource access and community empowerment reflect an attempt to democratize access to sustainable technologies, aligning with the food sovereignty approach. However, it remains necessary to discuss whether these measures are sufficient or if institutional barriers persist, limiting the empowerment of communities and producer associations.

4.3. CFS-RAI Principles and SDG in Healthy Food Production

The actions implemented by FCCC for each type of innovation, along with their alignment with CFS-RAI principles and the achievement of the relevant SDG in ongoing projects, are presented in Table 1.

From the perspectives of nutrition, environmental sustainability, social well-being, and governance, healthy food production has been addressed by the FCCC through various actions aligned with CFS-RAI Principle 1, which focuses on availability, acceptability, and quality of food, ensuring that it is safe and healthy. Essentially, the goal is to consolidate the right to food and health. Increasing the consumption of these foods stimulates agroecological production among female producers, ensuring both their health and that of the population. Likewise, these actions contribute to raising awareness about food security and nutrition, providing consumers with evidence-based knowledge about food types, nutritional content, and the socio-organizational processes behind healthy food production. While nutraceutical products are currently trending, the approach is not necessarily to introduce new products but rather to rescue and promote traditional foods with these properties, such as quinoa, mashua, and melloco.

Table 1. SDG and CFS-RAI principles in healthy food production in FCCC projects.

<table>
<tr><th rowspan="2">Type of Innovation</th><th colspan="2">Healthy Food Production</th><th rowspan="2">Actions Implemented in the FCCC Project</th><th colspan="2">Relation with</th></tr>
<tr><th>Scope</th><th>Elements</th><th>CFS-RAI</th><th>SDG</th></tr>
<tr><td rowspan="5">Social</td><td rowspan="5">Nutrition—Health</td><td rowspan="2">Nutritional Biodiversity</td><td>10 traditional products offered at bio-markets: morocho, wheat, barley, fava beans, machica, uchujaku, melloco, mashua, sweet potato, jicama.</td><td>RAI 1
RAI 8</td><td>SDG 2</td></tr>
<tr><td>Consumer access to healthy food options at 7 bio-markets.</td><td>RAI 1
RAI 8</td><td>SDG 2</td></tr>
<tr><td>Consumer Rights</td><td>Promotion of healthy and responsible consumption: 5 posts on Facebook, 5 on TikTok, 7 videos on YouTube, 4 posts on X (Twitter), 1000 fabric bags, and 1000 recipe booklets.</td><td>RAI 10</td><td>SDG 16</td></tr>
<tr><td>Producer Health</td><td>Family self-consumption: 60% of production is sold at bio-markets, 28% for self-consumption, and 12% is sold on farms.</td><td>RAI 1
RAI 8</td><td>SDG 8</td></tr>
<tr></tr>
<tr><td rowspan="10">Economic</td><td rowspan="10">Social Well-being</td><td rowspan="2">Economic Accessibility</td><td>Each bio-market has a steady flow of 10–20 permanent consumers.</td><td>RAI 1</td><td>SDG 2</td></tr>
<tr><td>Variety of fresh or minimally processed foods: mixed vegetables, toasted grains (savory and sweet), fried fava beans, toasted fava beans, bread.</td><td>RAI 1</td><td>SDG 8</td></tr>
<tr><td>Working Conditions</td><td>Women producers dedicate 4 to 6 h daily to farming activities on their plots.</td><td>RAI 2
RAI 3</td><td>SDG 5
SDG 8</td></tr>
<tr><td rowspan="2">Training</td><td>28 training events on organic production, agroforestry, soil and water management, post-harvest handling, livestock farming, cooperative marketing, family bioeconomy, organizational strengthening, and citizenship-rights-gender topics.</td><td>RAI 3
RAI 4</td><td>SDG 4
SDG 5</td></tr>
<tr><td>Knowledge exchange visits to agroecological farms, fruit farms, certified organic livestock farms, and guinea pig breeding and processing facilities.</td><td>RAI 3
RAI 4
RAI 7</td><td>SDG 4
SDG 5</td></tr>
<tr><td>Organizational Structures</td><td>Strengthening women's producer groups: 2 organizational capacity-building workshops, 1 workshop on citizenship-rights-gender topics.</td><td>RAI 3
RAI 4</td><td>SDG 5</td></tr>
<tr><td>Fair Trade</td><td>Local agroecological production: 8% cultivate 50–100 m^2; 48% cultivate up to 500 m^2; 26% up to 1000 m^2; 11% up to 5000 m^2; 7% over 10,000 m^2.</td><td>RAI 2</td><td>SDG 13</td></tr>
<tr><td>Inclusive Finance</td><td>55 microcredits granted, with an average loan amount of $2500 per producer.</td><td>RAI 2
RAI 3</td><td>SDG 5</td></tr>
<tr><td>Job Creation/Gender</td><td>72% of women producers receive help from their children or spouses for agroecological farming activities.</td><td>RAI 2
RAI 3</td><td>SDG 5
SDG 8</td></tr>
</table>

Table 1. *Cont.*

Type of Innovation	Healthy Food Production		Actions Implemented in the FCCC Project	Relation with	
	Scope	Elements		CFS-RAI	SDG
Technological	Environmental Sustainability	Use of Natural Resources	Production of bio-inputs: 45% of producers use 4 types of biofertilizers, 27% use 3, 22% use 2, and 6% use 1 type.	RAI 6	SDG 12
			Irrigation systems: 69% of women use sprinkler irrigation, 21% use drip irrigation, and 10% use flood irrigation.	RAI 5	SDG 12
		Pollution Reduction	Training: 2 workshops on integrated pest management and ethnoveterinary practices.	RAI 8	SDG 12
		Biodiversity	Diversified production: 13% cultivated 6–12 products, 26% cultivate 13–18, 30% cultivate 19–24, 23% cultivate 25–30, and 6% cultivate 31–36 products.	RAI 8	SDG 12
			Seed conservation and exchange: 1 annual seed exchange fair.	RAI 8	SDG 12
		Animal Welfare	Producción de especies menores:77%—Cuyes; 61%—Pollos; 40%—Gallinas ponedoras; 53%—Porcinos, 42%—Ovinos; 51%—Bovinos.	RAI 2 RAI 8	SDG 12
		System Resilience	Production of small livestock: 77% raise guinea pigs, 61% raise chickens, 40% have laying hens, 53% raise pigs, 42% raise sheep, and 51% raise cattle.	RAI 5	SDG 11
Governance	Food Systems Governance	Multi-level and Decentralized Governance	Strengthening the Producers' Council and Cantonal Women's Movement: 2 organizational capacity-building workshops, 1 workshop on citizenship-rights-gender topics.	RAI 9	SDG 17
			Project activity discussions with DAG Pichincha, DAG Cayambe, DAG Pedro Moncayo, DAG San José de Ayora, and the Ministry of Agriculture and Livestock.	RAI 9 RAI 10	SDG 16
		Local Development	Local policy (ordinance): granting public spaces for agroecological fairs, 39% of producers meet Participatory Guarantee System Green Card standards.	RAI 2	SDG 8

CFS-RAI Principle 2 aims to contribute to sustainable and inclusive economic development and poverty eradication. In this regard, FCCC strengthens women's agricultural production, with 90% of them dedicated exclusively to agroecology, 7% being homemakers, and 3% involved in small-scale flower production, livestock farming, and midwifery. The daily work on the chakras ensures that producers' health conditions are safeguarded, as they are not exposed to agrochemicals, and the food they consume is safe. Additionally, these women engage in embroidery, childcare, elderly care, and household responsibilities. Their agroecological activity generates monthly incomes of $250–300, used for personal expenses and family contributions. However, productive, reproductive, and community work is still undervalued within their families, and many do not have decision-making power over their income—only 31% do.

Bio-markets serve as employment spaces for these women, generating essential household income. The innovation in governance within the food system, facilitated through cooperation between international financial institutions and NGOs, supports dignified employment for women, aligning with their family and social dynamics. This innovation highlights how synergies can improve livelihoods in rural communities.

It is worth noting that 56% of these women have primary education, 17% have secondary education, and only 2% have higher education. Despite having basic education, many have undergone training in various topics, boosting their self-esteem. Initially, they considered themselves inferior to men (heads of households), but now they recognize themselves as heads of their own homes and productive agents. The FCCC's actions contribute to gender equality and women's empowerment (CFS-RAI Principle 3) through training processes, enabling them to access theoretical knowledge that strengthens their empirical practices, leading to technical innovations. These actions also align with CFS-RAI Principle 2.

Women have also fought for the recognition and visibility of their ancestral knowledge, integrating it with innovation to develop better production techniques, manage zoogenetic resources and biodiversity within farms and chakras, and maintain traditional diets while using medicinal plants (CFS-RAI Principle 7). Through their experience, they pass agroecological knowledge and chakra farming practices on to their children, not only as a means of ensuring food sovereignty but also as a potential source of employment for young people (CFS-RAI Principle 4). This has been reinforced through the training programs within the project. Leveraging their innovativeness and technology skills, young people could establish thriving businesses in emerging value chains, keeping them in rural areas, enhancing agricultural value, ensuring employment, and reducing rural-to-urban migration [57].

Sixty percent of women have a savings account, and 40% also belong to a community savings or credit bank. Fifty-eight percent access loans averaging $2555–5000, making this a mechanism for maintaining continuous production. Fifty-eight percent invest in small livestock, 24% invest in greenhouses, and 5% invest in irrigation infrastructure. This has significantly contributed to improving their productive capacity and consolidating their economic autonomy. As for individual productive assets, 71% own land, 75% have irrigation infrastructure, 48% have equipment and tools, 60% raise small livestock, and 40% raise large livestock. It can be said that they have a varying level of economic autonomy, conditioned by asset ownership. However, regarding their control over these assets, there is no evident need for individual decision-making, as they are considered family property, even if acquired by the women themselves. The financial assets obtained through the organization have fostered economic, social, and political autonomy, allowing them to build collective empowerment.

In the context of the climate crisis, access to water is becoming increasingly critical, which is why water use efficiency in food production is one of the project's objectives. While sprinkler and drip irrigation systems optimize water use, assessing the water footprint in food production remains pending to identify new ways to reuse water and prevent contamination. These practices are related to CFS-RAI Principle 6, which focuses on conserving and sustainably managing natural resources, increasing resilience, and reducing disaster risks. Additionally, in agroecosystems incorporating small livestock farming, women apply ethnoveterinary practices to ensure the health of both animals and themselves, preventing zoonotic diseases and significantly reducing the use of conventional medicines. Similarly, extensive farming systems guarantee animals' freedom to enjoy comfort, increasing productivity.

Regarding governance (CFS-RAI Principles 9 and 10), the FCCC promotes training in rights so that women, by knowing them, can propose public policies and advocate for them based on their interests, making their social contributions visible and participating in decision-making spaces at the local level as organized political subjects. The internalization of rights, both individually and collectively, strengthens unity and common objectives. CFS-RAI Principle 9 is also reflected in the transparency mechanisms of the FCCC's inclusive finance system, where communities manage this mechanism, which they have strengthened and maintained for many years, serving as a replicable model in similar contexts.

With the goal of improving their community and family situations, women producers have organized themselves to ensure full and effective participation with equal leadership opportunities in economic, political, and public life. They acknowledge that it is an ongoing process that requires strengthening. With the creation of the Cantonal Women's Movement and their collective work, women have begun to generate their own economy with a more informed perspective, which empowers them and helps them revalue themselves, not only as producers but also in reclaiming rights, culture, and knowledge. All project actions aim to promote women's participation in broader organizational processes, where empowerment, linked to CFS-RAI Principle 3, is reflected in political advocacy, particularly in proposing local policies such as the ordinance on the use of public spaces for bio-markets, which is in the implementation phase and requires follow-up and evaluation.

Although CFS-RAI Principle 10 was not explicitly outlined in the project design, it will be incorporated into its evaluation. This is essential to assess the fulfillment of deliverables, effects, and impacts through data and compliance with baseline indicators. Complying with this principle is also directly related to general project planning meetings, where actions can be rescheduled if necessary to optimize resource use. This represents a social innovation in action planning through social learning. Finally, by sharing progress and projections with public institutions, CFS-RAI Principle 10 is applied, as transparency in securing contributions from these entities has strengthened the territorial agroecological proposal.

The CFS-RAI principles underpin the achievement of the SDG. In the FCCC project, they contribute to SDG 2 (Zero Hunger) by promoting investments that enhance food security and improve nutrition in rural populations. They also support SDG 5 (Gender Equality) by emphasizing the equitable participation of women in decision-making, access, and efficient use of resources (SDG 12). Furthermore, they strengthen SDG 17 (Partnerships for the Goals) by fostering cooperation between governments, the private sector, international organizations, and local communities, ensuring that investments are implemented inclusively and in a coordinated manner.

5. Conclusions

The research shows that the following components are essential for the rural development innovation system in the production of healthy foods implemented by the Fundación Casa Campesina Cayambe and their alignment with the CFS-RAI principles and SDG.

1. Community microcredit model in healthy food production. The microcredit system managed by the FCCC has proven to be a key tool in strengthening healthy food production within the Kayambi people's territory. This model, based on community co-responsibility, allows producers to access financing without relying on traditional financial intermediaries, significantly reducing delinquency rates and increasing resource efficiency. The loans are primarily used for acquiring agroecological inputs, implementing irrigation systems, building greenhouses, and purchasing equipment. Unlike conventional financing schemes, this model fosters self-management and collective decision-making, strengthening women's economic autonomy and ensuring financial stability in rural areas.

2. Innovation and diversification in healthy food production. The implementation of agroecological technologies and the recovery of ancestral knowledge have been fundamental in enhancing the sustainability of the local agri-food system. Key innovations include the production of bio-inputs using local resources, crop diversification through the reintroduction of traditional varieties, the integration of small livestock farming, and the adoption of efficient irrigation systems. Additionally, the strengthening of short supply chains, such as bio-fairs and direct sales through digital platforms, has improved market

access without intermediaries. These strategies have not only increased profitability but also ensured the availability of healthy food for the local population, aligning with the SDG and CFS-RAI principles.

3. Strengthening territorial governance and women's participation. The collaboration between various territorial actors, including academia, local governments, and community organizations, has contributed to consolidating a participatory governance model in healthy food production. The FCCC, in partnership with UPS and DAGs, has promoted dialogue spaces and joint planning, as well as the design and implementation of public policies favorable to agroecology. Women producers have played a central role in this process, improving access to productive resources through networks, strengthening community leadership, and actively participating in decision-making on rural development. These actions have driven women's empowerment, ensuring that female producers are recognized not only as economic agents but also as political actors with influence in territorial management.

4. Innovation for sustainable rural development. Innovation has become a central element in sustainable rural development, encompassing social, economic, technological, and food governance innovations. Strengthening cooperation networks between communities, universities, and NGOs is key for knowledge exchange and agroecological and financial training. The implementation of decentralized financing mechanisms, such as microcredit, ensures economic autonomy without dependence on traditional banking. Additionally, digitalization can optimize commercialization and training through virtual platforms. For its expansion, it is crucial to adapt the model to each territory through participatory planning that integrates ancestral knowledge and local practices.

5. International and local cooperation in sustainable rural development. The FCCC experience demonstrates that a decentralized cooperation model, based on the collaboration between local and international actors, can significantly impact the transformation of rural agri-food systems. Unlike traditional cooperation, which often imposes external agendas, ICD—centered on strengthening local capacities—has fostered greater autonomy in decision-making and better project adaptation to territorial needs. This approach has facilitated the implementation of innovative strategies in production, commercialization, and financing, ensuring long-term sustainability while addressing the economic, social, and environmental challenges of rural communities. Furthermore, aligning these initiatives with the SDG and CFS-RAI principles reinforces their impact on food sovereignty, gender equity, and climate resilience, consolidating a replicable model for similar contexts.

However, one of the biggest challenges to the sustainability of agroecological production and food sovereignty is the low participation of young people in agriculture. The lack of access to productive resources and the perception of agriculture as an unprofitable activity have reduced younger generations' interest in continuing these practices. Nevertheless, the FCCC experience demonstrates that when opportunities for training, financing, and market access are created, young people can play a key role in revitalizing the agri-food sector.

Finally, it is recommended to further investigate the viability of the ICD model in different contexts, the long-term sustainability of projects, and the integration of young people into agroecological production to ensure its continuity and evolution.

Author Contributions: Conceptualization, C.C. and N.R.; methodology, C.C. and C.S.; formal analysis, A.A.; investigation, C.C., C.S. and N.R.; writing—original draft preparation, C.C. and N.R.; writing—review and editing, C.C., N.R. and A.A. All authors have read and agreed to the published version of the manuscript.

Funding: This research received no external funding.

Institutional Review Board Statement: This study is non-interventional, as it relies exclusively on interviews/questionnaires that do not pose any physical or psychological risks to participants. According to our internal ethical guidelines and regulations, this type of research is exempt from review by the Research Ethics Committee.

Informed Consent Statement: Ethical review and approval were waived for this study because all interviewees provided their verbal consent prior to participating in the interviews.

Data Availability Statement: Data are contained within the article.

Conflicts of Interest: The authors declare no conflicts of interest.

Abbreviations

The following abbreviations are used in this manuscript:

CFS-RAI	Principles for Responsible Investment in Agriculture
SDG	Sustainable Development Goals
NGO	International Civil Society Organizations
ICD	International Cooperation for Development
FCCC	Foundation Casa Campesina Cayambe
UPS	Universidad Politécnica Salesiana
FAO	Food and Agriculture Organization of the United Nations
WWP	Working with People
DAG	decentralized autonomous governments

References

1. Turner, C.; Aggarwal, A.; Walls, H.; Herforth, A.; Drewnowski, A.; Coates, J.; Kalamatianou, S.; Kadiyala, S. Concepts and Critical Perspectives for Food Environment Research: A Global Framework with Implications for Action in Low- and Middle-Income Countries. *Glob. Food Sec.* **2018**, *18*, 93–101. [CrossRef]
2. Havemann, T.; Negra, C.; Werneck, F. Blended Finance for Agriculture: Exploring the Constraints and Possibilities of Combining Fi Nancial Instruments for Sustainable Transitions. In *Social Innovation and Sustainability Transition*; Springer: Cham, Switzerland, 2022; pp. 347–358, ISBN 978-3-031-18560-1.
3. Norton, G.W.; Alwang, J. Changes in Agricultural Extension and Implications for Farmer Adoption of New Practices. *Appl. Econ. Perspect. Policy* **2020**, *42*, 8–20. [CrossRef]
4. Merino, R. The Geopolitics of Food Security and Food Sovereignty in Latin America: Harmonizing Competing Visions or Reinforcing Extractive Agriculture? *Geopolitics* **2022**, *27*, 898–920. [CrossRef]
5. Pinto, L. Soberanía Alimentaria, Justicia Ambiental y Resistencia Campesina Territorial Frente a Los Cambios Metabólicos Del Libre Comercio: Apuntes Teóricos y Empíricos Desde La Experiencia Mexicana. *Razón Palabra* **2016**, *20*, 517–542.
6. Devine, J.A.; Ojeda, D.; Garzón, S.M. Formaciones Actuales de Lo Campesino En América Latina: Conceptualizaciones, Sujetos/as Políticos/as y Territorios En Disputa. *Antipoda* **2020**, *1*, 3–25. [CrossRef]
7. Quintero, G. Hacia Un Enfoque Social de Los Territorios Inteligentes: Una Primera Aproximación. *Territorios* **2020**, *42*, 250–266. [CrossRef]
8. Acosta, M.; De los Ríos, I.; Ávila, C.; Castañeda, R. Responsible Investment in the Processes of Substitution of Illicit Crops from the "Working With People" Model: Case Study Guaviare (Colombia). In Proceedings of the International Congress on Project Management and Engineering, Donostia-San Sebastian, Spain, 10–13 July 2023; pp. 1711–1725.
9. Serradilla, A. Los Principios CSA-IAR de Fao Como Instrumento Para Revitalizar La España Vaciada: Proyecto de Ingeniería Alimentaria En Golmayo (Soria). In Proceedings of the 25th International Congress on Project Management and Engineering Alcoi, Alcoi, Spain, 6–9 July 2021; Volume 87, p. 947.
10. Acosta, M.; De Los Ríos, I.; Ávila, C.; Castañeda, R. Inversión Responsable En Procesos de Sustitución de Cultivos Ilícitos Desde El Modelo "Trabajando Con Personas": Estudio de Caso Guaviare (Colombia). In Proceedings of the 27th International Congress on Project Management and Engineering, Donostia-San Sebastian, Spain, 10–13 July 2023; Volume 29, pp. 1711–1725.
11. Requelme, N.; Afonso, A. The Principles for Responsible Investment in Agriculture (CFS-RAI) and SDG 2 and SDG 12 in Agricultural Policies: Case Study of Ecuador. *Sustainability* **2023**, *15*, 15985. [CrossRef]
12. Jiménez Aliaga, R.; De Los Rios Carmenado, I.; Cazorla Montero, A.; Huamán Critóbal, A.E. Definition of Priority Objectives for the Implementation of the CFS-RAI Principles in the Central Highlands of Perú. In Proceedings of the Proceedings 27th International Congress on Project Management and Engineering, Donostia-San Sebastian, Spain, 10–13 July 2023; pp. 1696–1710.

13. Ortuño, M.; De los Ríos, I.; Sastre-Merino, S. The Development of Skills as a Key Factor of the Cooperative System: Analysis of the Cooperative of Artisan Women Tejemujeres-Gualaceo-Ecuador from the WWP Model. *Sustainability* **2022**, *14*, 16233. [CrossRef]
14. Pérez, R. Falacias de Suma Cero En Las Lógicas Del Despojo. La Continua Representación de "Lo Indígena" y "Lo Rural" Como Oposición a La Ideología Desarrollista En Ecuador y Bolivia. *Estud. Rural.* **2022**, *7*, 138–164. [CrossRef]
15. Requelme, N.; Carvajal, J.; Lizano, R.; Cachipuendo, C.; Patricia, Y. *Mujeres En Resistencia y Territorios Agroecológicos*; Abya Yala: Quito, Ecuador, 2019; ISBN 978-9978-10-386-.
16. Cazorla, A.; De los Ríos, I. Hacia Una Metauniversidad Para Un Desarrollo Sostenible: Inversión Responsable En Agricultura y Sistemas Alimentarios. *Soc. Environ. Dev.* **2024**, *2024*, 83–92. [CrossRef]
17. Nijnik, M.; Sarkki, S.; Melnykovych, M.; Miller, D. *Social Innovation to Address Sustainability Challenges and Enlarge the Opportunities Provided by Forests for a Green and Just Transition of Marginalised Mountain Areas*; University of Ljubljana: Ljubljana, Slovenia, 2023; pp. 13–23. Available online: https://www.researchgate.net/publication/374368132_Social_innovation_to_address_sustainability_challenges_and_enlarge_the_opportunities_provided_by_forests_for_a_green_and_just_transition_of_marginalised_mountain_areas (accessed on 17 February 2025).
18. Schermer, M.; Kroismayr, S. Social Innovation in Rural Areas. *Osterreichische Z. Soziologie* **2020**, *45*, 1–6. [CrossRef]
19. Van Der Ploeg, J.; Renting, H.; Brunori, G.; Knickel, K.; Mannion, J.; Marsden, T.; De Roest, K.; Sevilla-Guzmán, E.; Ventura, F. Rural Development: From Practices and Policies towards Theory. *Sociol. Rural.* **2000**, *40*, 391–408. [CrossRef]
20. Stein, V.; Pentzold, C.; Peter, S.; Sterly, S. Digitalization and Civic Participation in Rural Areas. A Systematic Review of Scientific Journals, 2010–2020. *Raumforsch. Und Raumordn. Spat. Res. Plan.* **2022**, *80*, 251–265. [CrossRef]
21. Crotta, C.V.; Cendón, M.L.; Bruno, M.P. Alimentación Saludable: Una Revisión de La Literatura Para Una Definición Integral En El Marco de Los Circuitos Cortos de Comercialización. *Estud. Rural.* **2024**, *14*, 4. [CrossRef]
22. FAO. *CFS Principles for Responsible Investment in Agriculture and Food Systems*; FAO: Rome, Italy, 2014.
23. Paniagua-Sánchez, A. La Cooperación Internacional Para El Desarrollo En América Latina: Una Perspectiva Histórica. *Relac. Int.* **2024**, *97*, 69–86. [CrossRef]
24. Ayllón, B. La Cooperación Internacional Para El Desarrollo: Fundamentos y Justificaciones En La Perspectiva de La Teoría de Las Relaciones Internacionales. *Carta Int.* **2007**, *2*, 25–40.
25. Zalaquett, J. La Declaración de Naciones Unidas Sobre Derechos de Los Pueblos Indígenas. *Anu. Derechos Hum.* **2008**. [CrossRef]
26. López, M.; Piñero, F. Development and International Scientific-Technological Cooperation in Latin America. *Tla-Melaua* **2019**, *13*, 98–119. [CrossRef]
27. Solares, B. La Teoría de La Acción Comunicativa de Jürgen Habermas: Tres Complejos Temáticos. *Rev. Mex. Cienc. Polit. Soc.* **1996**, *41*, 9–33. [CrossRef]
28. Cachipuendo, C.; Sandoval, C.; Sandoval, J. Sentinel-2A Imaging in Mapping Greenhouse Rose Production and Rainwater Harvesting for Agricultural Irrigation Use. *Environ. Res. Commun.* **2024**, *6*, 111005. [CrossRef]
29. Cachipuendo, C. La Gestión Comunitaria Del Agua y La Sustentabilidad de Los Sistemas de Riego: Una Práctica Del Buen Vivir. In *Agua para la Gente. Experiencias de Gestión Comunitaria de Agua en el Ecuador*; Editorial Universitaria Abya-Yala: Quito, Ecuador, 2021; pp. 139–168, ISBN 9789896540821.
30. Cazorla, A.; De los Ríos, I.; Salvo, M. Working with People (WWP) in Rural Development Projects: A Proposal from Social Learning. *Cuad. Desarro. Rural.* **2013**, *10*, 131–157.
31. De Los Ríos Carmenado, I.; Mur Nuño, C. Inclusive and Transparent Governance Structures from IRA Principles. Case Fesbal "Food Bank". In Proceedings of the Proceedings 27th International Congress on Project Management and Engineering, Donostia-San Sebastian, Spain, 10–13 July 2023; pp. 1737–1751.
32. Jasper, J.M. Social Movement Theory Today: Toward a Theory of Action? *Sociol. Compass* **2010**, *4*, 965–976. [CrossRef]
33. Garzón, B.; Cárdenas, J.; Herrán, J.; Ortiz-T, P.; Altamirano, J.; Pesantez, F.; Flores, Á.; Juncosa, J.; Cachipuendo, C.; Requelme, N.; et al. *Incidencia de Los Proyectos de Vinculación Con La Sociedad de La Universidad Politécnica Salesiana*; Editorial Abya Yala, UPS: Cuenca, Ecuador, 2022; ISBN 978-9978-10-623-5.
34. Mancini, N.; Rodriguez, H.; Tadeo, D.; Prado, J.P.; Hourcade, O.; Niño, J.L. Una Aproximación a Las Relaciones Internacionales Subnacionales. *Trab. Investig. En. Parad.* **2011**, *1*, 180.
35. Herrán, J. El Microcrédito Como Gestión Compartida: La Experiencia de La Casa Campesina Cayambe. *Core UK* **2015**, *22*, 47–68. Available online: https://dspace.ups.edu.ec/bitstream/123456789/11062/1/2%20El%20microcredito%20como%20gestion%20compartida%20la%20experiencia%20de%20la%20Casa%20Campesina%20Cayambe.pdf (accessed on 30 January 2025).
36. Semenova, T. Political Mobilisation of Northern Indigenous Peoples in Russia. *Polar Rec.* **2007**, *43*, 23–32. [CrossRef]
37. Wilson, E. What Is Benefit Sharing? Respecting Indigenous Rights and Addressing Inequities in Arctic Resource Projects. *Resources* **2019**, *8*, 74. [CrossRef]
38. Melucci, A. Teoría de La Acción Colectiva. In *Acción Colectiva, Vida Cotidiana y Democracia*; El Colegio de Mexico: Ciudad de México, Mexico, 2010; pp. 25–54, ISBN 968-12-0869-2. Available online: https://muse.jhu.edu/pub/320/oa_monograph/chapter/2571434/pdf (accessed on 31 January 2025).

39. Easdale, M.; Conti, S.; Nuñez, P. El Desafío de Integrar Investigación y Extensión Rural En Procesos de Innovación Tecnológica Orientados al Desarrollo Territorial. *Rev. Fac. Agron.* **2017**, *116*, 51–60.
40. Villalba-Eguiluz, U.; Arcos-Alonso, A.; Pérez, J.; Urretabizkaia, L. Social and Solidarity Economy in Ecuador: Fostering an Alternative Development Model? *Sustainability* **2020**, *12*, 6876. [CrossRef]
41. García, A. El Comercio Justo: ¿una Alternativa de Desarrollo Local? *Polis* **2011**, *7*, 105–140.
42. Sotomayor, O.; Ramírez, E.; Martíne, H. *Digitalización y Cambio Tecnológico En Las Mipymes Agrícolas y Agroindustriales En América Latina*; Naciones Unidas: Santiago, Chile, 2023.
43. Muñoz, M.; Serra, A.; Rogozinski, A. Cómo La Bioeconomía Circular Es Una Solución Integral Para Cuidar La Casa En Común. *Rev. Lib.* **2021**, *4*, 30. [CrossRef]
44. Geissdoerfer, M.; Savaget, P.; Bocken, N.M.P.; Hultink, E.J. The Circular Economy—A new sustainability paradigm. *J. Clean. Prod.* **2017**, *143*, 757–768. [CrossRef]
45. Altieri, M.; Nicholls, C.; Montalba, R. Technological Approaches to Sustainable Agriculture at a Crossroads: An Agroecological Perspective. *Sustainability* **2017**, *9*, 349. [CrossRef]
46. Sengupta, R.; Aubuchon, C. The Microfinance Revolution: An Overview. *Fed. Reserve Bank. St. Louis Rev.* **2008**, *90*, 9–30. [CrossRef]
47. Geier, U.; Fritz, J.; Greiner, R.; Olbrich-Majer, M. Agricultura Biodinámica, Una Síntesis Científica. *Rev. Grupo Intercomunitario Ibérico Con Respeto* **2018**, *1*, 101–123.
48. Cachipuendo, C. The Technification of Irrigation as a Strategy of Community Resilience. Case Study: Pisque River Basin, Ecuador. In *Smart Innovation, Systems and Technologies*; Springer Science and Business Media Deutschland GmbH: Berlin, Germany, 2022; Volume 252, pp. 207–216, ISBN 9789811641251.
49. Pieper, C.; Grundy, M. Respite Care Program. *J. Gerontol. Nurs.* **1977**, *3*, 49. [CrossRef]
50. Živojinović, I.; Ludvig, A.; Hogl, K. Social Innovation to Sustain Rural Communities: Overcoming Institutional Challenges in Serbia. *Sustainability* **2019**, *11*, 7248. [CrossRef]
51. Merrick, L.; Lyon, S.; Balow, K.; Murphy, K.; Jones, S.; Carter, A. Utilization of Evolutionary Plant Breeding Increases Stability and Adaptation of Winter Wheat across Diverse Precipitation Zones. *Sustainability* **2020**, *12*, 9728. [CrossRef]
52. Lindqvist-Kreuze, H.; Bonierbale, M.; Grüneberg, W.; Mendes, T.; De Boeck, B.; Campos, H. Potato and Sweetpotato Breeding at the International Potato Center: Approaches, Outcomes and the Way Forward. *Theor. Appl. Genet.* **2024**, *137*, 12. [CrossRef]
53. Winkel, T.; Bommel, P.; Chevarría-Lazo, M.; Cortes, G.; Del Castillo, C.; Gasselin, P.; Léger, F.; Nina-Laura, J.P.; Rambal, S.; Tichit, M.; et al. Panarchy of an Indigenous Agroecosystem in the Globalized Market: The Quinoa Production in the Bolivian Altiplano. *Glob. Environ. Change* **2016**, *39*, 195–204. [CrossRef]
54. Sengupta, P. Can Precision Agriculture Be the Future of Indian Farming?—A Case Study across the South-24 Parganas District of West Bengal, India. *Biol. Life Sci. Forum* **2023**, *30*, 3. [CrossRef]
55. Rivera-Mercado, C.; Zahn-Muñoz, C.; Díaz-López, G.; Gallard-Gallardo, J.; Braüning-Wistuba, M.; Garrido-Álvarez, O. Gobernanza Colaborativa Multinivel En Territorios Que Buscan Descentralización: Una Experiencia Innovadora Para La Sostenibilidad En La Región de Los Lagos, Chile. *Estado Gob. Gestión Pública* **2024**, *22*, 16–44. [CrossRef]
56. Altieri, M.; Toledo, V. The Agroecological Revolution in Latin America: Rescuing Nature, Ensuring Food Sovereignty and Empowering Peasants. *J. Peasant. Stud.* **2011**, *38*, 587–612. [CrossRef]
57. Jiménez Aliaga, R.; De los Ríos-Carmenado, I.; San Martín Howard, F.; Calle Espinoza, S.; Huamán Cristóbal, A. Integration of the Principles of Responsible Investment in Agriculture and Food Systems CFS-RAI from the Local Action Groups: Towards a Model of Sustainable Rural Development in Jauja, Peru. *Sustainability* **2022**, *14*, 9663. [CrossRef]

sustainability

MDPI

Article

Good Practices of Food Banks in Spain: Contribution to Sustainable Development from the CFS-RAI Principles

María Leticia Acosta Mereles [1,*], Carlos Mur Nuño [1,2], Ricardo Rubén Stratta Fernández [3] and Manuel Enrique Chenet [4]

[1] Planificación y Gestión Sostenible del Desarrollo Rural-Local (GESPLAN), Cátedra Banco de Alimentos, Universidad Politécnica de Madrid, Avda Puerta de Hierro 2, 28040 Madrid, Spain; carlos.mur@alumnos.upm.es

[2] Facultad de Ciencias Económicas, Empresariales y de la Comunicación, European University of Madrid, Calle Tajo s/n, 28670 Madrid, Spain

[3] Departamento de Desarrollo Rural, Facultad de Ciencias Agrarias y Forestales, Universidad Nacional de La Plata, Av. 7 N° 776, Buenos Aires 1900, Argentina; rrs.ricardo.rs@gmail.com

[4] Facultad de Ciencias Económicas y Empresariales, Universidad Ricardo Palma, Jr. Morro Solar 500, Santiago de Surco 15039, Peru; manuel.chenet@urp.edu.pe

* Correspondence: marialeticia.acosta@alumnos.upm.es

Abstract: The Principles for Responsible Investment in Agriculture and Food Systems (CFS-RAI) are suitable standards for contributing to the Sustainable Development Goals (SDGs) in the area of sound consumption and sustainable food. In this context, food banks have demonstrated their significant role in supporting vulnerable groups and reducing food waste through the implementation of various projects and activities. This study identifies and classifies the good practices of 54 food banks that comprise the Spanish Federation of Food Banks (FESBAL). The methodology applied was based on the Working with People model, integrating a social and collaborative learning process based on the accumulated experience of food banks over 35 years. The analysis was carried out based on four dimensions of sustainability, namely social, economic, environmental, and governance, in alignment with the CFS-RAI Principles. The results obtained show the good practices of food banks, highlighting their positive effects on the dimensions of sustainability, consistent with the CFS-RAI Principles, and the SDGs, evidencing improved food security and a holistic contribution to sustainable development.

Keywords: food bank; sustainable development goals (SDGs); principles for responsible agricultural investment (CFS-RAI); food waste; good practices; food security; sustainable development

1. Introduction

The FAO believes that meeting the 2030 Agenda requires improving food security and facilitating sustainable livelihoods [1]. The Principles for Responsible Investment in Agriculture and Food Systems (CFS-RAI Principles), endorsed by the Committee on World Food Security in October 2014, contribute to food security, nutrition, and improved sustainable livelihoods with a holistic, responsible, and environmentally friendly vision [2]. Therefore, these principles are seen as a means of contributing to the achievement of Sustainable Development Goals (SDGs) by promoting sustainable development and food security through a holistic approach [3–7].

Food insecurity is defined as a socioeconomic inability to access food of sufficient quality and quantity [8] and affects millions of people around the world [9]. According to the FAO [10], 29.6% of the world's population is moderately or severely food-insecure. Moreover, it is estimated that 13.3% of the world's food is lost in the supply chain from the post-harvest to the retail level, and an additional 17% is wasted in households, food services, and retail [11], generating environmental, economic, and social consequences that affect sustainable development [12–15].

Reducing food waste leads to the more efficient use of land, water, energy, and labor resources, benefiting the global agri-food sector [16] and contributing to the fight against malnutrition [17]. The donation of surplus food, second in the food waste management hierarchy after source reduction [17–19], is increasingly recognized for its importance in preventing waste, alleviating hunger, and contributing to food security [20–24]. Among the organizations that carry out these actions are food banks (hereafter referred to as FBs).

FBs were first founded in 1967 in Arizona, the United States, and soon were adopted worldwide [25]. During their evolution, they were grouped into federations. In Europe, this idea was initiated in France in 1984. Later, in 1986, the European Federation of Food Banks (FEBA) was created. It is now based in Brussels and has an influence on European institutions and stakeholders, developing a network of food banks across the continent [16].

In Spain, the first FBs were founded in Barcelona in 1987; in 1996, the Spanish Federation of Food Banks (FESBAL) was formed, which currently groups together 54 banks throughout the country. FESBAL, a member of FEBA, fights hunger and poverty by reducing food waste in society [26,27]. In recognition of its social work, FESBAL was awarded the Prince of Asturias Award for Concord in 2012 [27].

FBs are non-governmental organizations dedicated to fighting poverty, reducing inequality, supporting vulnerable groups, and reducing food waste [28]. They carry out different activities to achieve their mission: on the one hand, they search for and obtain food from surpluses and donations, and on the other hand, they fight to raise awareness and educate people about waste and social inequality [29].

The multifaceted nature of food waste and food safety problems [30] requires the participation of different actors to address issues through different activities and at different levels [31]. Food security is not only a question of increasing production, but also of distributing food more effectively and achieving sustainable consumption [32]. It also emphasizes the importance of technological innovation, awareness, and public policies in addressing this issue [31].

Globally, and in the European Union, strategies and legislations are being implemented to prevent and reduce food waste in line with the SDGs [33,34]. In Spain, the first draft law on the "prevention of food losses and waste" was approved in 2021, which prioritizes donations to non-profit organizations or food banks [33]. This legal framework is aligned with the Spanish Circular Economy Strategy "Spain 2030" (EEEC), which develops policies and instruments for reducing food waste and promoting sustainable consumption, highlighting the collaboration and participation of companies, civil society, and FBs [35].

Numerous studies have confirmed the significant role of FBs in reducing food waste, food insecurity, and food poverty [36–42], and in contributing to sustainable consumption [30,33,43]. At a general level, they are efficiently organized and well established in their territories as civil society structures [29,44–46], which allows them to generate impacts in multiple areas and achieve the SDGs [47]. However, in the face of increasingly complex and changing food security situations, there are a number of challenges [48–50], and the need to improve their strategic vision, coordination, and resource management is evident [29,43,51–54].

Over the last 37 years, FESBAL and the 54 FBs have accumulated extensive experience and good practices in order to make them more efficient and achieve continuous improvements [55]. Although the people involved frequently reflect on and discuss how to improve their activities, the knowledge of good practices has not been evaluated and analyzed from a sustainability perspective in relation to social, economic, environmental, and governance dimensions. This article addresses this gap by proposing a comprehensive analysis that links these practices to the CFS-RAI Principles and SDGs, consolidating them as valid references for positively influencing food security and contributing to sustainable development, in order to make them useful for FBs and other related socioeconomic agents. In addition, it provides a novel theoretical framework that enriches the existing literature in the fields of food security and sustainable development. At the XXV Annual Congress of FESBAL in 2022, this weakness and the importance of sharing experiences and good practices was highlighted.

Previous research has collected isolated experiences of good practices by FBs to contribute to the transition towards sustainable food systems by articulating relationships with civil society and economic sectors and promoting solidarity [44,46,55,56]. However, it is necessary that these good practices be organized, evaluated with the participation of stakeholders, and analyzed from a sustainability perspective in line with the CFS-RAI principles and SDGs, so that they can be consolidated as valid references to positively influence food security.

1.1. Good Practices of Food Banks

Good practices (GPs) are considered valuable contributions because they have an impact on the quality of life of individuals and the community [57]. These practices are considered positive examples that can serve as a reference or model for others, as they have been proven to produce desirable, efficient, and effective results [58] and are based on empirical evidence or accumulated experience [59]. Their identification promotes learning, continuous improvement, replication, and transfer in different settings, entities, or geographical areas [58,60].

It should be borne in mind that good practice does not refer to perfect, flawless actions but rather to initiatives that meet a series of requirements that make them valuable experiences because of their capacity to guide the present and the future [61].

A set of principles, standards, or criteria are usually considered for the identification of good practices [62] that are able to satisfy needs and expectations in accordance with the mission and values of the entity that promotes them and that are successful examples that allow the improvement and addition of value [63].

At a general level, frequent criteria for identifying and analyzing good practices relate to effectiveness [64], the transformative effect [65], replicability [66,67], sustainability [68], and legitimacy [69,70].

UNESCO identifies four basic criteria for identifying "good practices": (1) innovation, when the practice introduces creative ways to improve, grow, and provide solutions to problems and perfect the internal or external functioning of a given action or organization; (2) effectiveness, when it demonstrates a positive and tangible impact on society, on the territory, and on the community in which it is developed; (3) sustainability, when it operates harmoniously over time and generates sustained impacts with lasting effects; and (4) replicability or transferability, when it allows for the repetition of its essential elements in a context other than that of its creation, with a high chance of success.

1.2. Committee on World Food Security's CFS-RAI Principles

In this context of good practices (GPs), the Committee on World Food Security's Principles for Responsible Investment in Agriculture and Food Systems (CFS-RAI), approved in 2014, are accepted as an appropriate standard to contribute to food security, nutrition, and sustainable livelihoods, with a holistic, responsible, and sustainable vision [2] from the perspective of the different actors involved. They support good practices in the context of sustainable food and food security [71].

Their objective is to promote "responsible" projects in the field of food security and nutrition, supporting the fulfillment of the right to adequate food [4,72], and allow actors to interrelate and complement actions and projects that incorporate these principles in their consideration as "good practices".

These principles seek to guide the actors involved in food systems, providing a guideline to achieve sustainable development that promotes food security, nutrition, and the right to food [7].

By following these CFS-RAI Principles (Table 1), the aim is to ensure the availability, access, and consumption of food in a sustainable, equitable, and safe manner. The CFS-RAI Principles also integrate four basic features to identify what they call good practices: innovation, effectiveness, sustainability, and replicability of initiatives.

Table 1. CFS-RAI Principles.

Principle 1: Contribute to food security and nutrition.
Principle 2: Contribute to sustainable and inclusive economic development and the eradication of poverty.
Principle 3: Foster gender equality and women's empowerment.
Principle 4: Engage and empower youth.
Principle 5: Respect tenure of land, fisheries, and forests, and access to water.
Principle 6: Conserve and sustainably manage natural resources, increase resilience, and reduce disaster risks.
Principle 7: Respect cultural heritage and traditional knowledge, and support diversity.
Principle 8: Ensure responsible governance mechanisms.
Principle 9: Promote safe and healthy agriculture and food systems.
Principle 10: Strengthen transparency, accountability, and commitment to sustainable development.

Source: [2].

Agriculture and food systems encompass a full range of activities related to the production, processing, distribution, marketing, consumption, and disposal of products derived from agriculture, including food and non-food products.

Food banks are part of the food chain, occupying a link between transformation and distribution through food distribution entities, which are retail organizations that deliver food to beneficiaries as final consumers.

Through an agreement between the UPM and FAO, since September 2016, the Research Group on Planning and Sustainable Management of Local Rural Development (GESPLAN) has been working to analyze good practices that integrate the CFS-RAI Principles to help achieve food and nutrition security, facilitating access to knowledge and concrete experiences related to development cooperation in the fields of agriculture, food, the environment, and climate change [73]. From this work, experiences and good practices have been demonstrated at the international level as successful examples of including or respecting these principles.

1.3. Dimensions of Good Practices from the SDGs and CFS-RAI

Good practices from the SDGs and CFS-RAI are classified into four dimensions (social, economic, environmental, and governance) [73], allowing for a more complete and detailed understanding of the interconnections and synergies between initiatives and different objectives. The classification of good practices was based on the dimensions and pillars of the SDGs [74–78] and CFS-RAI [5], as shown in Table 2.

Table 2. Good practices and SDG dimensions: CFS-RAI.

Dimension	Pillars	SDGs [1]	CFS-RAI [2]
Social	People	1, 2, 3, 4, 5	3, 4
Economic	Prosperity	7, 8,9,10	1, 2, 7, 8
Environmental	Planet	6, 12, 13, 14, 15	5, 6
Governance	Peace—Partnership	16, 17	9, 10

[1] SDGs: 1. No poverty 2. Zero hunger 3. Good health and well-being 4. Quality education 5. Gender equality 6. Clean water and sanitation 7. Affordable and clean energy 8. Decent work and economic growth 9. Industry, innovation, and infrastructure 10. Reduced inequality 11. Sustainable cities and communities 12. Responsible consumption and production 13. Climate action 14. Life below water 15. Life on land 16. Peace, justice, and strong institutions 17. Partnerships for goals. [2] Principles of CFS-RAI: 1. Contribute to food security and nutrition 2. Contribute to sustainable and inclusive economics 3. Foster gender equality and women's empowerment 4. Engage and empower youths 5. Respect tenure of land, fisheries, and forests, and access to water 6. Conserve and sustainably manage natural resources, increase resilience, and reduce disaster risk 7. Respect cultural heritage, traditional knowledge, and support diversity 8. Ensure responsible governance mechanisms 9. Promote safe and healthy agriculture and food systems 10. Strengthen transparency, accountability, and commitment to sustainable development. Source: Adapted from [4,76,77].

The SDGs are characterized by their holistic vision, recognizing the interdependence and synergy between them. Each SDG is intrinsically related to the others, forming an interconnected system in which progress in one objective can have an impact on other objectives [79,80]. Likewise, the CFS-RAI Principles support the achievement of the SDGs by promoting sustainable development and food security through a holistic approach [3–6].

The good practices of FBs from the social dimension include activities or projects aimed at raising awareness of food waste and people's food assets [81–83]. These initiatives are aligned with SDGs 1, 2, 3, 3, 4, and 5 as well as with CFS-RAI Principles 3 and 4, promoting participation and equality through volunteering and solidarity actions aimed at beneficiaries, students, and the public.

Several studies have analyzed awareness campaigns to reduce food waste [84–86]. These initiatives include lectures, exhibitions, and training [87] with the objective of raising awareness of how to make a significant impact on food waste issues. They concluded that educational programs can be effective tools to raise awareness and support social change [81], highlighting incentives and collaboration among stakeholders as key factors for success [86].

The environmental dimension of FBs relates to the positive impact of their activities on the environment. By reducing food waste, FBs contribute to mitigating the environmental impacts of the agri-food sector [33,88]. These practices include activities related to the fight against climate change [89], the reduction in greenhouse gas emissions [90], reducing carbon footprints [91], and promoting the use of renewable energies. The SDGs linked to this dimension are 6, 11, 12, 13, 14, and 15, and are also related to CFS-RAI Principles 5 and 6.

The economic dimension refers to initiatives that benefit those who access the services of the FBs and are related to cost reduction, sustainable economic growth, decent employment, the creation of new jobs [22,92], and the reduction in economic inequalities. The SDGs highlighted in this dimension are 7, 8, 9, and 10 and are also linked to CFS-RAI Principles 1, 2, and 7.

The governance dimension is related to the management of the organization of FBs, its transparency, and the independence of the management bodies [93,94]. Good practices in this dimension cover different initiatives to strengthen the efficiency, effectiveness, and transparency of FBs' functions. The SDGs most closely related to the good governance criteria are SDGs 16 and 17. The CFS-RAI Principles corresponding to the governance dimensions are 9 and 10, respectively.

This framework is applicable to the good practices carried out by food banks to generate sustainability impacts with their activities [92,95], contributing to the SDGs [28] and the

CFS-RAI Principles, and for FBs, experiences are highlighted in relation to Principle 1 for contributing to food security and nutrition and Principle 2 for contributing to sustainable and inclusive economic development and poverty eradication.

The purpose of this study is to identify, classify, and analyze the good practices carried out by FESBAL food banks in relation to the SDGs and CFS-RAI. It also seeks to reflect on these good practices in relation to the fight against hunger, poverty, and food waste, addressing these issues based on the four dimensions of the SDGs and CFS-RAI.

Addressing the capitalization of experiences of FBs, through the recognition and valuation of management, in relation to the SDGs and the CFS-RAI Principles will contribute to generating new knowledge to improve future interventions, promoting information to show the impact of their activities, and making decisions to continue improving their fight against hunger, poverty, and food waste through the use and distribution of food to the most needy people, and also contribute to protecting the environment.

2. Materials and Methods

The results of this study were based on a methodology that incorporated different tools and sources of empirical information (Figure 1). Quantitative and qualitative data were collected. The source of information included a series of collaborative virtual workshops, ArcGIS Survey123, interviews with food bank managers with relevant experience, and documentary analysis (FESBAL reports and the 54 FBs).

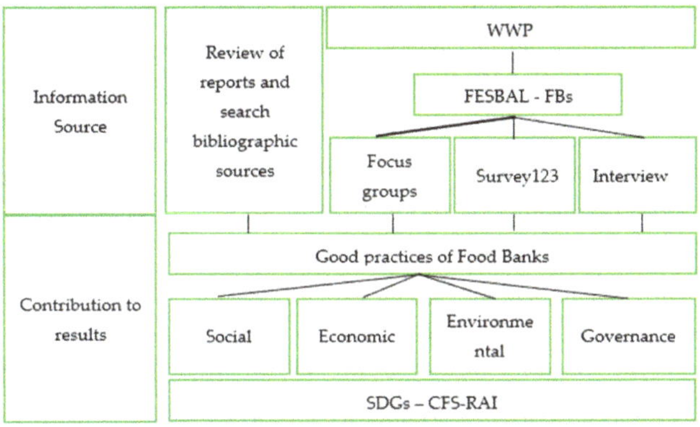

Figure 1. Methodological scheme of research.

The methodological process applied was "Working with People" (WWP), the result of collaborative experience from working with multiple stakeholders (public–public–private) in the implementation of projects related to sustainability, based on social learning [96], and this process has been validated as an innovative approach in multiple contexts [97–108], as well as in the context of the projects carried out by the Food Bank-UPM Chair [46,55], facilitating the analysis of the implementation of the CFS-RAI Principles [4].

The methodological approach of the WWP model has contributed to the sustainable development of territories, strengthening the organizational structure [103], and generating links between agents in the four spheres of the social relations system: the political, public–administrative, private–business, and social spheres [109].

Participatory processes with key informants to capture experienced knowledge constitute a methodological tool for obtaining primary information in working with people, according to the WWP model [96].

To guide the analysis, the following research questions were posed: how do the projects and good practices carried out by food banks contribute to the SDGs and the

CFS-RAI principles? How can the integration of teaching, research, and connection with civil society enhance the implementation and sustainable impact of food bank projects aligned with the CFS-RAI Principles and the SDGs? Finally, what role do partnerships between the public, private, and civil society sectors play in enhancing the impact and sustainability of good practices? Based on these questions, it was hypothesized that such projects generate significant impacts on sustainability based on the contribution to the SDGs and the CFS-RAI, enhanced by innovation and strategic collaboration.

2.1. Case Study

This research work was developed in the context of a collaboration between the University and FESBAL, which materialized through the Food Bank Cathedra—Polytechnic University of Madrid [110]. Since 2013, the agreement has been in place with the aim of raising awareness and promoting rational food consumption through training activities, knowledge transfer, and the promotion of R&D&I [111].

The target population focuses on 54 food banks in Spain grouped by the FESBAL. These FBs operate as non-profit organizations based on volunteerism, recover surplus food, and redistribute it through charities [43]. They manage the logistics of storage, classification, and distribution of food, collaborating with a network of charities and aid centers to reach those most in need. The sustenance that banks obtain comes from the contributions of individuals, companies, and governmental entities (supermarkets, food producers, wholesalers, and civil society). FESBAL acts as a key organization that coordinates the efforts of banks throughout the country, working with partner distribution and delivery organizations to ensure efficient distribution of food to recipients [26,112].

Table 3 gives an idea of the relative importance of food banks in Spain from 2018 to 2023, where it can be seen that they recovered on average 152,945.18 tons of food with the help of 3222 volunteers, and they serve more than one million beneficiaries—that is, 9.9% of the population at risk of poverty [113].

Table 3. Food assistance from FESBAL.

Year	N° FBS	Volunteers	Tn food	Charities	Beneficiaries
2023	54	3238	138,046	6493	1,187,976
2022	54	3214	151,684	6919	1,246,106
2021	54	3100	173,000	7497	1,353,276
2020	54	3327	158,746	7955	1,560,000
2019	54	3211	144,551	7216	1,050,684
2018	55	3247	151,645	7402	1,195,167

Source: [27].

2.2. Participatory Process: Focus Meetings

To obtain the initiatives and successful experiences of the FBs, a participatory process was carried out by means of focus groups through video conference systems using the Zoom platform. Five workshops were held in 2023, with the participation of 23 FBs.

To facilitate the participatory process and contextualize the workshops, the FBs were subdivided according to 4 main criteria: (1) the existence of an awareness-raising area or department; (2) contributions to the educational resource guides; (3) awareness-raising activities; and (4) participation in the CORAL Program (Rational Food Consumption), a program implemented by the CBA. Based on these indicators, the FBs were organized into levels A, B, C and D, where level A corresponds to outstanding participation in most of the criteria, while level D includes FBs with no record of activities in the aspects evaluated. In this way, the participants of the banks presented diversity according to activity and participation in projects in different areas, which will greatly enrich our reflection on the topics to be discussed. In view of the aforementioned diversity, the participants presented homogeneity in the issues dealt with as main actors of the FBs, facilitating both the dialog and the understanding of the topics to be discussed.

In all workshops, the agenda included an introduction in which the conceptual framework was presented, framing the objectives and methodology. The implementation and convening of the workshops were jointly coordinated between FESBAL and CBA-UPM, where they were provided with information on the day, time, and place. The workshop was coordinated according to the following steps.

First, all participants were introduced to each other. This was followed by a presentation of the conceptual framework that provided a basis for discussion and reflection on several key issues: the current problems faced by and opportunities arising from food banks were explored. The third point of discussion focused on actions and projects that can be considered good practices carried out by food banks to fulfill their work. Strategies to improve the dissemination and scope of these practices were also explored. Finally, the results obtained were shared and integrated, and the conclusions reached were shared

The same methodology was applied to the five virtual collaborative workshops held to ensure the greatest possible participation of food banks.

2.3. *ArcGIS Survey123*

Complementarily, using ArcGIS Online software (version 3.20.69) developed by the Environmental Systems Research Institute (ESRI), a survey was conducted using ArcGIS Survey123, which stands out as an application for collecting, analyzing, and reporting data and can be used flexibly on a smart device (e.g., smartphone, computer, tablet), with or without a field Internet connection. Survey123 is useful for collecting survey data of various types, including text, numbers, images, and location [114]. The survey was based on the same key questions raised in the collaborative workshops. Thus, the scope of this research included 54 food banks.

2.4. *Analysis of Results*

Data analysis combined qualitative [115–117] and quantitative [118] approaches to ensure a comprehensive and robust interpretation of the results. The various sources of information were integrated, allowing the triangulation of data to strengthen the validity of the findings.

The data obtained in the five focus groups were analyzed using a content analysis approach. Before each session, informed consent was obtained from the participants for the recording of the sessions, guaranteeing confidentiality. During the workshops, handwritten notes were taken as a complement to the main recording, and the recordings were transcribed in full to ensure the inclusion of all interventions. The responses were coded and organized in a Microsoft Excel database, including variables such as the code of the good practice, autonomous community, province, food bank, SDG-CFS-RAI dimension, type of good practice, description of the activity, persons in charge, agents involved, and web information.

The data from the survey conducted using ArcGis Survey123 were exported, verified, and integrated into the same database, which made it possible to combine quantitative and qualitative information in a systematic way. The interviews, on the other hand, provided greater detail on the issues addressed.

In addition, good practices were evaluated, according to the criteria described (innovation, effectiveness, sustainability, and replicability), based on a Likert scale [119] between 1 (very low) and 5 (very high), which is commonly used in social sciences to measure perceptions and quantitative aspects [120].

The assessment described was qualitative in nature, since it did not correspond to quantities or natural units of measurement but to ranges subject to value judgment. Therefore, it was a subjective valuation resulting from the perception of the authors of this study.

2.5. Ethical Considerations

This study was conducted in collaboration with the Spanish Federation of Food Banks (FESBAL), which encompasses 54 food banks and the Food Bank Chair UPM, ensuring compliance with established ethical principles. All participants provided informed consent after being adequately informed about the study's objectives, procedures, and potential implications.

3. Results and Discussion

The results of the research after the integration and analysis of the information are presented below. The problems, opportunities, and good practices identified by the 54 food banks and agents involved are presented. Good practices are analyzed based on the four dimensions of the SDGs and CFS-RAI (social, economic, environmental, and governance). Finally, the practices are evaluated according to the criteria of innovation, effectiveness, sustainability, and replicability

Through their activities, food banks generate impacts in different areas and achieve the SDGs [47]. Food rescue provides many positive benefits [121]; however, humanitarian organizations continually face challenges in meeting their objectives and must be increasingly innovative [122].

3.1. Main Problems of Food Banks

Table 4 shows the main problems identified by Spanish FBs. They highlighted that they face three main problems or challenges: the lack of human resources and volunteers, the lack of dissemination of activities, and the change in the model of the European food aid system of the FEAD program (funds for European aid to the most deprived persons).

Table 4. Main problems of FBs.

Main Problems	Total (%)
Lack of human and volunteer resources	30.77
Lack of dissemination of activities	20.51
Change in the European food aid system of the FEAD program	15.38
Lack of sharing of experiences and good practices among FBs	7.69
Lack of team training	5.13
Lack of communication between FBs	5.13
The new law on food waste	2.56
Lack of information on recipient entities	2.56
Lack of involvement of public administrations	2.56
Old technological equipment	2.56
Lack of access routes to warehouses	2.56
Inadequate valuation of beneficiaries	2.56

The lack of volunteers (permanent and temporary) for activities and food collection campaigns is a problem for many food banks. According to the FB representative of Salamanca, the number of volunteers has reduced in recent years, while previously it was much higher. This shortage of volunteers is a major challenge facing food banks [123], as volunteers play a crucial role in their activities. In addition to operational activities, volunteers also contribute to promoting social interactions and building more supportive community networks. Likewise, active participation benefits the people who serve and enriches the volunteers' experience, providing them with the opportunity to empower themselves and develop their skills and knowledge [124].

Meanwhile, the lack of dissemination of FBs' activities was also highlighted. This limits their capacity to generate awareness of their needs, attract volunteers, and obtain additional donations. According to Kruse et al. [125], food banks' lack of dissemination of good practices and innovative initiatives may hinder their efforts to address the current challenges. The representative of the Melilla FBs mentions the "little participation in social networks of the activity; only the Great Collection and Kilo operations have been published. There is a lack of much more information flowing from the global point of view of all the banks through networks and media, all the work should be publicized more at the national level".

Finally, the change in the model of the European food aid system of the FEAD program, planned for 2024, is another major concern for FBs. This new system proposes the implementation of wallet cards for direct purchase in supermarkets. Some food banks warn that this new system could reduce the number of beneficiaries and jeopardize the operation of those food banks that do not receive sufficient private donations and collaboration from volunteers

Other problems mentioned were the need to share experiences, a lack of training, insufficient communication between food banks, limited involvement of public entities, and a lack of technological equipment, among others.

3.2. Food Bank Opportunities

Although FBs experience problems, they also have several opportunities for improvement (Table 5), which can contribute to addressing society's needs.

Table 5. Main opportunities for FBs.

Main Opportunities	Total %
Cooperation with entities (education centers, university, business, government, etc.)	29.03
Nationwide network for joint projects and assistance to vulnerable people	19.35
Social and vocational training and volunteer training	16.13
Volunteering and social responsibility	12.90
Dissemination of the FBs and their actions; observation of good practices	12.90
The new waste law	9.68
Total, general	100.00

Strategic alliances, networking, and cooperation with entities from the public and private sectors and non-profit organizations (educational centers, universities, government agencies, NGOs, companies, etc.) are crucial for food banks. As in Shaheen et al. [122] and Santini and Cavicchi [126], these collaborations provide access to additional resources, expertise, and expanded support networks. Likewise, Dodd and Nelson [127] reinforce that collaboration among stakeholders is a strength that translates into trusting relationships among the members of the association.

The alliances and partnerships of FBs in Spain mainly focus on opportunities related to the social dimension, stressing the importance of raising awareness, providing education on food waste, and helping vulnerable people. They are also linked to volunteering and solidarity. In this regard, the representative of FBs in Cantabria mentions that "We believe that food banks have a great opportunity to offer the population and society, infrastructure, and means to help all vulnerable families in Spain, through efficient and transparent work".

Another opportunity refers to socio-labor training projects for volunteers and beneficiaries. These actions have an impact on the economic dimension, since they facilitate the incorporation of labor of people in need, contributing to reduced economic inequality. For example, the Malaga FB representative highlights this opportunity: "We have been working for eight years, and with our model we have managed to incorporate more than 1100 people into the labor market, who depended on food plans".

From an environmental perspective, cooperation with universities to demonstrate the positive impact of FBs' activities on the environment, the fight against climate change, the reduction in greenhouse gas emissions, and the use of renewable energy is also considered an important opportunity.

Another great opportunity is the existence of FESBAL as a national network that brings together 54 food banks with provincial structures, demonstrating great humanitarian work in the fight against hunger. In addition, this network is part of FEBA, extending its reach to the European level. These opportunities are related to the governance dimension of FBs, as the union provided by FESBAL is considered a benefit for social organizations that play a relevant role in sustainable development. It also highlights the opportunity to build a network for the exchange of experiences and food surpluses that allows collaboration between various agencies

Cooperation is essential to implement sustainability-based management models, enabling organizations to access the resources and capabilities needed to develop dimensions of sustainability that they cannot achieve individually [128]. Thus, Porter and Kramer [129], in their "Creating Shared Value" approach, mention that entities can create shared value, in which economic, social, and environmental value are mutually reinforcing. In this context, SDG 17 highlights the importance of strengthening partnerships as a key mechanism for achieving sustainable development [130].

3.3. Types of Food Bank Good Practices

A total of one hundred and fifty-two GPs carried out by the fifty-four FBs in Spain were identified and grouped into nine types (Table 6). Of particular note are awareness-raising actions, food collection initiatives, and the development of social and labor insertion activities.

Table 6. Types of good practices of FBs.

Types of Good Practices	Number of Projects	Percentage (%)
Awareness actions	54	35.5
Food collection	44	28.9
Socio-labor insertion	18	11.8
Reduced impact on the environment	13	8.6
Promote dissemination and visibility	7	4.6
Acknowledgments to collaborators	6	3.9
Recruitment of personnel and interns	5	3.3
Solidarity food distribution and nutrition	3	2.0
Digitization of processes	2	1.3
Total, general	152	100.0

Changes are now taking place in the way banks are providing food assistance [131,132]. Some offer a variety of services such as nutrition education, job training opportunities, and health and social support services, in addition to food assistance [131–135]. These changes coincide with the findings of this study, which show that FBs in Spain carry out a variety of activities and projects that go beyond the food industry.

The nine types of GP identified are shown, all of which are related to the fight against hunger, poverty, and food waste through different activities.

Moreover, these GPs are classified into the four dimensions of sustainability (Table 7), as the contribution of food banks to the SDGs is increasingly recognized [55] as benefitting the environment and adding economic and social value, particularly for vulnerable people [136].

Table 7. Dimensions of good practices of FBs.

Dimension and Type of GP	Number of Projects	%
Social	63	41.4
Awareness actions	54	35.5
Acknowledgments to collaborators	6	3.9
Solidarity food distribution and nutrition	3	2.0
Governance	14	9.2
Dissemination and visibility	7	4.6
Recruitment of personnel and interns	5	3.3
Digitization of processes	2	1.3
Economic	62	40.8
Food collection	44	28.9
Socio-labor insertion	18	11.8
Environmental	13	8.6
Reduced impact on the environment	13	8.6
Total, general	152	100.0

Most good practices (4.45%) corresponded to the social and economic dimensions (40.79%). The remaining 20% were distributed between the governance dimension (9.21%) and the environmental dimension (8.55%).

If the results are analyzed according to the agents involved in the GPs, as shown in Table 8, the participation of public, private, and other agents, such as universities, in the four dimensions of sustainability is identified. These agents establish the social capital necessary to conduct ethical research and effectively disseminate findings [127]. This demonstrates the significant capacity of FBs to establish joint projects with different partners.

Table 8. Agents involved in the good practices of FBs.

GP Dimension	% Publics Agents	% Private Agents	% Other Agents
Social	54.95	50.00	31.33
Economic	36.26	36.89	43.37
Governance	3.30	9.84	12.05
Environmental	5.49	3.28	13.25
Total	100	100	100

As mentioned, the integration of diverse actors is critical to the success of food banks. This collaboration between different actors strengthens interventions, enriches initiatives, and maximizes the community impact. Edge and Meyer [137] argued that the integration of actors allows for a comprehensive approach to the needs of vulnerable populations, developing more effective and sustainable solutions to the problems of hunger and poverty.

3.3.1. Social Dimension

Sixty-three GPs (41% of the total) were identified in the social dimension, interacting with different agents of society. The GPs that stood out in this dimension were grouped into awareness-raising actions, the recognition of collaborators, and the solidarity distribution of food and nutritional counseling (Table 9).

Table 9. Best practices identified within the social dimension.

GP Practices Within the Social Dimension	N°	%
Awareness actions	54	35.5
Awareness-raising talks and workshops on waste in schools	44	28.9
Itinerant exhibition of the activities carried out by the FBs	1	0.7
Agreements with entities and institutions	2	1.3
Interactive audiovisual awareness	1	0.7
Workshops on nutrition and responsible consumption	6	3.9
Solidarity food distribution and nutritional counseling	3	2.0
Awareness actions with nutritional counseling	1	0.7
Supermarket food delivery with nutrition counseling	1	0.7
Food distribution in sports facilities	1	0.7
Acknowledgments to collaborators	6	3.9
Thanks to donors and volunteers	6	3.9

These relate to CFS-RAI Principles 3 and 4, which promote participation and equality through volunteering and solidarity actions aimed at beneficiaries, students, and the public. Education and public participation change behavior and reduce food waste [82].

Raising awareness is key to promoting individual and collective responsibility in reducing food waste [84–86]; increasing awareness of the problem encourages greater reflection and action on the part of students [81]. According to Redman and Redman [83], a change in attitude, behavior, and culture is essential to achieving any substantial change.

Awareness-raising activities include talks and workshops on food waste and responsible consumption in educational centers. Particularly noteworthy is the "CORAL Program" drawing contest carried out by the CBA-UPM in collaboration with food banks. Universities play a fundamental role as agents of change in raising awareness and promoting more responsible behaviors in food management [81,138].

Another innovative GP is the interactive audiovisual awareness-raising campaign called "Aventura Interactiva" (Interactive Adventure) with a pedagogical base "Ayúdanos a ponerle freno al desperdicio" (Help us put a stop to waste), carried out by the FBs of Navarra in collaboration with a private company, which uses gamification to awaken the interest of children and adolescents. Similarly, the FBs of Tarragona are carrying out a "traveling exhibition" with the support of the Ministry of Agriculture with the aim of raising awareness among the public, especially students, about food wastage. Through their nutrition area, Lleida's FBs offer training in food safety, preferential consumption, solidary cooking, cooking for profit, and nutritional advice through agreements with entities and institutions aimed at volunteers and beneficiaries. These activities seek to improve the nutritional and health status of people living in food poverty, providing services to improve their diet and health.

Solidarity of food distribution and nutrition: Some FBs (Valencia and Lleida) carry out food distribution activities outside their facilities, using sports clubs, supermarkets, and soup kitchens. An outstanding project is "El Menjador" of the FBs of Lleida, which offers daily meals to the neighbors of the historical center and the whole city, in addition to providing an educational and awareness-raising space on food waste.

Recognition of collaborators: This recognition is given to donors and volunteers as a token of gratitude and appreciation for their solidarity. This good practice is identified in six FBs (Jaen, Soria, Valladolid, Rias Altas, Madrid, and Melilla).

3.3.2. Economic Dimension

There are 62 good practices identified in the economic dimension, carried out by 35 FBs, grouped into two main types: social and labor insertion and food collection (Table 10). Although they have a significant social impact, they are considered in this dimension because they generate economic benefits.

Table 10. Dimensions of good practices of FBs.

Economic Dimension	Number of Projects	%
Food collection	44	28.9
Fresh food recovery	14	9.2
Donation for tickets	10	6.6
Food collection campaign	8	5.3
Recovery of processed food	2	1.3
Monetary donation	5	3.3
Donation for participation	2	1.3
Commutation of fines for food donation	1	0.7
Exchange of products with other FBs	1	0.7
Socio-labor insertion	18	11.8
Training courses for beneficiaries	15	9.9
Volunteer training courses	3	2.0

Socio-labor insertion consists of training opportunities for volunteers and people in vulnerable situations to improve their employability and integration into the labor market. These training programs contribute to improving the economic situation by providing skills and work experience, facilitating access to employment opportunities, and integrating into society [22].

Food collection is the second type of good practice, with the second highest number of GPs identified as performing this type of activity, with several initiatives that provide economic benefits by recovering food and supporting the food security of vulnerable families. These initiatives are linked to sporting, musical, and cultural activities, with original names such as "Concierto de la Leche", "Por un cacho de turrón", "Subasta solidaria", "Acción sin desperdiciar", "Ayúdanos a Ayudarlos", and "Tradicional Solidaria", among others. All these GPs are related to CFS RAI principles 1 and 2, contributing to food security and economic development.

3.3.3. Environmental Dimension

Numerous studies have highlighted the positive impact of food waste reduction on the environment [33,89,139]. They emphasize that donations of surplus food from the food chain can contribute to mitigating environmental impacts such as climate change [89], reduce greenhouse gas emissions [90], reduce carbon and water footprints [33,89,91], and the use of renewable energies.

Thirteen good practices related to environmental improvements were identified. The activities focus on the carbon footprint, water footprint, and photovoltaic panel installation projects (Table 11). These initiatives reduce the company's dependence on non-renewable energy sources, in addition to reducing fixed costs, while reducing food waste contributes to reducing greenhouse gas emissions and using water more efficiently, thereby reducing the water footprint.

Table 11. Good practices from the environmental perspective.

Environmental Dimension	Number of Projects	%
Reduced impact on the environment	13	8.6
Cleaning of waste on beaches	1	0.7
Carbon Footprint Project	6	3.9
Photovoltaic panels project	5	3.3
Water Footprint Projects	1	0.7

Navarra stands out as the only group of FBs with a project on the "Water Footprint" as a continuation of the Carbon Footprint report, in collaboration with the FBs and the Institute for Innovation and Sustainability in the Agri-Food Chain (IS-FOOD) of the University of Navarra. The FBs of Barcelona, Córdoba, Huelva, Málaga, Madrid, Murcia, Navarra, Rías Altas, Tarragona, and Toledo also implemented environmental GPs.

These actions are closely related to CFS-RAI principles 5 and 6, which call for respecting forests, land, and access to water and conserving natural resources, contributing to environmental sustainability and environmental protection.

3.3.4. From the Governance Dimension

Finally, within the governance dimension, 14 good practices were identified, representing 9.2% of the total (Table 12), in ten food banks (Alicante, Almería, Asturias, Cáceres, Cantabria, Las Palmas, Granada, Murcia, Navarra, and Toledo).

Table 12. Good practices from the governance perspective.

Governance Dimension	Number of Projects	%
Dissemination and visibility	7	4.6
"LA CAIXA Foundation" auction exhibition	1	0.7
Food bank newsletter	1	0.7
Letters of invitation to schools	1	0.7
Broadcasting in radio station "La Voz de Almería"	1	0.7
Broadcasting in interviews on radio San Vicente	1	0.7
Media coverage (from interviews)	1	0.7
Broadcasting in the media (Regional television)	1	0.7
Recruitment of personnel	5	3.3
Hiring of professional experts	3	2.0
Scholarship holders from various professions	2	1.3
Digitization of processes	2	1.3
Digitization of processes	1	0.7
"ERUS" management program	1	0.7

These practices focus on promoting the dissemination and visibility of FBs through the media (television, radio, interviews, newsletters, etc.). Improved access to information fosters transparency and effective decision making, ensuring that information is available and accessible [140]. Also noteworthy are projects that hire personnel and digitize processes. The advent of technology has provided new opportunities for food banks to develop web platforms and/or mobile applications to support food distribution, information management, and logistics processes [45,141].

These GPs have a direct impact on the good governance of food banks by increasing transparency and efficiency in the management of resources, optimizing internal processes, and generating trust in society, which are essential elements of good governance [142]. Food banks, in their continuous improvement efforts, present inclusive governance initiatives applying the CFS-RAI principles and can replicate this knowledge to other organizations in the tertiary sector [93]. These activities are contemplated within CFS-IAR principles 9 and 10, with the purpose of promoting knowledge of and respect for the laws in force, fostering transparency in the organization in terms of decision making and monitoring and control processes, and enabling accountability for good governance.

3.4. Assessment of Good Practices

The 152 good practices identified were evaluated according to the four criteria described (innovation, effectiveness, sustainability and replicability) based on a Likert scale ranging from 1 (very low) to 5 (very high) (Table 13).

Table 13. Overall assessment of best practices by dimension.

GP Dimension	Inn.	Effec.	Sust.	Rep.	Average
Environmental	4.00	4.50	4.80	4.60	4.69
Reduced impact on the environment	4.85	4.46	4.85	4.62	4.69
Economic	3.00	4.30	4.70	4.70	4.18
Food collection	3.02	4.00	5.00	5.00	4.26
Socio-labor insertion	3.00	5.00	4.00	4.00	4.00
Governance	3.60	4.50	4.00	4.00	4.04
Dissemination and visibility	3.00	4.00	4.00	5.00	4.00
Recruitment of personnel and interns	4.00	5.00	4.00	3.00	4.00
Digitization of processes	5.00	5.00	4.00	3.00	4.25
Social	3.00	4.30	4.60	4.90	4.21
Awareness actions	2.80	4.17	4.69	4.96	4.15
Acknowledgments to collaborators	4.00	5.00	5.00	5.00	4.75
Solidarity food distribution and nutrition	5.00	5.00	3.33	3.67	4.25
Total, general	3.23	4.32	4.63	4.72	4.23

Inn.: Innovation. Effec.: Effectiveness. Sust.: Sustainability. Rep.: Replicability.

At the global level, the results are very positive (with an average of 4.22), with the GPs of the environmental dimension having the highest average (4.69), followed by the GPs of the social (4.21) and economic dimensions (4.18).

In general, innovation is rated very high (4.8) in the GPs belonging to the environmental dimensions and high (3.6) in the GPs related to governance, as it introduces creative ways to improve the services provided, increasing the holistic approach of the FBs as organizations capable of providing solutions to problems and improving their functioning.

In relation to the criterion of effectiveness, the GPs identified in the environmental (4.5) and governance (4.5) areas also stand out as they demonstrate positive and tangible impacts on society and in the contexts of the FBs where they are developed.

Sustainability is highly valued in the GPs belonging to the environmental (4.8), economic (4.7), and social (4.6) dimensions, which shows that the FBs that implement them are capable of implementing initiatives that allow them to function harmoniously over time because they have a social, technical, and economic organizational structure. Their capacity to adapt is important for maintaining a service offered through volunteering, making visible a sustained impact with long-lasting effects.

Furthermore, the results suggest that environmental performance not only directly contributes to the sustainability of food banks but also acts as a catalyst for the adoption of more robust public policies in waste management and the transition to renewable energy. These findings are consistent with recent studies highlighting the transformative role of public–private partnerships in maximizing the environmental and social impact of sustainable initiatives [36,52]. Similarly, the economic and social dimensions reveal significant interconnections, particularly in labor inclusion projects that not only generate economic value but also strengthen social cohesion and opportunities for vulnerable populations.

A comparative analysis of the dimensions revealed noteworthy contrasts and areas for targeted improvement. While the environmental dimension excels in sustainability and replicability (average scores of 4.80 and 4.60, respectively), the governance dimension lags in innovation (3.60), suggesting a potential gap in adopting cutting-edge technologies or novel strategies to enhance transparency and operational efficiency. Addressing this gap could significantly elevate governance performance, particularly by integrating digital tools for decision making and data management.

Moreover, the social and economic dimensions demonstrate a high degree of synergy, particularly in labor inclusion projects that address economic inequalities while fostering social cohesion. This dual impact highlights the importance of cross-sector collaborations, in which public and private partnerships can amplify the reach and effectiveness of food bank initiatives. For instance, targeted training programs for volunteers and beneficiaries not only improve employability but also build a resilient support network within vulnerable communities.

Interestingly, while replicability scores are consistently high across all dimensions, the capacity to scale up environmental initiatives, such as renewable energy integration, may face operational and financial barriers. Developing standardized frameworks or securing dedicated funding streams could mitigate these challenges, enabling the broader adoption of these practices across food banks in different regions.

Finally, at the global level, replicability is highly valued in the social (4.9), economic (4.7), and environmental (4.6) dimensions, with numerous projects and actions serving as a reference for other FBs, in other places, and in different contexts.

3.5. Interpretation and Implications of the Research Results

The findings presented in this article provide a comprehensive perspective on the contribution of FBs to sustainability [28,92,95], addressing social [81,82,84], economic [22,92], environmental [33,88], and governance [93,94] dimensions and allowing for an analysis aligned with the CFS-RAI principles [4,72] and the SDGs. These results not only reinforce the relevance of FBs in achieving global sustainability goals but also provide a theoretical framework that enriches the existing literature in the field of food security and sustainable development.

Furthermore, this theoretical integration opens up new opportunities to investigate how other organizations can adopt similar strategies to address social issues.

Through the identification of good practices and innovative strategies, FBs have proven to be key actors in the fight against food insecurity and the promotion of sustainable development.

While these documented practices show high potential for sustainability, this study also reveals significant challenges that limit their implementation. One of the main challenges is the lack of stable human and financial resources, a concern shared by Kruse [125], who highlighted how reliance on volunteerism hinders the operational sustainability of FBs. Similarly, some sustainable initiatives, while effective, can be costly and difficult to replicate in different settings.

Similarly, the analysis identifies FB governance as an area for improvement, as there are numerous tools that can be implemented, such as the use of information technology [29], which, although it has certain limitations, can be adapted to the needs of users. Training and education are critical for food banks to effectively use technology tools [133].

In this context, the establishment of public–private partnerships emerges as a crucial strategy to mitigate financial, logistical, and technical support barriers. Recent studies, such as those by Santini and Cavicchi [126] and Dodd and Nelson [127], have pointed out that these collaborations not only strengthen support networks but also generate positive impacts on society. This study reinforces this view by demonstrating how FESBAL and its 54 FBs have articulated effective networks that enhance operational sustainability.

Future research could focus on developing specific tools that strengthen the capacity of FBs to operate sustainably. This includes the creation of replicable governance models, the use of accessible technologies, and the identification of new sources of funding.

4. Conclusions

The purpose of this study was to capitalize on experiences from food banks aligned with the SDGs and the CFS-RAI principles from the perspective of sustainability (social, environmental, economic, and governance), which is of great importance, since these experience have largely gone unexplored so far.

The results showed 152 good practices carried out by food banks in Spain, reflecting their contribution to the SDGs and CFS-RAI through a more comprehensive approach to food security and sustainable development.

Within the social dimension, awareness-raising activities against food loss and waste are highlighted; within the economic dimension, job placement opportunities are identified; within the environmental dimension, activities to reduce the impact on the environment are demonstrated; and from the governance dimension, improvements in management efficiency and transparency are highlighted.

This study also shows a change in the way food banks provide food assistance, demonstrating constant innovation and creativity in attracting volunteers and resources as well as raising awareness and transmitting their relevant message to society. It also highlights FBs' partnerships with public and private entities to achieve their mission and to strengthen their impact on and contribution to food security.

This research highlights that the 152 documented good practices not only represent immediate solutions to issues such as food insecurity and waste but also offer a framework that can be adapted and replicated across various organizations within the tertiary sector. By creating structured methods for sharing and documenting these initiatives, food banks can facilitate wider adoption and ensure that their practices have lasting and scalable impacts in diverse contexts.

Furthermore, the integration of food banks' efforts with broader sustainability policies, such as the Spanish Circular Economy Strategy (Spain 2030) and the European Green Deal, could significantly enhance their role in systemic change. Strengthening these connections would enable food banks to position themselves as essential contributors to national and international goals, particularly in addressing food waste, promoting responsible consumption (aligned with SDG 12), and reducing hunger (aligned with SDG 2).

The identified environmental initiatives, including renewable energy projects and reductions in carbon emissions, illustrate the transformative potential of such practices. However, limited financial resources and logistical hurdles often challenge their implementation. Establishing stronger collaborations with private companies and public institutions focused on sustainability could help to address these challenges by providing technical expertise and funding, which could accelerate the adoption of these environmental solutions.

In addition, this study identifies governance as an area with room for improvement. While steps have been taken to enhance transparency and efficiency, adopting advanced management practices, such as digital tools for monitoring and accountability, could further refine organizational processes. Enhanced governance structures would not only optimize the internal operations of food banks, but also increase trust among donors, volunteers, and beneficiaries.

Finally, community engagement remains an integral part of the impact of food banks. Expanding partnerships with schools, universities, and community organizations could help deepen public awareness of food waste issues and promote long-term cultural shifts toward sustainability. These educational efforts, combined with collaborations at the local level, would solidify the role of food banks as key drivers of sustainable development and social cohesion.

Author Contributions: Conceptualization: M.L.A.M., C.M.N., R.R.S.F. and M.E.C.; Methodology: M.L.A.M. and C.M.N.; Validation: M.L.A.M., C.M.N., R.R.S.F. and M.E.C.; Formal analysis: M.L.A.M. and C.M.N.; Investigation: M.L.A.M., C.M.N. and R.R.S.F.; Resources: M.L.A.M.; Data curation: M.L.A.M.; Writing—original draft preparation: M.L.A.M.; Writing—review and editing: M.L.A.M., C.M.N. and R.R.S.F.; Visualization: M.L.A.M. and C.M.N. and M.E.C.; Supervision: R.R.S.F.; Project administration: M.L.A.M.; Funding acquisition: M.L.A.M. All authors have read and agreed to the published version of the manuscript.

Funding: This research received no external funding.

Institutional Review Board Statement: Not applicable.

Informed Consent Statement: Informed consent was obtained from all subjects involved in this study.

Data Availability Statement: All data regarding the results of this research are available.

Conflicts of Interest: The authors declare no conflicts of interest.

References

1. Afonso, A. Hacia una Conducta Empresarial Coherente con los Principios CSA. In *Avanzando en los Principios de Inversión Responsable en la Agricultura y los Sistemas Alimentarios*; Afonso, A., Cazorla, A., Echevarría, E., Eds.; Working Paper; GESPLAN Group, Technical University of Madrid: Madrid, Spain, 2021.
2. Food and Agriculture Organization of the United Nations. FAO and the Sustainable Development Goals: Achieving the 2030 Agenda Through Empowerment of Local Communities. 2022. Available online: https://www.fao.org/3/cc2063en/online/cc2063en.html (accessed on 10 October 2023).
3. Committee on World Food Security. Principles for Responsible Investment in Agriculture and Food Systems. 2014. Available online: http://www.fao.org/3/au866s/au866s.pdf (accessed on 13 October 2023).
4. Requelme, N.; Afonso, A. The Principles for Responsible Investment in Agriculture (CFS-RAI) and SDG 2 and SDG 12 in Agricultural Policies: Case Study of Ecuador. *Sustainability* **2023**, *15*, 15985. [CrossRef]
5. Jiménez Aliaga, R.; De los Ríos-Carmenado, I.; San Martín Howard, F.; Calle Espinoza, S.; Huamán Cristóbal, A. Integration of the Principles of Responsible Investment in Agriculture and Food Systems CFS-RAI from the Local Action Groups: Towards a Model of Sustainable Rural Development in Jauja, Peru. *Sustainability* **2022**, *14*, 9663. [CrossRef]
6. Buele, I.; Zúñiga, D.; Tobar, L. Principles for Responsible Investement in Agriculture and Food Systems and Their Social Impact: Application to Universitary Projects. *Acad. Entrep. J.* **2021**, *27*, 1–13.
7. Jordana, J. Towards business conduct consistent with CSA principles. In *Advancing the Principles of Responsible Investment in Agriculture and Food Systems. International Seminars to Share and Capitalize on University and Agribusiness Experiences in Latin America and Spain*; Afonso, A., Cazorla, A., Echeverría, E., Eds.; Working Paper; GESPLAN Group: Madrid, Spain, 2021; pp. 37–39.
8. Cazorla, A.; De los Ríos, I.; Afonso, A. *Principles for a Responsible Investment in Agriculture and Voluntary Guidelines on the Responsible Governance of Tenure of Land: Links with University*; Working Paper; GESPLAN Group: Madrid, Spain, 2018. Available online: https://oa.upm.es/50898/ (accessed on 20 October 2023).
9. Trząskowska, M.; Łepecka, A.; Neffe-Skocińska, K.; Marciniak-Lukasiak, K.; Zielińska, D.; Szydłowska, A.; Bilska, B.; Tomaszewska, M.; Kołożyn-Krajewska, D. Changes in Selected Food Quality Components after Exceeding the Date of Minimum Durability—Contribution to Food Waste Reduction. *Sustainability* **2020**, *12*, 3187. [CrossRef]

10. Reusken, M.; Cruijssen, F.; Fleuren, H. A Food Bank Supply Chain Model: Optimizing Investments to Maximize Food Assistance. *Int. J. Prod. Econ.* **2023**, *261*, 108886. [CrossRef]
11. FAO; FIDA; OMS; PMA; UNICEF. *The State of Food Security and Nutrition in the World 2023: Urbanization, Agrifood Systems Transformation and Healthy Diets Across the Rural–Urban Continuum*; FAO: Rome, Italy, 2023. Available online: https://openknowledge.fao.org/handle/20.500.14283/cc3017en (accessed on 23 October 2023).
12. Food and Agriculture Organization of the United Nations. *Tracking Progress on Food and Agriculture-Related SDG Indicators 2022*; FAO: Rome, Italy, 2022. Available online: https://openknowledge.fao.org/handle/20.500.14283/cc1403en (accessed on 27 October 2023).
13. Marimuthu, S.; Saikumar, A.; Badwaik, L.S. Food Losses and Wastage within Food Supply Chain: A Critical Review of Its Generation, Impact, and Conversion Techniques. *Waste Dispos. Sustain. Energy* **2024**, *6*, 661–676. [CrossRef]
14. Gonçalves, C.; Saraiva, S.; Nunes, F.; Saraiva, C. Food Waste in Public Food Service Sector—Surplus and Leftovers. *Resources* **2023**, *12*, 120. [CrossRef]
15. De Boni, A.; Ottomano Palmisano, G.; De Angelis, M.; Minervini, F. Challenges for a Sustainable Food Supply Chain: A Review on Food Losses and Waste. *Sustainability* **2022**, *14*, 16764. [CrossRef]
16. Roodhuyzen, D.M.A.; Luning, P.A.; Fogliano, V.; Steenbekkers, L.P.A. Putting Together the Puzzle of Consumer Food Waste: Towards an Integral Perspective. *Trends Food Sci. Technol.* **2017**, *68*, 37–50. [CrossRef]
17. Farcas, A.C.; Galanakis, C.M.; Socaciu, C.; Pop, O.L.; Tibulca, D.; Paucean, A.; Jimborean, M.A.; Fogarasi, M.; Salanta, L.C.; Tofana, M.; et al. Food Security during the Pandemic and the Importance of the Bioeconomy in the New Era. *Sustainability* **2021**, *13*, 150. [CrossRef]
18. Stanciu, M.; Popescu, A.; Antonie, I.; Sava, C.; Nistoreanu, B.G. Good practices on reducing food waste throughout the food supply chain. *Amfiteatru Econ.* **2022**, *24*, 566–582. [CrossRef]
19. Lahiri, A.; Daniel, S.; Kanthapazham, R.; Vanaraj, R.; Thambidurai, A.; Peter, L.S. A Critical Review on Food Waste Management for the Production of Materials and Biofuel. *J. Hazard. Mater. Adv.* **2023**, *10*, 100266. [CrossRef]
20. Facchini, F.; Silvestri, B.; Digiesi, S.; Lucchese, A. Agri-Food Loss and Waste Management: Win-Win Strategies for Edible Discarded Fruits and Vegetables Sustainable Reuse. *Innov. Food Sci. Emerg. Technol.* **2023**, *83*, 103235. [CrossRef]
21. Aloysius, N.; Ananda, J.A. A Circular Economy Approach to Food Security and Poverty: A Case Study in Food Rescue in Sri Lanka. *Circ. Econ. Sustain.* **2023**, *3*, 1919–1940. [CrossRef]
22. Arcuri, S. Food Poverty, Food Waste and the Consensus Frame on Charitable Food Redistribution in Italy. *Agric. Hum. Values* **2019**, *36*, 263–275. [CrossRef]
23. Vittuari, M.; De Menna, F.; Gaiani, S.; Falasconi, L.; Politano, A.; Dietershagen, J.; Segrè, A. The Second Life of Food: An Assessment of the Social Impact of Food Redistribution Activities in Emilia Romagna, Italy. *Sustainability* **2017**, *9*, 1817. [CrossRef]
24. Bucknum, M.; Bentzel, D. Chapter 13—Food Banks as Local Food Champions: How Hunger Relief Agencies Invest in Local and Regional Food Systems. In *Institutions as Conscious Food Consumers*; Thottathil, S.E., Goger, A.M., Eds.; Academic Press: London, UK, 2019; pp. 285–305.
25. Mook, L.; Murdock, A.; Gundersen, C. Food Banking and Food Insecurity in High-Income Countries. *Voluntas* **2020**, *31*, 833–840. [CrossRef]
26. Tenuta, N.; Barros, T.; Teixeira, R.A.; Paes-Sousa, R. Brazilian Food Banks: Overview and Perspectives. *Int. J. Environ. Res. Public Health* **2021**, *18*, 12598. [CrossRef]
27. Spanish Federation of Food Banks (FESBAL). Available online: https://www.fesbal.org.es/ (accessed on 3 February 2024).
28. Raimo, N.; Rella, A.; Vitolla, F.; Sánchez-Vicente, M.-I.; García-Sánchez, I.-M. Corporate Social Responsibility in the COVID-19 Pandemic Period: A Traditional Way to Address New Social Issues. *Sustainability* **2021**, *13*, 6561. [CrossRef]
29. Akkerman, R.; Buisman, M.; Cruijssen, F.; de Leeuw, S.; Haijema, R. Dealing with Donations: Supply Chain Management Challenges for Food Banks. *Int. J. Prod. Econ.* **2023**, *262*, 108926. [CrossRef]
30. Coque, J.; González-Torre, P. Adapting Non-Profit Resources to New Social Demands: The Food Banks in Spain. *Sustainability* **2017**, *9*, 643. [CrossRef]
31. Thapa Karki, S.; Bennett, A.C.T.; Mishra, J.L. Reducing Food Waste and Food Insecurity in the UK: The Architecture of Surplus Food Distribution Supply Chain in Addressing the Sustainable Development Goals (Goal 2 and Goal 12.3) at a City Level. *Ind. Mark. Manag.* **2021**, *93*, 563–577. [CrossRef]
32. Närvänen, E.; Mesiranta, N.; Mattila, M.; Heikkinen, A. Introduction: A Framework for Managing Food Waste. In *Food Waste Management*; Närvänen, E., Mesiranta, N., Mattila, M., Heikkinen, A., Eds.; Springer International Publishing: Cham, Switzerland, 2020; pp. 1–24. ISBN 978-3-030-20560-7.
33. Shabanali Fami, H.; Aramyan, L.H.; Sijtsema, S.J.; Alambaigi, A. The Relationship between Household Food Waste and Food Security in Tehran City: The Role of Urban Women in Household Management. *Ind. Mark. Manag.* **2021**, *97*, 71–83. [CrossRef]
34. Penalver, J.G.; Aldaya, M.M. The Role of the Food Banks in Saving Freshwater Resources through Reducing Food Waste: The Case of the Food Bank of Navarra, Spain. *Foods* **2022**, *11*, 163. [CrossRef] [PubMed]

35. Patel, S.; Dora, M.; Hahladakis, J.N.; Iacovidou, E. Opportunities, Challenges and Trade-Offs with Decreasing Avoidable Food Waste in the UK. *Waste Manag. Res.* **2021**, *39*, 473–488. [CrossRef]
36. Quintanero, S.; De los Ríos, I.; Zuluaga, C.; Roncancio, M.; Acosta, L. Analysis of the Legal Framework on Food Waste in Spain: Role of Food Banks. FESBAL—Cátedra Banco de Alimentos UPM. 2022. Available online: https://drive.google.com/file/d/1HoVC_AZqJmLPj53OJ_AxCriBUiO2nAgc/view (accessed on 10 November 2023).
37. Safayet, M.; Casellas Connors, J.P.; Watson, M. Measuring Access to Food Banks and Food Pantries: A Scoping Review of Spatial Analysis Approaches. *Health Place* **2024**, *88*, 103251. [CrossRef]
38. Mossenson, S.; Giglia, R.; Pulker, C.E.; Chester, M.; McStay, C.; Pollard, C.M. Evidence for Initiating Food Safety Policy: An Assessment of the Quality and Safety of Donated Food at an Australian Food Bank. *Food Policy* **2024**, *123*, 102589. [CrossRef]
39. Rivera, A.F.; Smith, N.R.; Ruiz, A. A Systematic Literature Review of Food Banks' Supply Chain Operations with a Focus on Optimization Models. *J. Humanit. Logist. Supply Chain Manag.* **2023**, *13*, 10–25. [CrossRef]
40. Wetherill, M.S.; White, K.C.; Rivera, C.; Seligman, H.K. Challenges and Opportunities to Increasing Fruit and Vegetable Distribution through the US Charitable Feeding Network: Increasing Food Systems Recovery of Edible Fresh Produce to Build Healthy Food Access. *J. Hunger Environ. Nutr.* **2019**, *14*, 593–612. [CrossRef]
41. Vågsholm, I.; Arzoomand, N.S.; Boqvist, S. Food Security, Safety, and Sustainability—Getting the Trade-Offs Right. *Front. Sustain. Food Syst.* **2020**, *4*, 16. [CrossRef]
42. Cloke, P.; May, J.; Williams, A. The Geographies of Food Banks in the Meantime. *Prog. Hum. Geogr.* **2017**, *41*, 703–726. [CrossRef]
43. Watuleke, J. *The Role of Food Banks in Food Security in Uganda: The Case of the Hunger Project Food Bank, Mbale Epicentre*; Nordiska Afrikainstitutet: Uppsala, Sweden, 2015; p. 92.
44. González-Torre, P.L.; Coque, J. How Is a Food Bank Managed? Different Profiles in Spain. *Agric. Hum. Values* **2016**, *33*, 89–100. [CrossRef]
45. Berti, G.; Giordano, C.; Mininni, M. Assessing the Transformative Potential of Food Banks: The Case Study of Magazzini Sociali (Italy). *Agriculture* **2021**, *11*, 249. [CrossRef]
46. Michelini, L.; Principato, L.; Iasevoli, G. Understanding Food Sharing Models to Tackle Sustainability Challenges. *Ecol. Econ.* **2018**, *145*, 205–217. [CrossRef]
47. De los Ríos Carmenado, I.; Polo, C.; Otero, P.; Castaño, J. Madrid Food Bank: 20 Years Working on Volunteering Projects in Favor of Needed People. In Proceedings of the 20th International Congress on Project Management and Engineering, Cartagena, Spain, 13–15 July 2016; AEIPRO: Cartagena, Spain, 2016.
48. Capodistrias, P.; Szulecka, J.; Corciolani, M.; Strøm-Andersen, N. European Food Banks and COVID-19: Resilience and Innovation in Times of Crisis. *Socioecon. Plan. Sci.* **2022**, *82*, 101187. [CrossRef]
49. Eicher-Miller, H.A. A Review of the Food Security, Diet and Health Outcomes of Food Pantry Clients and the Potential for Their Improvement through Food Pantry Interventions in the United States. *Physiol. Behav.* **2020**, *220*, 112871. [CrossRef]
50. Middleton, G.; Mehta, K.; McNaughton, D.; Booth, S. The Experiences and Perceptions of Food Banks amongst Users in High-Income Countries: An International Scoping Review. *Appetite* **2018**, *120*, 698–708. [CrossRef]
51. Bazerghi, C.; McKay, F.H.; Dunn, M. The Role of Food Banks in Addressing Food Insecurity: A Systematic Review. *J. Community Health* **2016**, *41*, 732–740. [CrossRef]
52. Ataseven, C.; Nair, A.; Ferguson, M. The Role of Supply Chain Integration in Strengthening the Performance of Not-for-Profit Organizations: Evidence from the Food Banking Industry. *J. Humanit. Logist. Supply Chain Manag.* **2020**, *10*, 101–123. [CrossRef]
53. Galli, F.; Cavicchi, A.; Brunori, G. Food Waste Reduction and Food Poverty Alleviation: A System Dynamics Conceptual Model. *Agric. Hum. Values* **2019**, *36*, 289–300. [CrossRef]
54. Holmes, E.; Black, J.L.; Heckelman, A.; Lear, S.A.; Seto, D.; Fowokan, A.; Wittman, H. "Nothing Is Going to Change Three Months from Now": A Mixed Methods Characterization of Food Bank Use in Greater Vancouver. *Soc. Sci. Med.* **2018**, *200*, 129–136. [CrossRef] [PubMed]
55. Iafrati, S. "We're Not a Bottomless Pit": Food Banks' Capacity to Sustainably Meet Increasing Demand. *Volunt. Sect. Rev.* **2018**, *9*, 39–53. [CrossRef]
56. Afonso, A.; Sastre, S. Good Practices for the Management of Food Banks: Capitalization of Experiences. FESBAL—Cátedra Banco de Alimentos—UPM. 2017. Available online: https://drive.google.com/file/d/1hi-UPxffxQF0TUU5OwCtscP4cMPjKpLD/view (accessed on 15 October 2023).
57. Hebinck, A.; Galli, F.; Arcuri, S.; Carroll, B.; O'Connor, D.; Oostindie, H. Capturing Change in European Food Assistance Practices: A Transformative Social Innovation Perspective. *Local Environ.* **2018**, *23*, 398–413. [CrossRef]
58. Slowinski, G.; Sagal, M.W. Good Practices in Open Innovation. *Res. Technol. Manag.* **2010**, *53*, 38–45. [CrossRef]
59. Vettoretto, L. A Preliminary Critique of the Best and Good Practices Approach in European Spatial Planning and Policy-Making. *Eur. Plan. Stud.* **2009**, *17*, 1067–1083. [CrossRef]
60. Coffield, F.; Edward, S. Rolling out 'Good', 'Best' and 'Excellent' Practice. What Next? Perfect Practice? *Br. Educ. Res. J.* **2009**, *35*, 371–390. [CrossRef]

61. Burijovich, J. El concepto de buenas prácticas en salud: Desde un enfoque prescriptivo a uno comprensivo. In *Coloquios de Investigación Cualitativa: Subjetividades y Procesos Sociales*; Universidad Nacional de Córdoba: Córdoba, Argentina, 2011; pp. 29–40.
62. Pernas, R.G.; Villar, M.B.C. Good Practices as a Resource for Community Action: Identification and Search Criteria. *Contextos Educ. Rev. De Educ.* **2016**, *19*, 75–88.
63. Koumalatsou, E. Conceptualizing the Social Best Practices at the International and European Level. *J. Community Posit. Pract.* **2011**, *4*, 126–144.
64. Bascón, M.; Rebollo, M.A.; Prados, M.M.; Saavedra, J.; Sala, A.; Ignacio, M.J. Analysis of Good Coeducational Practices within the Framework of the Teon XXI Project: A Sociocultural Approach. In Proceedings of the II National University Congress on Research and Gender, Sevilla, Spain, 17–18 June 2010.
65. Marquès, P. Good Teaching Practices. *Rev. DIM* **2002**, *19*, 1–14.
66. González Ramírez, T. The Concept of "Best Practices": Origin and Development. *Prim. Not. Comun. Y Pedagog.* **2007**, *222*, 32–35.
67. Epper, R.; Bates, A.W. *Teaching Faculty How to Use Technology: Best Practices from Leading Institutions*; UOC: Barcelona, Spain, 2004.
68. Escudero Muñoz, J.M. Good Practices and Extraordinary Programs of Attention to Students at Risk of Educational Exclusion. *Profesorado. Rev. Currículum Form. Profr.* **2009**, *13*, 107–141.
69. Rebollo Catalán, M.Á.; García Pérez, R.; Piedra de la Cuadra, J.; Vega Caro, L. Diagnosis of Gender Culture in Education: Teachers' Attitudes towards Equality. *Rev. Educ.* **2011**, *355*, 521–546.
70. Boza, A.; Toscano, M.O. Good Practices in ICT Integration in Education in Andalusia: Two Case Studies. In Proceedings of the VI Virtual Congress, AIDIPE. 2011. Available online: http://www.uv.es/aidipe/congresos/Ponencia_VIICongresoVirtual_AIDIPE.pdf (accessed on 15 October 2023).
71. Harrisson, D.; Laberge, M. Innovation, Identities and Resistance: The Social Construction of an Innovation Network. *J. Manag. Stud.* **2002**, *39*, 497–521. [CrossRef]
72. Gashu Adam, A.; Wenedem Agegnehu, A.; Tiruneh Dessie, B.; Sewunet Minale, A. Responsible Governance of Land Tenure in the Context of Agricultural Investment from Theory to Practice: Lessons from Ethiopia. *Res. Glob.* **2023**, *6*, 100131. [CrossRef]
73. GESPLAN. Research Group on Planning and Sustainable Management of Rural-Local Development. Available online: https://ruraldevelopment.es/index.php/es/ (accessed on 23 March 2024).
74. United Nations. *Transforming Our World: The 2030 Agenda for Sustainable Development*; United Nations: New York, NY, USA, 2015. Available online: https://sustainabledevelopment.un.org/content/documents/21252030%20Agenda%20for%20Sustainable%20Development%20web.pdf (accessed on 18 February 2024).
75. Barbier, E.B.; Burgess, J.C. Introduction to the SDGs. In *Economics of the SDGs*; Springer International Publishing: Cham, Switzerland, 2021; pp. 3–13, ISBN 978-3-030-78697-7.
76. Tremblay, D.; Fortier, F.; Boucher, J.-F.; Riffon, O.; Villeneuve, C. Sustainable Development Goal Interactions: An Analysis Based on the Five Pillars of the 2030 Agenda. *Sustain. Dev.* **2020**, *28*, 1584–1596. [CrossRef]
77. Dzhunushalieva, G.; Teuber, R. Roles of Innovation in Achieving the Sustainable Development Goals: A Bibliometric Analysis. *J. Innov. Knowl.* **2024**, *9*, 100472. [CrossRef]
78. Hansmann, R.; Mieg, H.A.; Frischknecht, P. Principal Sustainability Components: Empirical Analysis of Synergies between the Three Pillars of Sustainability. *Int. J. Sustain. Dev. World Ecol.* **2012**, *19*, 451–459. [CrossRef]
79. Zhu, J.; Zhai, Y.; Feng, S.; Tan, Y.; Wei, W. Trade-Offs and Synergies among Air-Pollution-Related SDGs as Well as Interactions between Air-Pollution-Related SDGs and Other SDGs. *J. Clean. Prod.* **2022**, *331*, 129890. [CrossRef]
80. Song, J.; Jang, C.-H. Unpacking the Sustainable Development Goals (SDGs) Interlinkages: A Semantic Network Analysis of the SDGs Targets. *Sustain. Dev.* **2023**, *31*, 2784–2796. [CrossRef]
81. Fraj-Andrés, E.; Herrando, C.; Lucia-Palacios, L.; Pérez-López, R. Informative Initiatives as a Useful Tool to Raise Awareness of Food Waste: An Application to Higher Education. *Int. J. Sustain. High. Educ.* **2023**, *24*, 840–858. [CrossRef]
82. Soares Pinto, R.; Machado dos Santos Pinto, R.; Silva Melo, F.; Santos Campos, S.; Marques-dos-Santos, C. A Simple Awareness Campaign to Promote Food Waste Reduction in a University Canteen. *Waste Manag.* **2018**, *76*, 28–38. [CrossRef] [PubMed]
83. Redman, E.; Redman, A. Transforming Sustainable Food and Waste Behaviors by Realigning Domains of Knowledge in Our Education System. *J. Clean. Prod.* **2014**, *64*, 147–157. [CrossRef]
84. Ellison, B.; Savchenko, O.; Nikolaus, C.J.; Duff, B.R.L. Every Plate Counts: Evaluation of a Food Waste Reduction Campaign in a University Dining Hall. *Resour. Conserv. Recycl.* **2019**, *144*, 276–284. [CrossRef]
85. Pearson, D.; Perera, A. Reducing Food Waste: A Practitioner Guide Identifying Requirements for an Integrated Social Marketing Communication Campaign. *Soc. Mar. Q.* **2018**, *24*, 45–57. [CrossRef]
86. Aschemann-Witzel, J.; de Hooge, I.E.; Rohm, H.; Normann, A.; Bossle, M.B.; Grønhøj, A.; Oostindjer, M. Key Characteristics and Success Factors of Supply Chain Initiatives Tackling Consumer-Related Food Waste—A Multiple Case Study. *J. Clean. Prod.* **2017**, *155*, 33–45. [CrossRef]
87. Zamri, G.B.; Azizal, N.K.A.; Nakamura, S.; Okada, K.; Nordin, N.H.; Othman, N.; MD Akhir, F.N.; Sobian, A.; Kaida, N.; Hara, H. Delivery, Impact and Approach of Household Food Waste Reduction Campaigns. *J. Clean. Prod.* **2020**, *246*, 118969. [CrossRef]

88. Damiani, M.; Pastorello, T.; Carlesso, A.; Tesser, S.; Semenzin, E. Quantifying Environmental Implications of Surplus Food Redistribution to Reduce Food Waste. *J. Clean. Prod.* **2021**, *289*, 125813. [CrossRef]
89. Penalver, J.G.; Armijos, A.; Soret, B.; Aldaya, M.M. Food Banks against Climate Change, a Solution That Works: A Case Study in Navarra, Spain. *Foods* **2022**, *11*, 3645. [CrossRef]
90. Albizzati, P.F.; Tonini, D.; Chammard, C.B.; Astrup, T.F. Valorisation of Surplus Food in the French Retail Sector: Environmental and Economic Impacts. *Waste Manag.* **2019**, *90*, 141–151. [CrossRef]
91. Eriksson, M.; Spångberg, J. Carbon Footprint and Energy Use of Food Waste Management Options for Fresh Fruit and Vegetables from Supermarkets. *Waste Manag.* **2017**, *60*, 786–799. [CrossRef]
92. Schneider, F. The Evolution of Food Donation with Respect to Waste Prevention. *Waste Manag.* **2013**, *33*, 755–763. [CrossRef] [PubMed]
93. De Los Ríos Carmenado, I.; Mur Nuño, C. Inclusive and Transparent Governance Structures from RIA Principles: Case FESBAL Food Bank. In Proceedings of the 27th International Congress on Project Management and Engineering, Donostia-San Sebastián, Spain, 10–13 July 2023; AEIPRO: Donostia-San Sebastian, Spain, 2023. [CrossRef]
94. Fernández Moscol, M.E.; Mogollón Calderón, A. Criterios ASG: Trends and some notes on the situation in Peru. *Thēmis* **2022**, *81*, 77–88. [CrossRef]
95. Bergström, P.; Malefors, C.; Strid, I.; Hanssen, O.J.; Eriksson, M. Sustainability Assessment of Food Redistribution Initiatives in Sweden. *Resources* **2020**, *9*, 27. [CrossRef]
96. Cazorla, A.; De Los Ríos, I.; Salvo, M. Working with People (WWP) in Rural Development Projects: A Proposal from Social Learning. *Cuad. Desarro. Rural* **2013**, *10*, 131–157. Available online: https://www.redalyc.org/pdf/117/11726378007.pdf (accessed on 9 February 2024).
97. Cazorla-Montero, A.; De Los Ríos-Carmenado, I.; Díaz-Puente, J.M. The LEADER Community Initiative as a Rural Development Model: Application in the Capital Region of Spain. *Agrociencia* **2005**, *39*, 697–708. Available online: https://www.cabidigitallibrary.org/doi/full/10.5555/20053201973 (accessed on 9 February 2024).
98. Cazorla, A.; Negrillo, X.; Montalvo, V.; De Nicolas, V.L. Institutional Structuralism as a Process to Achieve Social Development: Aymara Women's Community Project Based on the Working with People Model in Peru. *J. Sociol. Soc. Welf.* **2018**, *45*, 55–57. Available online: https://scholarworks.wmich.edu/jssw/vol45/iss4/5/ (accessed on 28 February 2024). [CrossRef]
99. Castañeda, R.; Avila Cerón, C.; De Los Ríos-Carmenado, I.; Domínguez, L.; Gomez, S. Implementing the Voluntary Guidelines on the Responsible Governance of Tenure of Land, Fisheries and Forests from the Working with People Model: Lessons from Colombia and Guatemala. *J. Peasant Stud.* **2023**, *50*, 1820–1851. [CrossRef]
100. Rivera, M.; Knickel, K.; De Los Rios, I.; Ashkenazy, A.; Pears, D.Q.; Chebach, T.; Šūmane, S. Rethinking the Connections between Agricultural Change and Rural Prosperity: A Discussion of Insights Derived from Case Studies in Seven Countries. *J. Rural Stud.* **2018**, *59*, 242–251. [CrossRef]
101. Herrera-Reyes, A.; De Los Ríos Carmenado, I.; Martínez-Almela, J. Project-Based Governance Framework for an Agri-Food Cooperative. *Sustainability* **2018**, *10*, 1881. [CrossRef]
102. Fontana, A.; De Los Ríos Carmenado, I.; Villanueva-Penedo, J.; Ulloa-Salazar, J.; Santander-Peralta, D. Strategy for the Sustainability of a Food Production System for the Prosperity of Low-Income Populations in an Emerging Country: Twenty Years of Experience of the Peruvian Poultry Association. *Sustainability* **2018**, *10*, 4035. [CrossRef]
103. De Los Ríos-Carmenado, I.; Cadena-Iñiguez, J.; Díaz-Puente, J.M. Creating Local Action Groups for Rural Development in Mexico: Methodological Approach and Lessons Learned. *Agrociencia* **2011**, *45*, 815–829. Available online: https://www.scielo.org.mx/scielo.php?pid=S1405-31952011000700007&script=sci_abstract&tlng=en (accessed on 22 January 2024).
104. Cazorla, A.; De Los Ríos, I. Rural Community Management in the Global Economy: A Planning Model Based on Innovation and Entrepreneurial Activity. In *Rural Communities in the Global Economy: Beyond the Classical Rural Economy Paradigms*; Nicolae, I., De Los Ríos, I., Vasile, A., Eds.; Nova Science Publishers, Inc.: New York, NY, USA, 2017; pp. 1–22.
105. Díaz-Puente, J.M.; Cazorla, A.; De Los Ríos, I. Empowering Communities through Evaluation: Some Lessons from LEADER. *Community Dev. J.* **2009**, *44*, 53–67. [CrossRef]
106. López, M.; Cazorla, A.; Panta, M. Rural Entrepreneurship Strategies: Empirical Experience in the Northern Sub-Plateau of Spain. *Sustainability* **2019**, *11*, 1243. [CrossRef]
107. Stratta, R.; De Los Ríos, I.; López, M. Developing Competencies for Rural Development Project Management through Local Action Groups: The Punta Indio (Argentina) Experience. In *International Development*; Appiah-Opoku, S., Ed.; IntechOpen: London, UK, 2017.
108. De Los Ríos, I.; Rivera, M.; García, C. Redefining Rural Prosperity through Social Learning in the Cooperative Sector: 25 Years of Experience from Organic Agriculture in Spain. *Land Use Policy* **2016**, *54*, 85–94. [CrossRef]
109. Friedmann, J. *Empowerment: The Politics of Alternative Development*; Basil Blackwell Publishers: Cambridge, UK, 1992.

110. Nole Correa, P.; Aliaga, G.; De Los Ríos, I. Service-Learning Projects for Responsible Consumption: 10 Years of Experience from the Food Bank Chair UPM. In Proceedings of the 27th International Congress on Project Management and Engineering, Donostia-San Sebastián, Spain, 10–13 July 2023; AEIPRO: Donostia-San Sebastian, Spain, 2023. [CrossRef]
111. De Los Ríos Carmenado, I.; Afonso Gallegos, A.; Quintanero Lahoz, S.; Ortega Rincón, R.; Nole Correa, P.; Zuluaga, C.; Roncancio Burgos, M. FESBAL-UPM Food Bank Chair and the Service-Learning Projects from the 'Working with People' Perspective. In *Crisis After the Crisis: Economic Development in the New Normal*; Chivu, L., De Los Ríos Carmenado, I., Andrei, J.V., Eds.; Springer International Publishing: Cham, Switzerland, 2023; pp. 79–90, ISBN 978-3-031-30995-3.
112. Carrillo-Álvarez, E.; Costa-Tutusaus, L.; Milà-Villaroel, R.; Shankar Krishnan, N.; Salinas-Roca, B. Food Insecurity and Dietary Diversity among Food Aid Recipients in Urban and Rural Areas at the Covid-19 Outbreak: An Exploratory Study. *World Food Pol.* **2024**, *10*, 33–47. [CrossRef]
113. National Institute of Statistics. Living Conditions Survey 2023. Available online: https://www.ine.es/dyngs/Prensa/ECV2023.htm (accessed on 10 January 2024).
114. Phantuwongraj, S.; Chenrai, P.; Assawincharoenkij, T. Pilot Study Using ArcGIS Online to Enhance Students' Learning Experience in Fieldwork. *Geosciences* **2021**, *11*, 357. [CrossRef]
115. Bender, D.E.; Ewbank, D. The Focus Group as a Tool for Health Research: Issues in Design and Analysis. *Health Transit. Rev.* **1994**, *4*, 63–80. Available online: https://www.jstor.org/stable/40652078 (accessed on 6 January 2025).
116. Rabiee, F. Focus-Group Interview and Data Analysis. *Proc. Nutr. Soc.* **2004**, *63*, 655–660. [CrossRef]
117. Timmermans, S.; Tavory, I. Theory Construction in Qualitative Research: From Grounded Theory to Abductive Analysis. *Sociol. Theory* **2012**, *30*, 167–186. [CrossRef]
118. South, L.; Saffo, D.; Vitek, O.; Dunne, C.; Borkin, M.A. Effective Use of Likert Scales in Visualization Evaluations: A Systematic Review. *Comput. Graph. Forum* **2022**, *41*, 3. [CrossRef]
119. Likert, R. A Technique for the Measurement of Attitudes. *Arch. Psychol.* **1932**, *22*, 55.
120. Barbero, M.I. *Psicometría II. Métodos de Elaboración de Escalas*; UNED: Madrid, Spain, 1993.
121. Hecht, A.A.; Neff, R.A. Food Rescue Intervention Evaluations: A Systematic Review. *Sustainability* **2019**, *11*, 6718. [CrossRef]
122. Shaheen, I.; Azadegan, A.; Davis, D. Humanitarian Supply Chains and Innovation: A Focus on US Food Banks. *Int. J. Oper. Prod. Manag.* **2023**, *43*, 1920–1942. [CrossRef]
123. Power, M.; Doherty, B.; Pybus, K.J.; Pickett, K.E. How COVID-19 Has Exposed Inequalities in the UK Food System: The Case of UK Food and Poverty. *Emerald Open Res.* **2023**, *1*, 10. [CrossRef]
124. Agostinho, D.; Paço, A. Analysis of the Motivations, Generativity and Demographics of the Food Bank Volunteer. *Int. J. Non-Profit Volunt. Sect. Mark.* **2012**, *17*, 249–261. [CrossRef]
125. Kruse, D.J.; Goeldner, M.; Eling, K.; Herstatt, C. Looking for a Needle in a Haystack: How to Search for Bottom-up Social Innovations That Solve Complex Humanitarian Problems. *J. Prod. Innov. Manag.* **2019**, *36*, 671–694. [CrossRef]
126. Santini, C.; Cavicchi, A. The Adaptive Change of the Italian Food Bank Foundation: A Case Study. *Br. Food J.* **2014**, *116*, 1446–1459. [CrossRef]
127. Dodd, W.; Nelson, E. Shifting Discourse and Practice on Food Banks: Insights from a Community–University Partnership. *Voluntas* **2020**, *31*, 881–893. [CrossRef]
128. Talavera, C.; Sanchis, J.R. Alliances between For-Profit and Non-profit Organizations as an Instrument to Implement the Economy for the Common Good. *Sustainability* **2020**, *12*, 9511. [CrossRef]
129. Porter, M.E.; Kramer, M.R. Creating Shared Value. How to Reinvent Capitalism—And Unleash a Wave of Innovation and Growth. *Harv. Bus. Rev.* **2011**, *89*, 62–77.
130. Florini, A.; Pauli, M. Collaborative Governance for the Sustainable Development Goals. *Asia Pac. Policy Stud.* **2018**, *5*, 583–598. [CrossRef]
131. Rizvi, A.; Wasfi, R.; Enns, A.; Kristjansson, E. The Impact of Novel and Traditional Food Bank Approaches on Food Insecurity: A Longitudinal Study in Ottawa, Canada. *BMC Public Health* **2021**, *21*, 771. [CrossRef]
132. Vitiello, D.; Grisso, J.A.; Whiteside, K.L.; Fischman, R. From Commodity Surplus to Food Justice: Food Banks and Local Agriculture in the United States. *Agric. Hum. Values* **2015**, *32*, 419–430. [CrossRef]
133. Martin, K.S.; Redelfs, A.; Wu, R.; Bogner, O.; Whigham, L. Offering More Than Food: Outcomes and Lessons Learned from a Fresh Start Food Pantry in Texas. *J. Hunger Environ. Nutr.* **2019**, *14*, 70–81. [CrossRef]
134. Dave, J.M.; Thompson, D.I.; Svendsen-Sanchez, A.; McNeill, L.H.; Jibaja-Weiss, M. Development of a Nutrition Education Intervention for Food Bank Clients. *Health Promot. Pract.* **2017**, *18*, 221–228. [CrossRef]
135. Tarasuk, V.; Eakin, J.M. Charitable Food Assistance as Symbolic Gesture: An Ethnographic Study of Food Banks in Ontario. *Soc. Sci. Med.* **2003**, *56*, 1505–1515. [CrossRef]
136. Sundin, N.; Bartek, L.; Persson Osowski, C.; Strid, I.; Eriksson, M. Sustainability Assessment of Surplus Food Donation: A Transfer System Generating Environmental, Economic, and Social Values. *Sustain. Prod. Consum.* **2023**, *38*, 41–54. [CrossRef]

137. Edge, S.; Meyer, S.B. Pursuing Dignified Food Security through Novel Collaborative Governance Initiatives: Perceived Benefits, Tensions and Lessons Learned. *Soc. Sci. Med.* **2019**, *232*, 77–85. [CrossRef]
138. Kowalewska, M.T.; Kołłajtis-Dołowy, A. Food, Nutrient, and Energy Waste among School Students. *Br. Food J.* **2018**, *120*, 1807–1831. [CrossRef]
139. Reynolds, C.; Piantadosi, J.; Boland, J. Rescuing Food from the Organics Waste Stream to Feed the Food Insecure: An Economic and Environmental Assessment of Australian Food Rescue Operations Using Environmentally Extended Waste Input-Output Analysis. *Sustainability* **2015**, *7*, 4707–4726. [CrossRef]
140. Cárcaba, A.; González, E.; Ventura, J.; Arrondo, R. How Does Good Governance Relate to Quality of Life? *Sustainability* **2017**, *9*, 631. [CrossRef]
141. Michel-Villarreal, R.; Vilalta-Perdomo, E.L.; Canavari, M.; Hingley, M. Resilience and Digitalization in Short Food Supply Chains: A Case Study Approach. *Sustainability* **2021**, *13*, 5913. [CrossRef]
142. Popielarski, J.A.; Cotugna, N. Fighting Hunger through Innovation: Evaluation of a Food Bank's Social Enterprise Venture. *J. Hunger Environ. Nutr.* **2010**, *5*, 56–69. [CrossRef]

sustainability

MDPI

Article

Strengthening Production Systems in Social Organizations: Application of IRA Principles and the WWP Model in the Tejemujeres Cooperative

Mauricio Ortuño [1,*], Ricardo Grunauer [2,*], Milagros Panta [3] and Xavier Negrillo [1]

1 GESPLAN Research Group, Escuela Técnica Superior de Ingeniería Agronómica, Alimentaria y de Biosistemas, Universidad Politécnica de Madrid (UPM), 28040 Madrid, Spain; xavier.negrillo@upm.es

2 Dirección de Extensión y Vinculación con la Sociedad, Universidad Bolivariana del Ecuador (UBE), Durán 092405, Ecuador

3 Faculty of Economics and Business Administration, Professional School of Administration and Management, Ricardo Palma University (URP), Av. Alfredo Benavides 5440, Santiago de Surco, Lima 15039, Peru; milagros.panta@urp.edu.pe

* Correspondence: m.ortuno@alumnos.upm.es (M.O.); rgrunauer@ube.edu.ec (R.G.)

Abstract

The overall objective of the study was to propose the application of the IRA principles and the WWP model in the Tejemujeres Cooperative, with the aim of strengthening its production and management system without compromising its social identity. To this end, a mixed descriptive and explanatory methodology was used. Surveys were conducted among the organization's 110 members, and focus groups were conducted with internal and external stakeholders, in addition to a review of documents and bibliographic sources. This revealed structural limitations in the production system, such as a shortage of raw materials, low innovation, marketing difficulties, and limited technical training. However, the perception of economic sustainability remained positive, thanks to the social and cultural cohesion of the cooperative. Likewise, most of the members expressed openness to incorporating IRA principles and the WWP model, highlighting training, active participation in decision-making, strengthening internal governance, and creating commercial networks as priorities. In conclusion, it was determined that Tejemujeres' main strength lies in its community identity and human capital, rather than in traditional economic indicators. The proposed theoretical frameworks were found to be relevant and adaptable to the context of the organization. Finally, a hybrid strategy is proposed that combines the participatory flexibility of the WWP model with the methodological rigor of the IRA principles, which will enable the cooperative to consolidate an innovative, sustainable, and culturally legitimate production system.

Keywords: cooperatives; IRA principles; WWP model; social identity; social organization

1. Introduction

Strengthening production systems in social organizations represents a crucial, yet structural, challenge in rural development efforts, due to a consistent dilemma between productive efficiency and the community values that guide their collective work, which limits their capacity to promote viable economic processes without abandoning their principles of solidarity and participation [1]. Given these tensions, add that consolidating productive strategies in these areas requires understanding that economic activity is intertwined with local identities and social relations that influence daily management; otherwise, a direct impact on sustainability and the capacity to generate economic and social value persists [2]. Similarly, Hennchen and Schäfer [3] agree that the absence of productive adaptability measures, such as internal cohesion and inclusiveness, limits development and success based on more sustainable systems.

In this regard, studies such as that by Jiménez et al. [4] highlight that organizational strengthening or productive sustainability requires the integration of key instruments that promote a balance between efficiency, equity, and participation. However, as another limitation, Jiménez et al. [5] demonstrated that a lack of management capacities, clear governance mechanisms, or articulated planning processes negatively impacts the production system, limiting its integration into broader value chains and hindering more resilient rural development. Thus, the Principles for Responsible Investment in Agriculture (RIA) have emerged as a benchmark for investing in more inclusive and environmentally sustainable actions, prioritizing respect for the rights of rural communities. It is evident that, in rural areas of Peru, their integration has led to stronger organizational resilience and concrete collaboration among local, public, and private actors [4].

While the Working With People (WWP) model has become established as a perspective focused on integrating local and technical knowledge for the design of collaborative processes [6], its focus on ethical, methodological, and technical dimensions has proven useful in boosting management capacities, energizing participation, and generating contextualized solutions to production challenges in rural communities [7].

Thus, the integration of both models, IRA and WWP, contributes to a more robust analysis and management of social organizations, guiding them to generate added value and consolidate more efficient productive structures [6]. Furthermore, Regalado et al. [7] highlighted that cooperative systems with a high level of application of both models exhibit greater levels of stability and sustainable development compared to those that still maintain traditional models. Likewise, Jiménez [8] affirms that this integration promoted an improvement in the sustainability and resilience of rural cooperatives in Andean contexts, strengthening their internal governance and increasing their productive capacity without compromising their community values, thereby fostering sustainable rural development.

Given the above, this paper examines how both models can contribute to understanding and strengthening production systems in social organizations, as well as their challenges and opportunities. Therefore, the Tejemujeres Cooperative, an organization made up of women artisans dedicated to weaving and solidarity microcredit mechanisms, is considered as a case study, operating within a logical community context. This constitutes a key space for analyzing, from a real-world environment, how the IRA and WWP approaches can be jointly operationalized in production processes linked to dynamics of mutual aid, time availability, and community participation, where they clearly seek to achieve a balance between efficiency and community. Based on the above, this article has the general objective of analyzing the integration of the IRA principles and the WWP model as an operational framework to strengthen production systems and the internal management of social organizations. To achieve this, three specific objectives are proposed: (a) to describe the structural and organizational limitations of the production system in the Tejemujeres Cooperative in relation to sustainable production systems; (b) to identify the elements of the IRA principles and the WWP model that are relevant for their application in the reality of the cooperative; (c) to propose an integrated IRA-WWP model for the productive strengthening of the Tejemujeres Cooperative without compromising its social identity. Consequently, the study is oriented towards answering the following question: how can the integrated application of the IRA principles and the WWP model contribute to strengthening the production system and the management of a social organization without compromising its community identity?

https://doi.org/10.3390/su18052661

Development

This model is based on the premise that collective action generates a flow of formal and informal knowledge, which, when exchanged, allows for the creation of more sustainable and prosperous rural areas. Therefore, models are needed that integrate the knowledge generated in social learning processes [1], emphasizing practical action, experience, the quality of social interaction and communication, from decision-making structures and the relationships between stakeholders, considering the input of external knowledge as a fundamental element for improving decision-making [9].

According to Mur et al. [10], it is a collaborative methodology oriented towards participatory management in sustainable development projects. It is based on co-responsibility, active participation, and the inclusion of diverse actors in decision-making, promoting the strengthening of local capacities and effective governance in social organizations such as the Tejemujeres Cooperative. Its application allows for the articulation of interdisciplinary knowledge, fosters equity, and guarantees long-term impacts in productive communities.

The WWP model emerged from studies on rural development and participatory project management with the aim of improving the relationship between communities, institutions, and productive sectors through active cooperation processes. Its application has been key in social impact, education, and governance projects, demonstrating its flexibility and effectiveness in different contexts [11]. Thus, Figure 1 describes the model as a constant interaction between organizations and individuals for the co-creation of solutions adapted to each context, always prioritizing four fundamental axes:

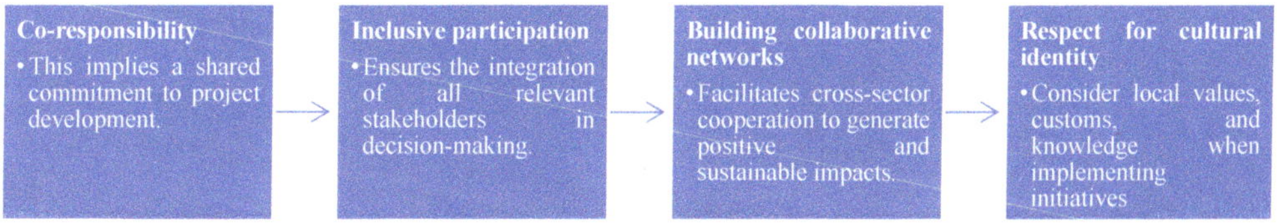

Figure 1. Key aspects of the WWP model. Note Jiménez et al. [5].

Furthermore, Vásquez [12] describes the principles of this model as fundamental to ensuring its effectiveness in any social and productive environment, where commitment implies shared responsibility among all project participants, promoting co-responsibility in decision-making and action execution. Transparency guarantees accessible communication throughout all implementation phases, allowing information to flow clearly and promptly.

Inclusion fosters equity in participation and decision-making, ensuring that all stakeholders have a voice in the process. Empowerment, for its part, focuses on developing capacities that allow those involved to achieve greater autonomy and leadership in project management. Finally, social innovation drives the creation of solutions tailored to local needs, promoting sustainable development and resilience in communities [12]. The implementation of the WWP model promotes the optimization of the production system through an efficient organizational structure, fostering equity in decision-making and strengthening community networks. Furthermore, it facilitates the adoption of innovative strategies for sustainable growth aligned with the principles of social responsibility, inclusive development, and, above all, based on its four components [13].

These principles were developed to ensure that investments in the agricultural sector contribute to sustainable development, respecting the rights of rural communities, promoting food security, and protecting natural resources [4]. These principles were established by the United Nations Conference on Trade and Development (UNCTAD), the Organization for Economic Co-operation and Development (OECD), the Food and Agriculture Organiza-

https://doi.org/10.3390/su18052661

tion of the United Nations (FAO), and the World Bank, to guide governments, investors, and agricultural stakeholders in making decisions that promote equitable and sustainable growth, as shown in Figure 2.

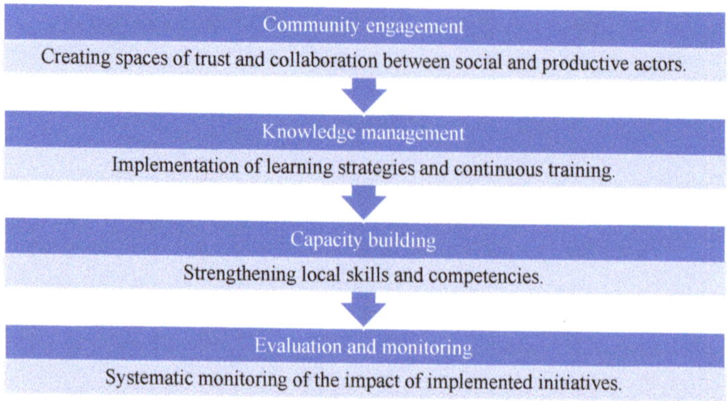

Figure 2. Components of the WWP model. Note. Adapted from Inquilla et al. [13].

The IRA principles are based on ten fundamental pillars, including respect for land tenure rights, strengthening food security, promoting sustainable agricultural practices, and generating economic and social benefits for the communities involved [14]. Furthermore, they seek to ensure that investments in agriculture are inclusive, transparent, and respectful of labor rights, fostering the participation of small-scale producers, women, and Indigenous communities in agricultural development. This allows for an equitable distribution of benefits and prevents forced displacement or the exploitation of natural resources without regard for the well-being of local populations.

2. Materials and Methods

The study was developed using a non-experimental, cross-sectional, and descriptive-explanatory design, based on a mixed-methods approach. This approach was deemed appropriate due to the need to understand the relevance of integrating the principles of Responsible Investment in Agriculture (RIA) and the Working with People (WWP) model as tools for strengthening the production system of the Tejemujeres Cooperative.

This social entity was selected for three reasons that make it a representative case of the general problem: (1) more than 14 years of consolidated organizational history under a horizontal structure and a solidarity economy approach, typical characteristics of rural social organizations in Latin America; (2) its intensive production activity based on artisanal labor, which reflects tensions between productivity, cultural identity, and sustainability, widely discussed in the literature on rural development, as described by Hennchen and Schäfer [3]; Finally, (3) availability of systematized institutional information to triangulate evidence according to IRA and WWP approaches [4,5]. Therefore, the case is not only an empirical unit, but also a suitable laboratory to operationalize the IRA principles and the dimensions of the WWP model in a real community context.

The study population consisted of the 110 active female members of the cooperative, all artisans dedicated to hand weaving and complementary activities. They varied in age, production experience, level of education, and degree of participation in the different internal committees, but were united by a common practice: artisanal production as a means of livelihood, a source of cultural identity, and a mechanism for a solidarity economy.

For this purpose, a non-probabilistic census sampling method was used, given the importance of the participation of all members for the quantitative component,

https://doi.org/10.3390/su18052661

while for the qualitative component, a small group of participants were selected using purposive sampling.

2.1. Techniques and Instruments

It was operationalized through three complementary components: a quantitative approach based on a survey, a qualitative approach developed through a focus group, and a documentary approach focused on the analysis of relevant information about the management of the cooperative.

2.2. Survey

The study involved administering a structured questionnaire to the 110 members, providing census data on key dimensions such as productive sustainability, internal governance, valuation of artisanal work, perception of the limitations of the current system, and preliminary knowledge of the IRA principles and the WWP model. The questionnaire included closed-ended questions, five-point Likert scales, and prioritization items, developed from the categories proposed by the IRA principles: participation, livelihoods, inclusion, and sustainability; and the dimensions of the WWP model: social interaction, capabilities, shared vision, and knowledge management. Before its final implementation, the instrument underwent expert validation and a subsequent pilot test with 12 members, allowing for adjustments to wording, cultural relevance, and conceptual clarity. The reliability of the scales was assessed using Cronbach's alpha, yielding values between 0.78 and 0.84 for the composite dimensions, which are adequate to ensure internal consistency.

Its processing and analysis were carried out using SPSS V.13 software, applying descriptive statistics, Chi-square tests, association analysis, and a K-Means cluster model to identify differentiated profiles around the implementation of the IRA–WWP model

2.3. Focus Group

The study was conducted through a focus group with 10 participants selected for their strategic roles within and outside the cooperative: artisan members, production and finance leaders, members of the Supervisory Board, designers involved in product development, and allied institutional stakeholders. This allowed for a deeper understanding of perceptions regarding the actual functioning of the production system, daily challenges, organizational sustainability, and the possibilities of integrating IRA principles and the WWP approach without compromising the organization's cultural identity.

The study was conducted in person, with audio recording and a written report, over two hours. The participants were asked guiding questions about a description of the current production system, operational and organizational challenges, economic sustainability, material and governance limitations, prior knowledge of IRA/PIRA, principles already present in daily practice, areas susceptible to improvement through IRA-WWP, external support needed for implementation, and cultural strategies for harmonizing changes without affecting community identity. Therefore, to ensure the analytical consistency of the qualitative process, the focus group transcripts were reviewed iteratively, identifying recurring units of meaning that were subsequently grouped into analytical categories. Open coding enabled the identification of expressions linked to production limitations, organizational dynamics, and perceptions of sustainability. Subsequently, through axial coding, these categories were articulated into thematic axes related to governance, innovation, social cohesion, and productive projection. Finally, selective coding allowed for the integration of these axes into interpretive patterns consistent with the IRA principles and the WWP model, ensuring that the findings did not reflect isolated observations, but rather systematic recurrences in the participants' discourse.

https://doi.org/10.3390/su18052661

2.4. Documental Analysis

The third component consisted of a comprehensive documentary analysis of the cooperative's main management and planning instruments: the Tejemujeres Final Report, the IEPS-Cuenca Production Process, the 2022–2030 Strategic Plan, and the 2024 Business Plan. These documents allowed for the reconstruction of the actual production flow, the identification of assigned roles, delivery times, quality control criteria, the marketing structure, support networks, and internal governance and sustainability challenges. The documentary analysis served as the basis for comparing quantitative and qualitative findings and for establishing alignment with the strategic dimensions of the IRA principles and the WWP model.

3. Results

3.1. Analysis of the Tejemujeres Cooperative's Production System

The Tejemujeres Women Artisans Cooperative is a leading example of a social and productive organization in Gualaceo. Its production system is based on artisanal weaving as both an economic activity and a cultural practice. Previous studies demonstrate that the cooperative's productive structure relies more on social and organizational dynamics than on strictly profitable indicators, highlighting strengths such as teamwork, communication, and ethical values, along with persistent weaknesses in leadership, planning, and digital skills. The results of a survey of 110 members show that membership growth (29.4%) coexists with a complete absence of young people under 25, confirming a generational shift issue. Participation remains concentrated in production (92%), while only 18% hold leadership roles. Time commitment is also limited by domestic responsibilities, as two-thirds of the members dedicate between 1 and 20 h per week to the cooperative.

This quantitative overview is complemented by focus groups, which identified perceptions that explain the structural limitations. Members indicated that low incomes discourage the incorporation of young people, and internal leaders mentioned resistance to innovation. External stakeholders added that limited infrastructure and a lack of productive diversification reduce growth capacity. Taken together, these elements show that the production system faces tensions that combine gender, age, organizational capacities, and market conditions.

Table 1 summarizes information from the survey, focus groups, and interviews with external stakeholders, integrating these three sources into an intersectional matrix focused on age, productive roles, management, time commitment, and main perceived strengths and weaknesses. This integration allows us to understand how social and organizational variables influence internal participation. Different profiles emerge between older women, middle-aged women, and a complete absence of young people. The former has the greatest tenure and experience but exhibit lower digital literacy and greater resistance to innovation. The second group simultaneously combines artisanal work with external activities, showing a greater willingness to seek training, although with family responsibilities. The absent group, comprised of those under 25, represents an untapped potential due to unattractive income and a lack of visibility of artisanal work as an economic opportunity. This implies that the future sustainability of the cooperative requires a generational succession strategy, as well as a redistribution of workloads and an expansion of leadership opportunities—elements consistent with IRA principles such as inclusion, recognition, and autonomy, and with WWP's focus on capacity building and governance.

https://doi.org/10.3390/su18052661

Table 1. Intersectional matrix on age, generation, role, time, strengths and weaknesses.

Group	Productive Rol	Management/Leadership	Time Spent	Strengths	Weaknesses/Challenges
60% = Older women (40–55 years old)	Artisanal production (high concentration)	Low participation in leadership (≤10%)	1–15 h per week	Experience and cultural cohesion, permanence (48% with more than 6 years in the cooperative)	Resistance to innovation, low digital literacy, physical fatigue; perception that the textile industry is more heritage than business.
40% = Middle-aged women (26–39 years old)	Handicraft production + external activities	Average management (30%)	10–20 h per week	Greater flexibility, use of social media (WhatsApp, Facebook), and willingness to train	Overburdened by dual roles (home + cooperative), limited income; they express distrust in opening new markets.
0% = Young people (<25 years)	Unrepresented	N/A	N/A	Potential for renewal and digitization if they were to enter	Current non-existence: low attraction due to lack of income and appreciation of artisanal work, as noted by members of the supervisory board.

Figure 3 summarizes the main factors limiting the productive sustainability of the Tejemujeres Cooperative and the proposed strengthening strategies. The survey identified that 69% of members report a shortage of raw materials, 58% a lack of innovation, 40% marketing problems, and 37% insufficient training. Despite these limitations, 88% positively evaluate the cooperative's economic sustainability, reflecting organizational resilience based on internal cohesion, member retention, and the cultural value placed on artisanal work. The qualitative analysis of the focus groups provides further insight. One member noted that "the poor quality of raw materials makes work slow and even demotivating," while a finance leader acknowledged that "there is a lack of confidence in developing new products."

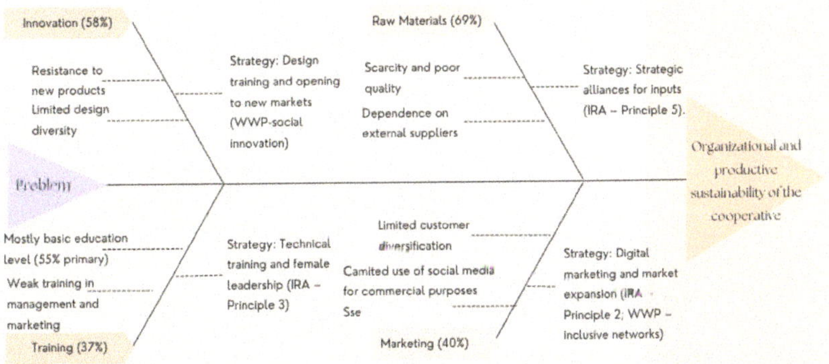

Figure 3. Key aspects of the WWP model.

External stakeholders, in turn, highlighted the need to "be open to innovation and secure strategic alliances" and to "strengthen training in management and digital marketing." These perceptions align with the survey results, which show interest in improving innovation and diversification (88%), collaborative work (79%), and digital marketing (76%).

Thus, the diagram not only organizes the production constraints but also connects their impacts with the improvement strategies suggested within the framework of the IRA principles and the WWP model, aimed at ensuring access to inputs, strengthening capacity building, diversifying markets, and fostering inclusive governance.

https://doi.org/10.3390/su18052661

3.2. Relationship Between the Limitations of the Production System and Sustainability

Table 2 shows that the chi-square tests for each limitation did not reveal a statistically significant association with the perception of sustainability, as the significance values (p-values) were greater than 0.05. Since several tables had cells with expected frequencies of five or fewer, these results should be interpreted with caution and only as a descriptive trend. The lack of significant associations suggests that perceived sustainability does not depend on a single factor, but rather on the interaction of multiple social and organizational dynamics within the cooperative.

Table 2. Chi-square test between limitations and sustainability.

Limitation	Chi-Square	Gl	p-Value	Exact Test
Lack of access to raw materials	4.568	4	0.335	ns
Low training	2.242	4	0.691	ns
Marketing difficulties	3.622	4	0.460	ns
Lack of innovation	5.178	4	0.270	ns
Financial management problems	0.670	4	0.955	ns
Communication/internal rules	0.993	4	0.911	ns

Note: All tests showed p-values > 0.05; furthermore, several tables showed more than 50% of cells with expected frequencies < 5, so these results should be interpreted with caution. "ns" = not significant. Data were generated from the analysis using SPSS V.13 statistical software. Prepared by the author.

The qualitative analysis of the focus groups provides greater depth, reflected in expressions such as: "Here we all support each other, even though things are lacking, we always find a way to move forward," or "we know we need to improve, but without losing who we are as artisans," which demonstrates a favorable disposition toward improvement processes as long as they respect the collective identity and traditional forms of organization.

In this context, it is essential to note that the limitations associated with marketing and access to raw materials do not operate in isolation, nor do they have homogeneous effects on the cooperative's sustainability. On the contrary, the qualitative findings show that these difficulties are partially mitigated by internal dynamics of mutual aid, cooperation, and collective organization. The members resort to informal support mechanisms, such as the exchange of supplies, the redistribution of workloads, or solidarity-based coordination during periods of lower production, which lessens the negative impact of structural constraints. This collaborative logic explains why, despite acknowledging persistent problems in innovation and marketing, the overall perception of sustainability remains relatively high. In this sense, social cohesion not only fulfills a cultural function but also acts as an operational factor that sustains the productive system in the face of adverse conditions, although it can also become a limitation if it replaces necessary modernization processes.

This finding is complemented by the results of the focus group, where members differentiated between everyday difficulties and overall sustainability. Low-quality inputs or a lack of trust in innovation affects processes, but do not necessarily alter the general perception of collective stability. This reinforces the importance of integrating both sources to interpret the production system.

The convergence between quantitative results and qualitative narratives allows us to affirm that the identified patterns do not respond to isolated perceptions, but to a collective construction of the meaning of sustainability, in which cooperation, cultural identity, and shared experience act as structuring axes of the productive system.

https://doi.org/10.3390/su18052661

3.3. Non-Hierarchical Cluster Analysis (K-Means) on Perceptions of the Implementation of the IRA-WWP Model

Cluster analysis was conducted to identify distinct patterns in how members understand and value the implementation of the IRA principles and the WWP model. A K-Means procedure was applied, with the number of clusters set at three. This decision was based on the criteria of substantive interpretability and a preliminary review of the model with k = 2 and k = 4, where the resulting structure showed less conceptual stability. The choice of k = 3 allowed coherent and comparable profiles, and the initial centroids showed sufficient distance to avoid overlapping. The variables used came from Likert-type scales with the same response range, so no additional standardization was applied. Even so, the exploratory nature of the exercise is acknowledged, especially considering the sample size (N = 110) and the type of variables used.

The consolidated results show the formation of three distinct profiles. The first cluster comprises 41 members who reported low familiarity with the IRA-WWP principles, but have a strong focus on internal participation, community work, and daily cooperation. The second cluster, made up of 25 participants, is characterized by a moderately greater understanding of both models and a stronger emphasis on training, professionalizing artisanal work, and organizational improvement. The third cluster includes 44 members who, despite reporting limited initial familiarity, express a high value for technical training and opportunities to connect with external markets.

Although the procedure did not incorporate sociodemographic variables into the clustering algorithm, comparing these profiles with the patterns recorded in the previously presented intersectional matrix allows for a more nuanced interpretation. The groups with greater openness to innovation, diversification, and commercial engagement tend to coincide with middle-aged artisans and those with more dynamic ties to management. Meanwhile, more conservative perceptions are more frequently observed among members with longer tenures within the organization.

Overall, the results suggest that the implementation of IRA-WWP does not depend on a single line of action, but rather is shaped by internal differences in knowledge, expectations, and levels of participation. Training emerges as the unifying element among the three profiles, while governance and productive sustainability require differentiated strategies that acknowledge the cooperative's internal heterogeneity and avoid homogeneous solutions that could reinforce existing gaps.

The differentiation of profiles obtained through cluster analysis reveals that resistance to or openness toward IRA-WWP principles is not solely determined by levels of knowledge, but also by organizational trajectories and accumulated experiences within the cooperative. While some groups value the stability and continuity of traditional practices, others see training and market access as opportunities for strengthening. This coexistence of perspectives explains why change processes require gradual strategies that acknowledge internal differences and avoid homogeneous approaches that could exacerbate existing tensions, as shown in Table 3.

Table 3. Final cluster centers and number of cases in each cluster.

	Cluster		
	1	2	3
Are you familiar with the IRA principles and the WWP model?	1.24	2.20	1.00
Promoting active participation of all members	0.90	0.48	0.57
Implementing best practices in sustainable production	0.54	0.20	0.05

https://doi.org/10.3390/su18052661

Table 3. *Cont.*

	Cluster		
	1	2	3
Strengthening connections with the market and consumers	0.34	0.36	0.45
Training and developing new skills	0.07	0.88	0.93
Strengthening internal governance	0.15	0.20	0.16
Number of cases	41	25	44

Note: Data was generated from analysis using SPSS V.13 statistical software. Prepared by the author.

3.4. Triangulation of Findings with IRA Principles, WWP Model, and GRI-OCB (IPMA)

The findings from the focus group and the intersectional matrix were clearly linked to the IRA principles, the WWP model's axes, and the GRI-OCB/IPMA governance guidelines. Beyond simply demonstrating mandatory alignment, this approach allowed the partners' perceptions regarding resource scarcity, commercial dependence, low youth participation, weaknesses in training and innovation, and limitations in internal management to be situated within widely recognized frameworks for sustainability and community governance. Table 3 illustrates that the challenges identified by the participants are not isolated but rather align with critical dimensions such as diversification, participation, transparency, innovation, and cultural resilience—central elements in both IRA and WWP, as well as in GRI-OCB/IPMA. Taken together, this triangulation serves as a conceptual framework that strengthens the proposed improvements, ensuring they simultaneously address Tejemujeres' organizational reality and international principles of sustainable management.

In this sense, as shown in Table 4, triangulation not only validates the empirical findings but also allows us to understand that the identified weaknesses—far from being isolated failures—are part of common structural tensions in community-based organizations. The alignment between the IRA principles, the WWP model, and the GRI-OCB standards underscores the importance of addressing sustainability from a holistic perspective, where governance, inclusion, and innovation are integrated with cultural identity and the dynamics of the territory.

Table 4. Triangulation of findings with IRA, WWP and GRI-OCB/IPMA.

Results	IRA Principles	Axes of the WWP Model	GRI-OCB Standard (ESG)
Shortage of raw materials and dependence on a buyer	P1 (food security), P2 (diversification), P5 (strategic alliances)	Social innovation	Risk management, sustainability (E)
Low representation of young people and limited leadership	P3 (participation), P4 (inclusion)	Inclusion	Leadership, teamwork, equity (S, G)
Deficiencies in training and product innovation	P6 (sustainable practices), P8 (responsible production)	Commitment, Social Innovation	Planning, innovation, customer focus (E, S)
Financial management problems and lack of transparency	P9 (participatory processes), P10 (accountability)	Transparency	Transparency, governance, accountability (G)
Social cohesion and cultural resilience	P7 (cultural preservation)	Commitment, Inclusion	Values management, cultural legitimacy (S)

3.5. Governance Model Based on the IRA Principles and the WWP Model

Based on the analysis developed using applied research techniques, it was possible to the analysis revealed that the perceptions of members, directors, and external stakeholders

https://doi.org/10.3390/su18052661

coincide with the need to strengthen internal cohesion, diversify productive activities, and consolidate more transparent governance processes. To guide these findings toward a structured proposal, a comparative perspective was adopted between the IRA principles and the relational approach of the WWP, two frameworks widely used in participatory rural development experiences. This comparison is not a circular exercise; rather, it allows for the interpretation of the results obtained considering international benchmarks that have demonstrated effectiveness in similar contexts. Upon reviewing the characteristics of each model, it was identified that the WWP stands out for its adaptability and applicability in community organizations, while IRA offers greater methodological rigor, especially in practices related to sustainability, diversification, and inclusion. The comparative assessment of both approaches is summarized in the following Table 5.

Table 5. Evaluation criteria for governance models with an ESG focus.

Criterion	WWP	IRA
ESG Coverage	4	5
Adaptability	5	4
Practical Feasibility	4	3
Methodological Rigor	4	5
Replicability	4	5
Average	4.2	4.4

The results showed that the WWP (Wholesale Value Project) scored highly in adaptability and practical feasibility, an aspect supported by experiences such as those of Ríos et al. [15], Castañeda et al. [16], and López et al. [17], who document its effectiveness in local participation processes and capacity building. However, these studies also indicate that the model has made more progress in the social sphere than in formal environmental monitoring mechanisms. For its part, IRA (Integrated Environmental Management) demonstrates greater methodological consistency and robustness in articulating ecological and economic sustainability, although it requires more technical resources and administrative capacity. This complementarity supports the decision to integrate both approaches into a hybrid model adaptable to the Tejemujeres context.

Based on this integration, in Table 6, a governance model is proposed that articulates three dimensions: ethical-social, technical-business, and political-contextual, consistent with the study's findings and the principles that guide sustainable associative practices. Each dimension incorporates organizational competencies, participation criteria, and expected outcomes.

The hybrid WWP-IRA model allows for a comprehensive measurement of effectiveness across all ESG dimensions, evaluating environmental, social, and governance aspects at the Tejemujeres Cooperative. Furthermore, it ensures strategic alignment with the Sustainable Development Goals (SDGs) of the 2030 Agenda, strengthening the social and environmental impact of its artisanal activities. In Tables 7–9, the incorporation of IPMA competencies contributes to professionalizing management, promoting leadership, planning, and collaborative work, while direct alignment with IRA principles guarantees social, environmental, and governance sustainability, consolidating a comprehensive and replicable governance framework for the organization.

https://doi.org/10.3390/su18052661

Table 6. Key elements of the governance model of the Tejemujeres Cooperative.

Description	IPMA Competencies (ICB4.0)	IRA Principles	Expected Result
Ethical-Social Dimension			
Focused on community cohesion, equity, and cultural transmission; it strengthens Tejemujeres' identity as an organization.	Teamwork Effective communication Management of values and ethical principles Social sustainability	P3: Consolidate the central role of members in decision-making. P4: Design strategies for generational succession. P7: Preserve the network as a cultural heritage and foster innovation. P9: Clear rules, participatory processes, and accessible accountability mechanisms.	An organizational structure based on cohesion, equity, and cultural legitimacy, where internal governance is strengthened through the inclusion of women and youth in leadership roles.
Technical-Business			
It addresses productive capacities, innovation and competitiveness in the face of a demanding market.	Planning Leadership Risk Management Customer Focus Innovation Digitalization	P1: Adapted to artisanal products as a contribution to community well-being. P2: Diversification of markets and reduced dependence on a single buyer. P6: Incorporation of sustainable raw materials and eco-efficient practices. P8: Guaranteeing safe and responsible processes in artisanal production.	An innovative, sustainable, and competitive production system that optimizes resources, diversifies supply, and ensures stable income without losing cultural identity.
Political-Contextual			
Focused on integration into territorial networks, fair markets, and local and global governance structures.	Negotiation, stakeholder management, strategic vision, long-term sustainability	P5: Strengthen alliances to guarantee sustainable raw materials. P9: Transparency in agreements with buyers and strategic partners. P10: Establish clear indicators of social, economic, and environmental sustainability.	Governance model connected with external actors legitimizing the cooperative as a benchmark for sustainable artisanal production in the region.

Table 7. Indicators of the ethical-social dimension.

General Objective: To Strengthen Social Cohesion, Inclusion, Equity and Participation of Members.					
Key Indicator	**Description**	**IRA Principle**	**IPMA**	**ESG**	**SDG**
Percentage of female members participating in strategic decisions	Measuring the effective participation of members in meetings and committees	P3, P4, P9	Leadership, communication, teamwork	Social, Governance	ODS5, ODS16
Percentage of members trained in ethical values and artisanal culture	Training in ethics, cultural identity and knowledge transmission	P7, P3	Values and ethical principles, teamwork	Social	ODS4, ODS5
Level of perceived inclusion and equity	Surveys on gender equality and youth participation	P3, P4	Communication, leadership	Social	ODS5, ODS10

https://doi.org/10.3390/su18052661

Table 8. Indicators of the technical-business dimension.

General Objective: To Strengthen Social Cohesion, Inclusion, Equity and Participation of Members.					
Key Indicator	**Description**	**IRA Principle**	**IPMA**	**ESG**	**SDG**
% of products with design innovation	Number of new or redesigned products annually	P8, P2	Innovation, planning	Social, Ambiental	ODS2, ODS9
% of sustainable production processes	Implementation of good practices, resource efficiency and eco-efficiency	P6, P1	Risk management, sustainability	Ambiental	ODS12, ODS13
Percentage of members trained in management and digital marketing	Business management training and online promotion	P2, P1	Leadership, planning	Social, Governance	ODS8, ODS9
Diversification of customers and markets	Number of buyers or marketing channels	P2	Strategic planning, leadership	Social, Governance	ODS8, ODS12

Table 9. Indicators of the political-contextual dimension.

General Objective: To Strengthen Social Cohesion, Inclusion, Equity and Participation of Members.					
Key Indicator	**Description**	**IRA Principle**	**IPMA**	**ESG**	**SDG**
Percentage of agreements with documented strategic allies	Formalization of alliances and commitments with foundations, municipalities and fair-trade networks	P5, P9, P10	Negotiation, stakeholder management	Governance	ODS16
% of processes with financial transparency	Clear and auditable accounting reports	P9, P10	Planning, Leadership	Governance	ODS16
Annual social and environmental impact assessment	Measuring impacts on community and ecosystem	P1, P6, P10	Risk management, sustainability	Ambiental, Social	ODS2, ODS12, ODS13

4. Discussion

The findings of the study at the Tejemujeres Cooperative align with previous studies argue that grassroots social organizations tend to rely more on their human capital and cultural identity than on profitability indicators, which explains Tejemujeres' resilience despite its business weaknesses.

The adaptability demonstrated by the members, who combine artisanal production with other economic activities, reinforces the findings, and it considers flexibility a key factor in strengthening sustainable production systems [18,19]. Along the same lines, it emphasizes that rural development projects should be built upon local perspectives, which is reflected in the members' willingness to integrate the IRA principles and the WWP model without losing their community identity [13].

However, a significant contrast is observed regarding innovation. While Caicedo and Grass [1] point out that agro-industrial sustainability depends largely on diversification and innovation, in the case of Tejemujeres, the lack of innovation was not perceived as a factor that compromises its economic sustainability. This divergence can be explained by the fact that the cooperative maintains a certain stability thanks to its social cohesion and the sustained demand from a majority buyer. However, as Vásquez [12] warns, dependence

https://doi.org/10.3390/su18052661

on a single development model without the capacity for real appropriation can limit future growth and increase vulnerability to external changes.

Regarding conceptual frameworks, the results confirm the relevance of the WWP model, which emphasizes active participation, inclusion, and co-responsibility in decision-making [10,11]. The high acceptance of this model by the members coincides with the findings of Jiménez et al. [5], who consider it a space for co-creating solutions adapted to the territory. In Tejemujeres, technical training, internal governance, and market access were identified as priority elements, in line with the findings of [20] regarding the role of association in organizational strengthening.

For their part, the IRA principles appear to be an essential complement, especially in the areas of environmental sustainability and governance. As evidenced by Jiménez et al. [4] in cooperative experiences in Peru, its implementation can improve resilience and sustainability, although it requires greater technical and financial capacities, an aspect already noted by Obando [18]. This contrast indicates that, while Tejemujeres recognizes the value of these principles, their full application could demand external support and strategic alliances for the development of training focused on its reality.

In turn, as methodological support for the proposal, the evaluation criteria for governance models with an ESG approach described in the study were used [20]. These criteria consider ESG coverage, adaptability, practical feasibility, methodological rigor, and replicability, thus validating the relevance of using the models within the context of Tejemujeres. Furthermore, the model is flexible and easily applicable to community projects, such as those in Romania analyzed by Ríos et al. [15], where the WWP approach facilitated the consolidation of competencies in regional project management, reinforcing inter-institutional cooperation and social participation as pillars of sustainability.

Castañeda et al. [16] demonstrated its application in Colombia and Guatemala, where the model was key to implementing the Voluntary Guidelines on the Responsible Governance of Tenure of Land, Fisheries and Forests, promoted by the FAO. In these cases, the WWP approach strengthened dialogue between rural and governmental actors, fostering more inclusive processes that were sensitive to the local context. However, the same authors caution that the model still has limitations in incorporating environmental and monitoring indicators aligned with international standards, which restricts its ability to systematically measure ecological impacts.

From an educational perspective, [21] the WWP model was adapted to improve evaluation systems in higher education, highlighting its potential to promote more collaborative learning processes oriented towards the holistic development of individuals. This shift of the approach from the rural sphere to university management demonstrates its flexibility and adaptability to diverse institutional contexts. Similarly, applied the model to rural entrepreneurship strategies in the northern plateau of Spain, finding that its application fosters autonomy and local leadership by linking human values with productive innovation [18]. Finally, the analysis by Cazorla and De los Ríos [21] confirms that the WWP model has evolved from Robert Chambers' participatory approaches in the 1970s toward a more relational vision, where trust, cooperation, and ethical management are pillars of sustainable development [2]. Therefore, the interpretation of the results obtained is supported, in which the social dimension and community linkage stand out over the environmental and regulatory aspects, showing that the model, although consolidated in its human dimension, requires further progress towards a more robust integration of ecological and global governance components.

Taken together, the results reinforce the arguments and maintain that sustainability in rural projects depends on a balanced integration of social, economic, and institutional factors [6,22]. In the case of Tejemujeres, the most viable strategy appears to be the con-

https://doi.org/10.3390/su18052661

struction of a hybrid model: leveraging the social proximity and flexibility of the WWP while gradually incorporating the methodological requirements of the IRA principles.

5. Conclusions

The integrated application of the IRA principles and the WWP model demonstrates that it is possible to strengthen the Tejemujeres Cooperative's production and management system without altering its social and community identity. Although limitations related to access to raw materials, innovation, and training persist, internal cohesion and the cultural value of weaving remain key pillars supporting its perception of sustainability. In this context, IRA provides structure, sustainability criteria, and transparency mechanisms, while WWP reinforces participation and community engagement. The complementarity between these two approaches allows for improved production practices, market diversification, and professionalized management, all while respecting the cooperative's core values. In this sense, the hybrid model offers a viable path to balancing efficiency, identity, and sustainability in Tejemujeres' future development.

Regarding the first objective, it was identified that the main limitations of the Tejemujeres Cooperative's production system are associated with a scarcity of raw materials, a lack of innovation in designs, low levels of technical training, and marketing difficulties. However, these restrictions have not directly impacted the members' perception of economic sustainability, as they value the organization's social and cultural cohesion as a fundamental asset for maintaining its stability. This finding reaffirms that the cooperative's strength lies more in its community identity and the shared responsibility of its members than in purely business indicators.

Regarding the second objective, the analysis revealed that both the IRA principles and the WWP model offer key elements that can be adapted to Tejemujeres' context. These include internal governance, active participation, technical training, social innovation, and links to fair markets, all aligned with the priorities identified by the members. Furthermore, the gradual incorporation of sustainable practices and financial transparency allows the organization to align its artisanal activities with international sustainability frameworks without losing its cultural essence.

Finally, in accordance with the third objective, it is proposed that the most appropriate strengthening strategy for the cooperative should combine the participatory flexibility of the WWP model with the methodological rigor and international reach of the IRA principles. This hybrid approach ensures a balance between respecting Tejemujeres' social identity and the need to improve its competitiveness in diversified markets. In this way, the organization will be able to consolidate an innovative, sustainable, and culturally legitimate production system, guaranteeing its long-term economic viability and strengthening its role as a benchmark for social development in rural areas.

From future lines of research, it is recommended that this analysis be replicated in other cooperatives and artisan organizations to test the validity of the IRA-WWP hybrid model in diverse contexts. It would also be pertinent to develop longitudinal studies to assess how governance, participation, and sustainability evolve, incorporating environmental variables and more precise cultural metrics. Furthermore, it is suggested that participatory methodologies that actively integrate young women be explored to strengthen generational renewal and better understand their motivations and expectations regarding artisan work. These approaches will contribute to refining the proposed model and expanding its potential application in sustainable rural development processes.

Author Contributions: Conceptualization, M.O., R.G., M.P. and X.N.; methodology, M.O., R.G., M.P. and X.N.; software, M.O., R.G., M.P. and X.N.; validation, M.O., R.G., M.P. and X.N.; formal analysis, M.O., R.G., M.P. and X.N.; investigation, M.O., R.G., M.P. and X.N.; resources, M.O., R.G., M.P. and

https://doi.org/10.3390/su18052661

X.N.; data curation, M.O., R.G., M.P. and X.N.; writing—original draft preparation, M.O., R.G., M.P. and X.N.; writing—review and editing, M.O., R.G., M.P. and X.N.; visualization, M.O., R.G., M.P. and X.N.; supervision, M.O., R.G., M.P. and X.N.; project administration, M.O.; funding acquisition, M.O., R.G., M.P. and X.N. All authors have read and agreed to the published version of the manuscript.

Funding: This research received no external funding. The previous studies were funded by the researchers, and the article processing charges (APC) were funded by Universidad Bolivariana del Ecuador (UBE).

Institutional Review Board Statement: The study was conducted in accordance with the Declaration of Helsinki and received formal ethical approval from the Board of the Cooperativa de Mujeres Artesanas Tejemujeres, which serves as the competent authority for social and organizational research within the institution. Approval was granted by the General Management and the Presidency of the Cooperative, acting as an Internal Ethics Committee in accordance with their institutional statutes (approval code: CT-ET/2024-01, approval date: 15 March 2024).

Informed Consent Statement: Informed consent was obtained from all subjects involved in the study.

Data Availability Statement: The data that support the findings of this study are available in Google Sheets at https://docs.google.com/spreadsheets/d/1wabWHpTQMht6Rqu55ZXDL8wtD2JlT2VQ/htmlview (accessed on 23 November 2025). The datasets were generated during the current study.

Acknowledgments: The authors would like to acknowledge Rosendo Ortuño Soledispa and the Tejemujeres Cooperative.

Conflicts of Interest: The authors declare no conflict of interest.

References

1. Caicedo, M.; Grass, J. Sostenibilidad en la Transformación Agroindustrial de la Costa Pacífica del Cauca: Modelos Organizativos, Retos y Oportunidades. *Cienc. Lat. Rev. Científica Multidiscip.* **2024**, *8*, 1228–1255. [CrossRef]
2. Fonseca, N.; González, M.; Narváez, C. Asociatividad para la administración los sistemas de producción campesina. *Rev. Estrateg. Organ.* **2020**, *9*, 1–7. [CrossRef]
3. Hennchen, B.; Schäfer, M. Do sustainable food system innovations foster inclusiveness and social cohesion? A comparative study. *Front. Sustain.* **2022**, *3*, 921169. [CrossRef]
4. Jiménez, R.; De los Ríos, I.; Huamán, A. Los principios CSA-IRA como instrumentos para la dirección de proyectos de desarrollo rural. El caso de sierra central de Perú. In Proceedings of the 26th International Congress on Project Management and Engineering, Terrassa, Spain, 5–8 July 2022; pp. 1632–1644.
5. Jiménez Aliaga, R.; De los Ríos-Carmenado, I.; Huamán Cristóbal, A.E.; Aliaga Balbín, H.; Marroquín Heros, A.M. Competencies and Capabilities for the Management of Sustainable Rural Development Projects in the Value Chain: Perception from Small and Medium Sized Business Agents in Jauja, Peru. *Sustainability* **2023**, *15*, 15580. [CrossRef]
6. Acosta Mereles, M.L.; Ríos Carmenado, I.D.L.; Ávila Cerón, C.A.; Castañeda Sepúlveda, R. Inversión responsable en procesos de sustitución de cultivos ilícitos desde el modelo "Trabajando con Personas": Estudio de caso Guaviare (Colombia). In Proceedings of the 27th International Congress on Project Management and Engineering, Donostia, Spain, 10–13 July 2023; pp. 1711–1725.
7. Regalado-López, J.; Maimone-Celorio, J.A.; Pérez-Ramírez, N. Strengthening of the Rural Community and Corn Food Chain Through the Application of the WWP Model and the Integration of CFS-RAI Principles in Puebla, México. *Sustainability* **2025**, *17*, 5442. [CrossRef]
8. Jiménez, R. Modelo de Desarrollo Rural Sostenible Desde los Principios de Inversión Responsable en Agricultura: Aplicación al Centro de Desarrollo Regional El Mantaro—Grupo de Acción Local Jauja, en Perú. Ph.D. Thesis, Universidad Politécnica de Madrid, Madrid, Spain, 2025.
9. García, J. *Modelo de Aprendizaje Basado en Proyectos Para el Desarrollo de Competencias de Contabilidad: Aplicación en la UPS*; Universidad Politécnica de Madrid: Madrid, Spain, 2021.
10. Mur Nuño, C.; Nole, P.; De Los Ríos, I.; Mejía, J. Buenas prácticas de gobernanza en los Bancos de Alimentos: Criterios ESG desde el modelo APS-WWP. In Proceedings of the VII Congreso Internacional sobre Aprendizaje, Innovación y Cooperación, Madrid, Spain, 18–20 October 2023; pp. 328–331.
11. De los Ríos Carmenado, I.; Nole Correa, P.; Zuluaga Zuluaga, C.L.; Roncancio Burgos, M. ApS projects from the Working with People–WWP–model in the framework of responsible food consumption: Experience from the UPM Food Bank Chair. In *Service-Learning Experiences at UPM: 2021 and 2022*; UPM: Kuala Lumpur, Malaysia, 2023.

https://doi.org/10.3390/su18052661

12. Vásquez, M. Las comunidades y organizaciones afro no se apropian del modelo de desarrollo: Caso Norte del Cauca. *Rev. Lumen Gentium* **2023**, *7*, 95–114. [CrossRef]
13. Inquilla, J.; Chaiña, F.; Sucapuca, M. Imaginarios sociales de desarrollo rural desde los actores locales: Caso de los regantes de Taraco y Pusi-Huancané-Perú. *Idesia* **2024**, *42*, 79–88. [CrossRef]
14. López, I. Governance and networking in the Third Sector of Social Action. *Hum. Rev. Int. Humanit. Rev./Rev. Int. Humanidades* **2022**, *15*, 1–14.
15. Ríos, I.; Rahoveanu, A.; Gallegos, A. Project Management Competencies for Regional Development in Romania: Analysis from "Working with People" Model. *Procedia Econ. Financ.* **2014**, *8*, 614–621. [CrossRef]
16. Castañeda, R.; Avila Cerón, C.; De los Ríos-Carmenado, I.; Domínguez, L.; Gomez, S. Implementing the Voluntary Guidelines on the Responsible Governance of Tenure of Land, Fisheries and Forests from the Working with People Model: Lessons from Colombia and Guatemala. *J. Peasant. Stud.* **2023**, *50*, 1820–1851. [CrossRef]
17. López, M.; Cazorla, A.; Panta, M. Rural Entrepreneurship Strategies: Empirical Experience in the Northern Sub-Plateau of Spain. *Sustainability* **2019**, *11*, 1243. [CrossRef]
18. Obando, M. *Análisis de los Estándares Internacionales de Gobernanza GRI y IPMA-OCB: Aplicación al Proyecto de la Gran Recogida de FESBAL*; Universidad Politécnica de Madrid: Madrid, Spain, 2025.
19. de los Ríos-Carmenado, I.; Sastre-Merino, S.; Fernández Jiménez, C.; Núñez del Río, M.; Reyes Pozo, E.; García Arjona, N. Proposals for Improving Evaluation Systems in Higher Education: An Approach from the Model "Working with People". *J. Technol. Sci. Educ.* **2016**, *6*, 104–120. [CrossRef]
20. Cazorla, A.; De los Ríos, I. "Putting the Last First" to "Working with People" in Rural Development Planning: A Bibliometric Analysis of 50 Years of Research. *Sustainability* **2023**, *15*, 10117. [CrossRef]
21. Cazorla, A.; Negrillo, X. Los Principios Ira Como Base de Investigación Entre Universidades y Empresas. *Contab. Mark. Empresa* **2022**, *7*, 41–54.
22. Aldaz, J.C.C.; Cortez, J.L.P.; López, M.C.; Jacome, S.S.I. Adaptabilidad en el sistema de producción agrícola: Una mirada desde los productos alternativos sostenibles. *Rev. Cienc. Soc.* **2020**, *26*, 308–327.

https://doi.org/10.3390/su18052661

MDPI AG
Grosspeteranlage 5
4052 Basel
Switzerland
Tel.: +41 61 683 77 34

Sustainability Editorial Office
E-mail: sustainability@mdpi.com
www.mdpi.com/journal/sustainability

www.ingramcontent.com/pod-product-compliance
Lightning Source LLC
Chambersburg PA
CBHW040801250926
51850CB00007BA/1051
9783725873388